Technician's Guide to Industrial Electronics
How to Troubleshoot and Repair Automated Equipment

Technician's Guide to Industrial Electronics

How to Troubleshoot and Repair Automated Equipment

Robert S. Carrow

TAB Books
Division of McGraw-Hill
New York   San Francisco   Washington, D.C.   Auckland   Bogotá
Caracas   Lisbon   London   Madrid   Mexico City   Milan
Montreal   New Delhi   San Juan   Singapore
Sydney   Tokyo   Toronto

©1995 by **McGraw-Hill, Inc.**
Published by TAB Books, a division of McGraw-Hill, Inc.

Printed in the United States of America. All rights reserved. The publisher takes no responsibility for the use of any materials or methods described in this book, nor for the products thereof.

hc   2 3 4 5 6 7 8 9 10 DOC/DOC 9 9 8 7 6

Product or brand names used in this book may be trade names or trademarks. Where we believe that there may be proprietary claims to such trade names or trademarks, the name has been used with an initial capital or it has been capitalized in the style used by the name claimant. Regardless of the capitalization used, all such names have been used in an editorial manner without any intent to convey endorsement of or other affiliation with the name claimant. Neither the author nor the publisher intends to express any judgment as to the validity or legal status of any such proprietary claims.

Library of Congress Cataloging-in-Publication Data
Carrow, Robert S.
    Technician's guide to industrial electronics : how to troubleshoot and repair automated equipment / by Robert S. Carrow.
        p.   cm.
    Includes index.
    ISBN 0-07-011273-8 (H)
    1. Industrial electronics. 2. Industrial electronics—Maintenance and repair.   I. Title.
TK7881.C37 1995
670.42'7—dc20   95-3240
                                                                                  CIP

Acquisitions editor: Roland S. Phelps
Editorial team: Joanne Slike, Executive Editor
                Lori Flaherty, Supervising Editor
                Theresa Cunningham, Book Editor
Production team: Katherine G. Brown, Director
                 Jeffrey Hall, Computer Artist
                 Wanda S. Ditch, Desktop Operator
                 Nancy K. Mickley, Proofreading
                 Jodi L. Tyler, Indexer
Design team: Jaclyn J. Boone, Designer   GEN1
             Katherine Stefanski, Associate Designer   0112738

To my wife, Colette,
my children, Ian and Justine,
and to my mother and father, Red and Nonie

Contents

Foreword *xiii*

Acknowledgments *xvi*

Introduction *xvii*

1 How we got here   *1*
What is automation?   *1*
The Industrial Revolution   *2*
The digital revolution   *3*
Controlling motion   *3*
Automating the process   *4*

2 Basic electricity and electronics   *7*
Standards organizations   *8*
Symbols   *10*
Electricity   *10*
Electrical formulas   *13*
Frequency and amplitude   *16*
Electrical hardware   *16*
Transformers   *21*
Power semiconductors   *23*
Wiring and cabling   *26*
Tools of the trade   *28*
Fusing and circuit protection   *29*
Electronics   *30*
Bibliography and suggested further reading   *31*

3 Power transmission basics   33
Power transmission   33
Horsepower   35
Torque   38
How to use speed, torque, and horsepower curves   41
Inertia   43
Friction   46
Motors, or prime movers   47
Stress   48
Gear reduction   48
Couplings   49
Gearboxes   50
Gears   51
V-belts   52
Chains and sprockets   53
Clutches and brakes   54
Variable-speed pulleys   55
Fluidized speed variators   55
Proportionately infinitely variators   56
Ball screws   56
Drives   57
Pumping applications   59
Fan and blower applications   60
Hydraulics   61
Pneumatics   62
Noise   63
Efficiency   64
Maintenance   65
Power transmission authorities and standards   66
Conclusion   66
Bibliography and suggested further reading   67

4 Industrial computers   69
Microprocessing basics   72
Software   78
Hardware   78

Memory 79
Peripherals 82
Online in the factory 82
Uploading and downloading 84
Industrial computers and workstations 85
CAD, CAM, and CAE 88
Industrial computer applications 91
Computers: today and tomorrow 94
Bibliography and suggested further reading 96

5 Process controllers and PLCs 97

PLC hardware 99
Digital, discrete, and analog I/O handling 107
Other modules 109
Man-machine interfaces 113
Remote I/O, I/O drops, peer-to-peer, and master/slaving 113
Networks 115
Cell control 116
Distributed control systems 117
PLC software and programming 118
Keeping track of all that data 123
Programming summary 125
PID loop control 125
PLC applications 130
Troubleshooting, fault handling, and diagnostics 133
Bibliography and suggested further reading 134

6 Electric motors 145

Motor cooling 147
Protecting the motor 149
Motors (ac) 152
Motors (dc) 157
Servomotors 163
Stepper motors 167
Conclusion 168
Bibliography and suggested further reading 170

7 Motion control   *171*

Motion control standards   *173*
Old technology   *173*
New technology   *175*
Drives (ac)   *176*
Drive selection (ac)   *191*
Flux vector drives (ac)   *196*
Drives (dc)   *198*
Specialty electronic drives and soft starts   *201*
Braking and regeneration   *210*
Harmonics   *212*
Radio frequency interference   *218*
Power factor   *219*
Electronic-drive applications   *220*
Troubleshooting drive systems   *222*
Conclusion   *226*
Bibliography and suggested further reading   *228*

8 Sensors and feedback devices   *229*

Open-loop control   *230*
Closed-loop control   *231*
Speed and position sensors   *232*
Resolvers   *238*
Process control devices   *240*
Temperature control   *242*
Pressure sensing   *243*
Flow and level control   *244*
Photoelectrics   *245*
Other sensors and feedback devices   *247*
Methods of transmitting sensor data   *248*
The future of sensor technology   *249*

9 Communications   *251*

Protocol   *254*
Communication media   *255*
Transmission distances   *256*
Transmission speeds   *257*
Serial and parallel communication   *257*

Networks *260*
Wide area networks *263*
Noise *264*
Harmonic distortion *265*
EMI and RFI *266*
Grounding *266*
Power fluctuations *267*
Electrical disturbances *268*
Shielding *270*
Isolation versus nonisolation *271*
Signal conditioners and filters *272*
Bibliography and suggested further reading *273*

10 Machine vision *275*
The five disciplines of machine vision *275*
Digitized video and actual processing *284*
Online applications *284*
Offline machine-vision systems *288*
Common offline machine-vision inspections *293*
Tools *295*
Lighting techniques *295*
Fixtures *297*
Troubleshooting and specifications *298*
The future of machine vision *299*
Bibliography and suggested further reading *302*

11 Machines and system integration *303*
Robotics *304*
Other machine systems and applications *315*
System integrators *320*
Why do machines stop running? *322*
Downtime analysis *322*
Training, spares, warranty, and field service *324*
Conclusion *325*
Bibliography and suggested reading *326*

12 Industrial safety *327*
Safety labeling *328*
Laboratory testing *330*

Field labeling *333*
Performance testing *335*
Hazardous locations *335*
Emergency and fail-safe *338*
Enclosures *340*
Industrial noise *342*
On-the-job safety *343*
Bibliography and suggested further reading *344*

13 Total quality management, statistical process control, and ISO-9000 *345*

Statistical process control *347*
Quality control and statistical quality control *354*
The history of quality *354*
What is quality control and quality assurance? *356*
TQM *360*
ISO-9000 *361*
Where to turn *367*
Why get ISO certified, anyway? *372*
Who is certified and who should be *374*
How to get certified *375*
Conclusion *376*
Bibliography and suggested reading *377*

14 Where we are headed *379*

Service and support *380*
Where are we now? *382*
Future developments *384*

Appendices

A Common acronyms *387*

B English to metric conversions *391*

Index *395*

Foreword

The future brings many challenges in how we, as technicians, can network the multitude of interdisciplinary pieces that constitute today's automation systems. Although historically unproved, automation brings the fear of lost jobs along with it. The truth, however, is that individuals have more often been moved into new job classifications that required different skills. These changing classifications quickly resulted in new educational and training programs; a recent example being industrial robots. Robotics was viewed as both a threat to employment and an automation revolution—technician training programs with promises of high-paying jobs sprang up overnight. Unfortunately, these graduates were labeled Robot Technician/Specialist, which limited the individuals to a narrowly focused segment of automation that soon became a flat industry. The 1984 media hype of robotics spawned over 300 U.S. manufacturers of robots, yet only a handful of these companies exist today. Currently, robots are primarily imported and are used, along with other types of automation, at an evolutionary rate rather than the revolutionary rate predicted during the 1980s.

In the 1980s, robotics offered a new application of existing technologies, rather than a new technology. Robot technicians were thus trained in basic interdisciplinary specialties that could be applied in almost any situation requiring motion to be controlled. Interdisciplinary applications ranged from complicated automation systems to copying machines.

Technological change has historically been a result of an immediate need more often related to national defense. A combination of defense and industrial needs are once again the driving force toward developing a new technology that can ultimately result in jobs requiring a whole new educational system.

Imagine, if you will, a machine with 200 axes of controlled motion or one with only four moving parts that has a thrust capability of up to 8000 pounds and accelerations of 1.5 G. There is no need to imagine these machines because they are here today; their features

will be dominant technologies of the near future. The technological changes on the horizon include the following:
- faster control response, in the sub-millisecond range without precompiling
- doubled or quadrupled servo-loop performance
- improved sensor technology with the ability to read and react within the controller
- simultaneous sensor feedback and trajectory correction
- hardware- and software-independent communication protocol architecture features

A greater opportunity for technological creativity has not existed in the history of humankind greater than exists today. In short, advances in technology create new jobs. New jobs create new wealth, and new wealth is what made America the power it is and gives Americans the ability to constantly invest in new technology. Service-oriented jobs often result in a redistribution of existing money, rather than the creation of new wealth and new jobs.

This book explores the interdisciplinary nature of the technologies that surround how we move automation-related components more intelligently. The topics are presented in an easy-to-understand format. The reader can gain an understanding and appreciation for the motion-control package that consists of the muscle (made up of actuators), the load (often described as mechanisms), and the brain (which controls the system digitally).

The American Institute of Motion Engineers describes motion control as

"the broad application of various technologies to apply a controlled force to achieve useful motion in fluid and/or solid electromechanical systems."

It further describes a motion control specialist as

"one who understands and can integrate the many disciplines required to solve motion-control application problems."

Is it not more useful for the individual with the narrow label of robot technician to be referred to as a motion-control specialist? After all, each does the same work. Although advances still need to evolve, the technology of muscle and control loops exists for our immediate application needs. We have come a long way since the early days of robotics and limited choices of motion control.

Yes, software improvements will continue to help sell our systems and system modeling might not. Understand, however, that the effort must continue, collectively. The most important feature of the system is the individual who can integrate the motion-control solutions that enhance the productivity of America. Although the reader of the book can become smarter and, in some cases, become the entrepreneur, the ultimate goal is for the reader to become more useful.

Professors Fred Sitkins and Dr. Frank Severance,
Codirectors of the American Institute of Motion Engineers,
Professors at the College of Engineering and Applied Sciences,
Western Michigan University, Kalamazoo, Michigan.

Acknowledgments

Many thanks go out to the following people:
- Shawn Daly of Intertek, an Inchcape Testing Services Company
- Umbereen Mustafa of ETL Testing Laboratories, an Inchcape Testing Services Company
- Steve Crang and Charles Seifert of View Engineering, Inc.
- Fred Sitkins and Frank Severance of the American Institute of Motion Engineers
- Eric Carlen of Carlen Controls
- Jerry Gerloff of Bailey-Fischer and Porter
- Russ Berstein of Amprobe Instrument, a Division of Core Industries, Inc.
- Terry Loftis of Micon, a subsidiary of Powell Industries
- Pete Cheuvront of Fisher Communications
- Roger Boss of Adept Technology, Inc.
- Joan Rickards of Two Technologies, Inc.
- My wife, Colette

Introduction

Walk into any factory, plant, or manufacturing facility and look around. What do you see and hear? Many machines and processes in noisy motion. If you are on a tour through the plant, you will most likely notice the final product and some of the components that constitute the final product. If, however, your purpose at the plant is because you are a technician, engineer, or new employee, this book is for you. This book is for the people who keep the factories and the plants producing.

When I saw my first machine, I was in awe. I had so many questions and so few answers. I felt embarrassed to ask stupid questions about the computers, mechanics, and electronics. I thought that the answers would come later, and they did, but it took time and experience. This book is the book I looked for 10 years ago to give me an edge, to help me better understand industrial electronics. The book did not exist then and, until now, is not in existence. This book is for all those practical, technical automation people and the unanswered questions they have.

Technician's Guide to Industrial Electronics: How to Troubleshoot and Repair Automated Equipment covers the many hot, important, and crucial topics and issues found in today's factories. This book is the automation glossary and dictionary for all the acronyms, hard-to-understand terms, and nebulous matters facing the factory technician.

Technician's Guide to Industrial Electronics is practical and written so anyone can understand. It contains few formulas (although some basic ones must be used), nor is it a "math-heavy" textbook. Rather, it is for the average industrial worker so he or she can keep the plant producing. It is for the plant electrician, the mechanic, the machine operator, the plant engineer, and the plant manager. Finally, this book is for the sales engineers who might know their particular product well but want to understand all the processes, machines, and peripherals that go into producing their product.

Each chapter includes its own charts and tables. Similarly, each chapter has its own bibliography and suggested reading list. Other tools included with this book are various forms and logs for charting machine downtime data, I/O logging, preliminary ISO-9000 auditing, and others. Some chapters include actual specifications and troubleshooting trees for various equipment types.

Chapter 1, "How We Got Here," analyzes early developments that contributed to automation today. Microprocessor and power semiconductor technologies are briefly discussed. This chapter also takes you through a typical automation project from concept to installation—step-by-step.

Chapter 2, "Basic Electricity and Electronics," covers ac and dc power circuit basics and explains Ohm's Law. Types of switches, relays, diodes, transistors, and other electrical components are discussed, as are transformers, fusing, and test equipment. The topics of frequency, amplitude, and power semiconductors are included, as well as a list of the standards organizations and their addresses is also furnished.

Chapter 3, "Power Transmission Basics," explains converting electrical energy into power and mechanical energy, otherwise known as power transmission. Horsepower versus torque is clarified. Hydraulics, inertia, and pneumatics are discussed, along with clutches and brakes. This chapter also covers ball screws, gears, geartrains and gearboxes, couplings, and motors. The concepts of gear reduction, friction, and efficiencies are explored, and applications of fans and pumps are reviewed. The chapter also compares different mechanical methods of variable speed control.

Chapter 4, "Industrial Computers," covers microprocessor basics, software, hardware (including surface-mount devices), computer memory (including flash ROM), and I/O devices. Uploading, downloading, and CAD/CAM basics are discussed. Applications for these concepts include hotspotting, client/server systems, CD-ROM, touchscreens, and small hand-held computers.

Chapter 5, "Process Controllers and PLCs," provides a complete look at programmable logic controller (PLC) hardware, configurations, and software used in process control. Proportional-integral-derivative (PID) loop control, data acquisition (SCADA), distributed control system (DCS), and work cell control are included. The chapter starts with a look at relays, relay logic, and ladder logic basics, and covers Boolean algebra. Applications and troubleshooting are also discussed.

Chapter 6, "Electric Motors," covers electric motor basics, including design, construction, heat, weight, slip, insulation, and protection.

Introduction

The ac induction motors, dc motors (shunt wound, series, and permanent magnet), steppers, and servomotors are covered in detail.

Chapter 7, "Motion Control," is the most significant chapter in the book because so many disciplines evolve around motion control. The ac and dc electronic drive controller types are compared extensively. Regen vs. non-regen, four quadrant, flux vector control, synchronous, and spindle drive techniques are covered. Drive harmonics (input and output) is addressed along with EMI, RFI, and noise. Braking techniques and reduced voltage starters are also discussed, along with many variable and constant torque applications. Speed control and servo position controllers are also included.

Chapter 8, "Sensors and Feedback Devices," shows how that without these components, automation cannot exist. Encoders, resolvers, and tachometers are covered regarding types, signals, and the hardware itself. Other topics, such as transducers and photoelectrics, are covered in detail.

Chapter 9, "Communications," covers such topics as talking to computers on the plant floor. This chapter begins with the communications basics of serial and parallel circuits. LANs and WANs and fiberoptic networks are compared. Electrical noise, ground loops, EMI, RFI, and isolation and related topics are defined along with recommendations on how to avoid them. Safe operating distances and filtering are also covered.

Chapter 10, "Machine Vision," discusses machine vision and image processing basics. Online and offline systems are fully defined, along with basic metrology and fixturing methods.

Chapter 11, "Machines and System Integration," explores the field of robotics. The articulated arm, SCARA, gantry, and Cartesian robots are discussed. In addition, new machines, old machines, retrofits, and various control/machine applications are reviewed. The question "Why do machines stop running?" is analyzed from the point of troubleshooting. Downtime analysis, training, and service are also mentioned.

Chapter 12, "Industrial Safety," covers safety labeling, certification, field labels, product testing, and performance testing. It defines emergency (fail-safe and intrinsically safe) and hazardous locations (Class I, II, and III) and discusses audible noise issues.

Chapter 13 covers total quality management and ISO-9000. TQM and ISO-9000 certification are currently very important issues for manufacturers. Companies all over the world are seeking certification. This chapter defines the standards, how and where to seek certification, what it costs, and what not to do. Statistical process control techniques are also covered.

Chapter 14, "Where We Are Headed," concludes with the everyday questions concerning the many disciplines and products involved in factory automation. Is ac or dc better? Will there be PLCs in the future? Why is good customer service hard to find?

The appendix lists most of the acronyms and their definitions found throughout the book, as well as the necessary charts and conversion rules of thumb to help make the technician's job easier.

Automation is happening all around us, not only in industry, but also in the home, office, schools, and hospitals. It is happening quicker than we would like to imagine. Now is the time to gain a basic understanding so you can grow with the technology rather than resist it.

1

How we got here

Since the beginning of time, humans have always tried to make things easier for themselves by doing routine tasks quicker with less personal labor. For example, the invention of the wheel made things much better—large objects could be moved with fewer people and less work. Perhaps the wheel was the very first automation-driven occurrence. It is certainly the basis for most motion today and is explored in later chapters. First, though, we must ask, what is automation?

What is automation?

Automation is industrial electronics. Automation is using electricity and computers, the essence of electronics, to operate a factory. Throughout this book, automation and industrial electronics are used interchangeably. In addition, as we look further into the elements that compose the electronic equipment, we find that much evolves around motion and motion control. First, let's look at the history and evolution of automation and electronics.

Automation can be defined many ways. The dictionary defines automation as "to make automatic or have certain parts act in a desired manner at a certain time." To control the event, it should be repeatable and consistent, which is the basis for industry—seeking the perfect system and bettering the elements that create the event. All this improvement is conducted in a capitalistic society for the betterment of productivity and profitability.

Automation, as a term, supposedly came from the automobile industry's efforts to build cars quicker and better. That industry's approach was to use the best available technology and equipment for the time. This approach still applies today. The automobile industry certainly wasn't the first to try its luck at automating, but it is the automation trendsetter to this day.

The Industrial Revolution

The automation of the car industry began in the 1940s, but we have actually been automating processes much longer than that. During the Industrial Revolution in the eighteenth century, machine and automation precedents were set. Eli Whitney's gun, made of interchangeable parts, and his cotton gin, which allowed for the increased production of cotton, are good examples. Although powered by man, animal, or water, the cotton gin was a step in the right direction: repeat the process, gain more output of product, and ultimately free valuable labor for other tasks. Eventually, as demands on production increased, machines came to be powered by other means, such as steam.

These mechanical machines existed until after World War II. At this time, a new industrial revolution took place. The baby boom was in full swing and Americans were ripe to produce a strong economy. This time, however, electricity played a big role, and one might call this era the time of electrical automation, the Electrical Revolution. Americans were embarking on a new frontier of electronics and control.

Major electrical companies emerged in the late 1940s and early 1950s. Alternator and generator devices were being developed to replace steam- and combustion-powered motion in plants. Electric motors were now coming into their own. After all, something had to move in any factory to make a product, and to move something, power had to exist behind the movement.

Some industries relied on manual labor and brute force to create their product. A prime example is the steel industry (its reinvestment of profits into automation happened too late). Many other industries, however, were looking to machines for better *throughput* (a key term) and a more competitive product.

As equipment became more sophisticated in the plant, so did the controlling of the equipment. First, raw materials came into the plant, then came fabrication, assembly, testing and inspection, finishing (with more inspection), packaging, and finally shipment of a finished good—thoroughly inspected. The process repeated many tasks. Industry then tried to control the process better and keep better record of all the happenings during that process, using crude electric relays, which controlled when an event happened, and rheostats, which monitored the speed of each step of the process. To record items in the process factories used more paper, more reports, and more physical inspections because it was the best technology available—until the 1970s and early 1980s.

The digital revolution

A couple of key components were created during the 1970s and 1980s that thrust the world into a new industrial/automation revolution. The first was the microprocessor; it changed everything then and is still changing everything today.

The microprocessor allowed more information to be processed quicker and more consistently. The microprocessor chip opened the door to a digital environment. No more large, bulky cabinets in the factory that housed countless relays and analog devices. The cabinets became smaller because the world of digital electronics was upon us. It was happening in the office, too! Less paper (at least in theory) was needed! New methods of tracking large amounts of information were emerging and so were the geniuses to create them. This small yet powerful component would prove to be the largest most important development of the century.

The microprocessor was used in computers, programmable logic controllers (PLCs), and motion controllers, just to name a few of the more important pieces of control equipment. One would be surprised at how many microprocessors are in the average factory today! They are everywhere and in almost every electrical product, many of which are explored in more detail later. The microprocessor chip spawned two complete industries: the computer software and hardware industries.

The second key component developed during this time was the transistor. Although actually invented in the late 1940s, the transistor's impact was not felt until much later. This semiconductor provided rectification and switching of power at higher current ratings. The issue of efficient use of energy, a huge concern in the 1970s, gave the transistor developmental "push." The transistor provided the means by which energy-saving products could be incorporated into the factory. Power-consuming processes were now looked at closely and plant efficiencies reached new heights as industries tried to get as close to 100 percent as possible.

Controlling motion

As companies looked to the microprocessor to aid in their production, they had to address motor control and power demands at the same time. Transistors, diodes, and integrated circuits, all using semiconductor technology, could convert, or rectify, power to gain control of motors (ac and dc) in the plant. Control meant controlling the speed without wasting energy when powering a motor full speed, as well as using other devices such as throttling valves or dampers to lower throughput when required.

Power semiconductors allowed for this phenomenon. Semiconductors are crystal-like solids that exhibit a "middle-of-the-road" posture as to electrical conductivity: they are neither a good conductor nor a good insulator. Once chemically treated, however, their conducting capabilities can change to whatever is required. For example, the silicon-controlled rectifier (SCR) can be manufactured to have a specific turn-off time when used as a switching device in a power circuit. As time went on, the transistor became the mainstay, creating a new industry for electrical drives and servo systems.

In industrial automation, the control and power components might be called the *primary elements*, but several *secondary elements* also exist, including the power transmission components, the feedback systems important to making a process self-contained or closed-loop, communications between physical devices, the software or programs that instruct the otherwise dumb devices, and many others. It takes many pieces to complete the industrial automation puzzle, most of which are detailed in later chapters.

Automating the process

The technology relative to industrial automation is changing so fast that equipment can become outdated as soon as it is put into service. For example, if a plant decides that it needs to automate to gain a competitive edge or manufacture a new product, the first step is to begin designing. Many individuals must contribute and be consulted for the design, from the operator of the old machine on how to do the task better to the utility company on how much and what type of electrical power is required. No single person has all the answers. What is needed is an industrial-electronics guru, someone with a degree in automation engineering with several years experience designing and implementing systems. These individuals are very rare.

Next, a thorough evaluation of present technology with respect to the equipment needed for the project occurs. This evaluation can take months, depending on how thorough the evaluation is. Finally, equipment decisions are made and the final design is completed. Orders are then given to vendors and subcontractors.

Each vendor and subcontractor has its own build and delivery cycle, which requires more time. Soon, equipment, componentry, hardware, and software start to arrive. If everything goes as planned, nothing arrives damaged or too late. Most often, however, one or the other happens to every precisely planned or coordinated project.

Automating the process 5

Finally, it is time to install the new equipment. Installing all this machinery and peripheral componentry also takes time. The installation process is scheduled around planned downtime within the factory (maybe over a holiday). More valuable time can be lost during the downtime. Once all the equipment is onsite, work can commence. Timing of contracted help (integrators, programmers, vendors, etc.) is crucial. If the decision is made to have something electronic tested on the machine or process, all wires must be landed. Also, if one piece of hardware is missing or if the other technicians are not ready, bringing in an outside specialist is futile.

Once installed, crucial phases of the process can be started, such as trial runs and equipment debugging. After the test phase comes the day when the first production item can be run off the assembly line, and everyone is relieved and happy. Unfortunately, their joy is short-lived if they seriously look at then and now. A couple of years have elapsed from the design phase, and many technological changes have been occurring during this time. The plant probably has an outdated machine even before it ever runs its first production cycle. In addition, this investment must continue to produce for several years to recoup the initial costs!

The microprocessor industry is changing so fast that it almost isn't fair. Once the decision is made to purchase computerized, electronic equipment, it is virtually outdated when it ships. The remaining elements for the machine and process have longer useful life expectancies as far as technology goes. The computer equipment can run for many years, but it will be notably deficient in capability within a few years, or possibly even months. This comparison can be seen in Fig. 1-1, which shows the useful life expectancy of many factory electrical products versus their initial costs. This comparison is tempered by technological changes and developments.

Nevertheless, this is the automation process. Many factory projects have more aggressive design and install schedules. Premiums can be paid for accelerated schedules. Some large projects still take longer. The scenario for industry is simple: make the decision to automate now or be forced to do it later.

At some point in time, every company must automate in some form. Almost all companies have been automating along the way in some form and probably do not even realize it. Equipment is purchased with electronic capabilities that are not even used. Many times, equipment is purchased with the intent to link it or integrate it later to a large-scale electronic investment. The key here is not to wait too long, or the technology will change dramatically and what once seemed like a good idea is not worth pursuing. In the proceeding

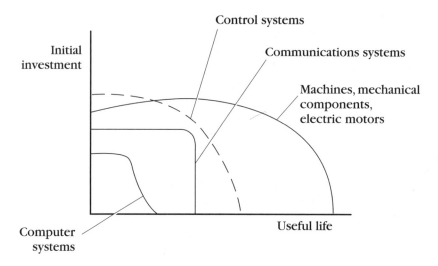

1-1 *The cost of electronic and industrial equipment versus its useful life.*

chapters, I look at most of the crucial electronic and electrical elements of industrial automation, because they all interact with one another in some way.

Industrial electronics is an exciting and ever challenging field. Technicians cannot rest on their laurels when it comes to a complete understanding of the field. Industrial electronics is one of those fields where the learning cannot stop. Keep your eyes and ears open, and all the processes and machines in today's factories will keep producing!

2

Basic electricity and electronics

Four years of dedicated college education can typically earn a person a bachelor degree in electrical or mechanical engineering. Depending on the school, the engineering curriculum might offer a focus on industrial automation. Some specialized schools might even have a dedicated program primarily for those interested in automation and robotics engineering. What is really needed is a degree program in automation engineering.

No matter the particular education received, the reality is that the engineer or technician fresh out of school must learn on-the-job, because technology changes daily. The engineer or technician must quickly become an automation specialist. The specialist becomes the company's expert on certain equipment and machines. For example, specific controllers, made by a specific manufacturer, are programmed and maintained by the one person in the plant most familiar with them. It is not believed to be worth another individual's time nor is it even practical to become trained on every complex piece of equipment.

This philosophy of training is changing, however, using the team fundamentals of total quality management (TQM). Certain plant people do become specialists (and get the phone call at home at 3:00 AM on a Sunday to troubleshoot a failing system). They also become fast learners in disciplines other than the one in which they were trained. For instance, an electrical engineer might be asked to size a motor for a machine because the machine is an electrically powered device. The electrical engineer must quickly determine the torque and inertia and what their relationship to speed means. The electrical engineer might also need to select a coupling and a gearbox, which is not really that difficult, but can be frightening to an inexperienced engineer.

On-the-job experience for the engineer does not end with an occasional and casual off-discipline assignment. Most electrical equipment has some microprocessor content to them. It might involve anything from the engineer programming the product to simply knowing that a chip is in the product. This chapter cannot begin to cover even the tip of the iceberg as to the amount of knowledge necessary to be proficient at one's job in the factory. The intent is rather to supply relevant facts and be used as a reference for those issues that occur from time to time in the workplace.

Standards organizations

Numerous authoritative sources provide information and standards on their products to provide a common ground on which industry can operate. Suppliers and users of electromechanical equipment need rules and standards to design and manufacture a product. Although standards exist for sizes and ratings of most hardware, it is difficult for plants to maintain standards on other equipment because technology is changing so rapidly.

The main organizations to which plant and factory personnel turn to for reference are large and all-encompassing, while others are smaller, specialized associates. For example, NEMA, the National Electrical Manufacturers Association, is probably one of the largest groups of its kind. IEEE, the Institute of Electronic and Electrical Engineers, is another. Both have large memberships and have existed a long time. Their guidelines are referred to daily by industrial and factory personnel. Still, other organizations must also be considered. ASME, the American Society of Mechanical Engineers, has its own set of standards. Even though ASME focuses on mechanical engineering issues, it has recognized that the electrical content of factory work has increased dramatically over the past decade.

The National Electric Code (NEC) and National Fire Protection Agency (NFPA) are both used extensively when engineers and technicians must know the electrically safe procedures within the plant. Which organization is ultimately correct? The largest? Which one has the final say? Often, these organizations merely provide recommended guidelines or practices. Sound judgment by previous decision makers who have the utmost concern for human safety is a necessary part of the formula. Agencies and organizations only provide the standards; they cannot make the day-to-day decisions. The standards have been established by members of the industry who meet periodically and discuss issues and set up guidelines.

Standards organizations

Many smaller specialized organizations are also good sources for information. One such smaller group is the American Institute of Motion Engineers (AIME), which focuses on motion control. Another is the Electrical Apparatus Service Association, or EASA. EASA focuses on service and repair of electrical products. The following is a list of all the organizations mentioned here, along with their respective addresses:

American Institute of Motion Engineers (AIME)
Kohrman Hall
Western Michigan University
Kalamazoo, MI 49008
(616) 387-6533

American National Standards Institute (ANSI)
11 West 42nd Street
New York, NY 10036
(212) 642-4900

Standards Council of Canada (CSA)
1200–45 O'Connor
Ottawa, Ontario K1P6N7
(613) 328-3222

Electrical Apparatus Service Association, Inc. (EASA)
1331 Baur Boulevard
St. Louis, MO 63132

Electronic Industries Association (EIA)
2001 Pennsylvania Avenue
Washington, DC 20006-1813
(202) 457-4919

ETL Testing Laboratories, Inc. (ETL)
Industrial Park, P.O. Box 2040
Cortland, NY 13045
(607) 753-6711

Institute of Electrical and Electronic Engineers (IEEE)
445 Hoes Lane
Piscataway, NJ 08854
(908) 562-3803

National Electrical Manufacturers Association (NEMA)
2101 L Street, NW
Washington, DC 20037
(202) 457-8400

National Fire Protection Association (NFPA)
1 Batterymarch Park
P.O. Box 9101
Quincy, MA 02269-9101

National Institute of Standards and Technology (NIST)
Building 221/A323
Gaithersburg, MD 20899
(301) 975-2208

National Standards Association (NSA)
1200 Quince Orchard Boulevard
Gaithersburg, MD 20878
(800) 638-8094

Robotic Industries Association (RIA)
900 Victors Way, P.O. Box 3724
Ann Arbor, MI 48106
(313) 994-6088

Society of Manufacturing Engineers (SME)
One SME Drive, P.O. Box 930
Dearborn, MI 48121
(313) 271-1500

Underwriters Laboratories (UL)
333 Pfingsten Road
Northbrook, IL 60062
(708) 272-8800

Symbols

Besides standards organizations, electrical symbols have also become standardized. A set of standard symbols allows people in the industry to graphically convey ideas. When showing an electrical configuration graphically, one typically uses a single-line drawing to convey a meaning or concept. Once presented, this single-line drawing can evolve into a complete set of drawings with wiring diagrams, all terminations marked, and schematics. *Schematics* are electrical diagrams and can include wiring drawings.

A symbol library in the electrical world consists of the basic elements: resistors, capacitors, inductors, etc. Beyond these symbols, standard symbols exist for items used in the microprocessor and power conversion industries. Figure 2-1 lists several common symbols and their appropriate names. These symbols can be used in a single-line diagram to describe a desired circuit, etc. They are used throughout this book to illustrate different circuits and electronic devices.

Electricity

Electricity and magnetism are the basic elements of the factory. They were once thought to be two separate forces, but Albert Einstein's

Electricity

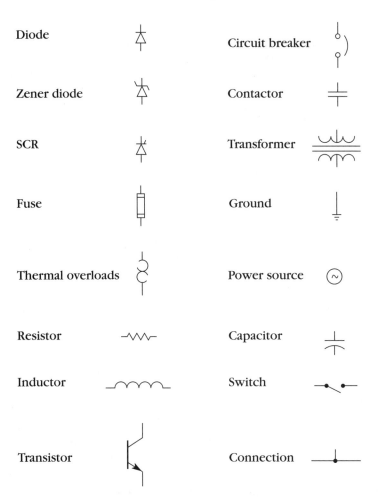

2-1 *Common electrical symbols used in single-line diagrams.*

theory of relativity showed that the two were very much interrelated. Electricity is produced from electric charges, whereas magnetism is created by charges in motion, reacting to moving charges. Electromagnetic forces and magnetic flux are all a result of the presence of electricity.

Most of what occurs in the factory is a direct result of electricity. Electricity is a form of energy that consists of mutually attracted protons and electrons (positively and negatively charged particles). Therefore, in every factory, electrons flow to make products. This electrical function might be apparent, but might not. Communications is a form of low-level electricity traveling over conductors to pass data. Motion, on the other hand, is caused by an electric prime mover, or a motor generated by higher levels of electrical energy. Everything can be traced back to the power-producing utilities.

Once the power enters the plant, it is distributed via *transformers*. These devices actually provide the usable voltages—460 V, 230 V, and 115 V being the most common. In addition, lower voltages, such as 24 V and lower, are used for most computer systems. Most of the equipment commercially available in the United States requires one of the supply voltages mentioned. Some voltages, such as 575 V, are used but are not as common. In Canada, 575-V power is very common. There is always some piece of equipment or odd supply that is an unusual voltage, which is where transformers come into play. Transformers are discussed further later in this chapter.

Electricity is either found in direct current (dc) or alternating current (ac). Electric motors and other pieces of equipment can run from a dc source or an ac source. Numerous discussions have taken place on which form of energy is better, but the answer is both have their advantages and disadvantages. Throughout this book, ac and dc power and equipment are discussed. Every application should be treated as an individual case when deciding which type of electrical supply is appropriate. Typical industrial power is provided as three-phase, and its frequency is normally 60 cycles, or hertz (Hz), in North America. Power of 50 Hz is typically seen in Europe.

An electrical circuit consists of three elements: voltage, amperage, and resistance. Voltage, usually shown as V or E, is the force that causes electrons to flow. Amperage, or current, is expressed as I or A and is the actual flow of electrons. It uses the unit of measure of amps. Resistance, R or ohms, is the opposition to current flow. These three elements constitute Ohm's Law, from which many basic electrical circuit calculations can be made. Ohm's Law is adequate for all dc circuit analysis and some ac circuit analysis. Three-phase power, however, tends to be a bit more complicated. Figure 2-2 shows the VAR pie chart, which is useful when trying to remember the following equations:

$$V = AR \text{ or volts} = \text{amps} \times \text{ohms}$$

$$A = \frac{V}{R} \text{ or amperage} = \frac{\text{volts}}{\text{ohms}}$$

$$R = \frac{V}{A} \text{ or ohms} = \frac{\text{volts}}{\text{amps}}$$

When using the pie chart, the V value is shown above the A and R values, which represents V being divisible by either A or R when

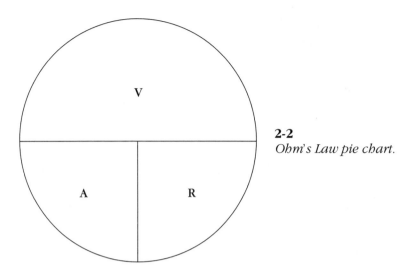

2-2
Ohm's Law pie chart.

solving for either A or R. Likewise, to find V, multiply A by R, as is true of the actual equation. The pie chart is an effective and easy way of remembering the aforementioned equations and their relationship to each other rather than memorizing the separate formulas. Ohm's Law allows the technician to quickly determine whether or not an immediate discrepancy exists.

Electrical formulas

Many derivations to Ohm's Law exist, along with other laws of electricity that must be understood. Electrical engineers are usually familiar with all facets of circuit design and analysis and, consequently, the plethora of calculations that can be performed involving higher levels of mathematics and calculus. Several analyses and transformations are used to prove and disprove certain issues. For our purposes, however, the basics and some rules of thumb are the tools by which the factory machines are kept running. I do look at certain analyses, postulates, and transformations, however, to help you understand how and why they are important to the average plant employee.

One such transformation is called the *La Place transformation*. It is a mathematical integration used in harmonic analyses and other issues related to electricity, magnetic, and gravitational concerns. Another is the *Fourier analysis*, which is also useful in harmonic and waveform investigation. The names for these analyses come from famous mathematicians, and many engineers and scholars refer to these theorems. Another electrical law is *Kirchoff's Law* of current and voltage for ac and dc circuits.

Electrical power (P), or the rate of doing work, is measured in watts and is expressed as a formula for dc circuits as

$$P \text{ (watts)} = V \times A$$

We need to differentiate between equations for dc and equations for ac circuits mainly because of steady-state conditions of dc and single-phase ac and the nonsteady-state conditions (such as three-phase ac power). Circuits that are ac must be analyzed differently when three phases of power are involved: ac power must be averaged and root mean square (RMS) values provided to get proper results. An example of determining power for a light bulb is as follows: A lightbulb is nothing more than a resistor, so its current can be quickly determined using the P = VA equation. If the light bulb is 100 W and the supply voltage is 120 V, then

$$I = \frac{100 \text{ W}}{120 \text{ V}} \text{ or } 0.83 \text{ A}$$

Power on a larger, different scale is discussed in terms of horsepower. Horsepower for a given dc circuit is equal to

$$\frac{\text{volts} \times \text{amps} \times \text{efficiency}}{746}$$

Because production is a direct relationship to work in the industrial plant, electrical power usage is also a gauge. Factories today, however, try to achieve the absolute best output for the electricity they purchase. Older mills and factories have enormous electrical bills. The goal now is to have equipment as close to 100 percent efficiency as possible. This goal is factored into automating any plant, new or old. The equation

$$\text{efficiency} = \frac{746 \times \text{output horsepower}}{\text{input watts}}$$

can be used for ac circuits and is a good indicator of where a particular process or piece of equipment is relative to its cost and productive output.

The incoming power to the plant is usually three-phase ac. Table 2-1 lists common formulas for power in ac circuits. Single-phase power is also available, especially at lower voltages, and is used for lighting and office circuits. The bulk of the work is powered by three-phase systems. Motors are designed and built around three-phase supplied power, although many single-phase motors are in use. One good question for thought is from a design standpoint: why is three-phase

Table 2-1. Common electrical formulas

Single-phase horsepower = $\dfrac{\text{volts} \times \text{amperes} \times \text{efficiency} \times \text{power factor (PF)}}{746}$

Single-phase power factor = $\dfrac{\text{input watts}}{\text{volts} \times \text{amperes}}$

Single-phase efficiency = $\dfrac{746 \times \text{horsepower}}{\text{volts} \times \text{amperes} \times \text{PF}}$

Single-phase amperes = $\dfrac{746 \times \text{horsepower}}{\text{volts} \times \text{efficiency} \times \text{PF}}$

Single-phase kilowatts = $\dfrac{\text{volts} \times \text{amperes} \times \text{PF}}{1000}$

Three-phase horsepower = $\dfrac{\text{volts} \times \text{amperes} \times 1.732 \times \text{efficiency} \times \text{PF}}{746}$

Three-phase power factor = $\dfrac{\text{input watts}}{\text{volts} \times \text{amperes} \times 1.732}$

Three-phase efficiency = $\dfrac{746 \times \text{horsepower}}{\text{volts} \times \text{amperes} \times \text{PF} \times 1.732}$

Three-phase amperes = $\dfrac{746 \times \text{horsepower}}{1.732 \times \text{volts} \times \text{efficiency} \times \text{PF}}$

Three-phase volt-amperes = volts × amperes × 1.732

Three-phase kilowatts = $\dfrac{\text{volts} \times \text{amperes} \times \text{PF} \times 1.732}{1000}$

Efficiency = $\dfrac{746 \times \text{output horsepower}}{\text{input watts}}$

used over single-phase? One important factor is conductor size. If single-phase motors in the hundreds of horsepower sizes had only one wire running to it, the actual diameter of that wire would be immense. Large wires are very difficult to work with and route in a factory.

Thus, three-phase systems, which incorporate three individual wires, are prevalent. The current is spread out, or averaged, over all three wires. The three-phase incoming power can be rectified into dc power to run dc motors and such, which is discussed in more detail later in this book. Also, dc is often the only power choice in remote locations when oil or diesel generators are producing the power.

Obviously, there is much to know about electricity and electrical calculations. The end of this chapter contains several good reference sources for more information on this subject. The discussion in this chapter is more of basic understanding, terminology, and what aids are available when applying and troubleshooting electrical systems.

Frequency and amplitude

The power that is supplied to the industrial plant is single- or three-phase and has a voltage rating. The other factor is frequency. *Frequency* is the amount of electrical pulses transmitted over a given period of time. Frequency is expressed in hertz (Hz). For example, most power in the United States is 60 Hz, which means that 60 pulses of electricity go through a given point in a wire every second. Figure 2-3 shows a typical waveform depicting the frequency and the *amplitude* of the wave. This wave is actually a sine wave. (Along with the sinusoidal wave form shown in Fig. 2-3 are examples of square waves and triangular waves.) For every electrical wave, there must be a corresponding amplitude to provide any usable power. This amplitude is often in the form of voltage. This wave has two time-based components, *periods* or *cycles*. A half-period is one of the halves of the wave and is sometimes called a *half-cycle*. It is typical for ac.

Outages are sometimes expressed in cycles, or portions of a second. If the lights flicker, most likely either the amplitude for a given cycle dropped or a complete cycle or wave was nonexistent. Not many look at the situation in the light. All they know is that the lights flickered.

As mentioned before, some remote locations might depend on generators or even battery power (dc). In other instances, three-phase ac power must be converted to dc to run dc motors and dc equipment. In yet other instances, dc power must be inverted back to ac. There is much to know about power conversion, or rectification, prior to any real automation occurrences.

Electrical hardware

This section looks at some of the components that might be found in the factory today, especially in the common electrical circuit. These devices all perform a specific function and are commonly seen throughout the automation process.

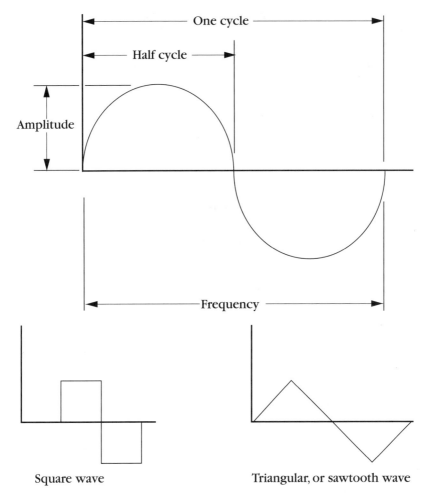

2-3 *The frequency and amplitude of a sine wave. A square wave and triangular shaped wave are also shown.*

The most prevalent electrical component seen in the factory is the resistor. A resistor's ratings are in ohms, or R, and can take on many shapes and sizes because electricity is flowing through so many different devices. Resistance exists in many areas of the plant we do not often consider. Lights, motors, and most other electrically driven components have some resistance value. On a smaller scale, one can find hundreds of small resistors on the common printed circuit board (PCB). Dynamic braking resistors, which might be housed in an enclosure because they are so much larger and have higher voltages flowing through them, are also commong in plants. Even an electric motor has a resistive value in the overall plant electrical circuit scheme. Over time, some of these resistive values change with repeated heating and cooling of the components. Resistors play a big role in making electrical and electronic devices perform as desired because impeding the flow of electricity is sometimes necessary.

Parallel and series

This is a good time to look at *parallel* and *series* circuitry. Many electrical devices, such as resistors, must be strategically located in an electrical circuit. Subtle differences between ac and dc circuits exist when using series or parallel schemes. Basically, the placing of resistor valves in a circuit can provide solutions for the electrical personnel in the plant. For example, Fig. 2-4 shows a series circuit. If $R_1 = 6\ \Omega$ and $R_2 = 8\ \Omega$, the total resistance in that circuit is $R_1 + R_2$, or $14\ \Omega$.

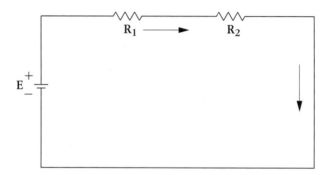

2-4 *A typical series circuit. Shown are two resistor values in series with one another.*

Conversely, in the parallel circuit shown in Fig. 2-5, the same values for R_1 and R_2 exist. Electrically, current flows differently here (the paths of least resistance), and we calculate the total resistance accordingly:

$$R = \frac{(6)(8)}{6 + 8}$$

$$= \frac{48}{14}$$

$$= 3.43\ \Omega$$

In one scheme, we can get $14\ \Omega$ of resistance, and in another $3.43\ \Omega$.

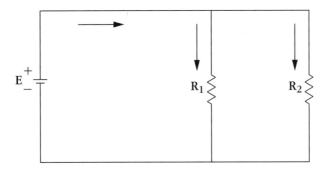

2-5 *A typical parallel circuit. Shown are two resistor values in parallel with one another.*

Capacitors, relays, diodes, and inductors

Other electrical components can be incorporated into either type of circuit. One of these is the *capacitor*. This component is capable of storing electrical energy. It is sometimes referred to as a *condenser*, as it consists of two conducting materials separated by a dielectric, or insulating, material. The conducting materials get charged, one positively and one negatively, thus creating a potential between them. The size and type of material of the conductors, the distance between them, and the applied voltage determine the capacitance in farads (one coulomb/volt). A *coulomb* is the basic unit of electrical charge. One farad is a large amount of capacitance; common capacitances are in themicrofarad range. Capacitors are used in circuits for tuning and as batteries to store electrical power.

When discussing electronics, it is necessary to include relays, diodes, and inductors, as they all play an integral role in the automated factory. Relays have been used for a very long time to control machines and processes. Relays work on the principle of electromagnetism, allowing the relay to function as a device that can either automate or control any electrical flow from a remote location. There are basically two types of circuits, the *relay circuit* and the *energizing circuit*. These work together to open and close a switch. Enough energy is provided via the relay circuit to magnetize the appropriate element and complete the other circuit. The other circuit can be and usually is a higher voltage and carries more current than the relay circuit. The operation of the common contactor can be compared to that of the conventional relay.

The relay is still very much an integral part of the factory automation process. Its switching is an important function in most every process, and much previous automation was by relay logic. Relay logic used interlocking relays and contactors to control machines and processes. If a machine was to start, permissive relays were used and then energized to allow that start. Today, equipment tends to be controlled by computers and microprocessors which, in effect, do the switching via software.

A *diode* is a solid-state rectifier that has an anode (the positive electrode) and a cathode (the negative electrode). These nodes allow electricity to flow in one direction only. Diodes are commonly used to convert alternating voltage to direct current. The diode, in effect, acts as a valve for electricity. Figure 2-6 shows the accepted symbol for a basic diode. This diode might be used for voltage regulation.

A reverse operating version of the standard diode is the *Zener diode*, whose symbol and characteristics are shown in Fig. 2-7. Zener diodes are sometimes called *breakdown diodes*, which allow reverse currents under breakdown conditions. This condition can be desirable to regulate voltage.

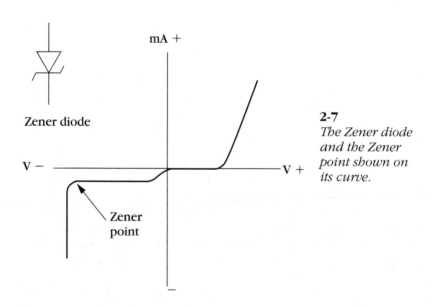

2-6
The diode with its discharging component (the anode) and its cathode component.

2-7
The Zener diode and the Zener point shown on its curve.

We have all gazed upon a light-emitting diode (LED). These red or green lit displays are very common in much of the electronic componentry in factories and in our homes. LEDs are made from materials that provide energy bands of different wavelengths for each color. The materials are made from semiconductors such as gallium-arsenide phosphide or gallium phosphide. These materials produce the red or green colored LEDs. A new LED exists, which is made from silicon carbide and is blue. When combined with red and green colors, the blue completes the red, green, blue (RGB) package, producing white light. In addition, varying current through RGB LEDs allows for many other colors, making the possibilities almost endless.

Another term relative here is *optical isolation*. Because certain diodes emit infrared light, it has become an electronics industry practice to use that capability to isolate expensive electrical components. Using the light emitter in conjunction with a light receptor requires no physical connection at all among components. Logic components in low-voltage control circuitry are thus fully separated, or isolated. The light receptor is sometimes known as a *phototransistor*. Separating components is a safe and effective means of enabling or disabling a circuit and is used frequently in controllers, optically isolating 0- to 10-volts dc (Vdc) and 4- to 20-milliamps (mA) reference signals. When interfacing the logic of one manufacturer's product with another's, isolation techniques that are compatible must be incorporated to avoid incompatibility between the equipment.

Another component often found in a factory's electrical circuitry is the inductor. *Inductors* are devices used to control current and the associated magnetic fields relative to the currents for a given period of time. Inductance is measured in a unit called *henry*, which equates to one volt per amp per second. Typically, an inductor consists of a coil around a conducting material of specific size and shape. This material is coiled around a core, most often of soft iron. For this reason, the inductor is sometimes called a *choke*. Most often, however, the inductor is called a *transformer* or *reactor*. An inductor is used to slow the rate of rising current or to suppress noise.

Transformers

Transformers, a specialized type of inductor, are an integral part of the factory and industrial electronics. Without these specialized inductors, usable electricity in lower voltage levels would be unavailable. A transformer's simplified construction and symbol is shown in Fig. 2-8. Transformers work on the principle of mutual inductance

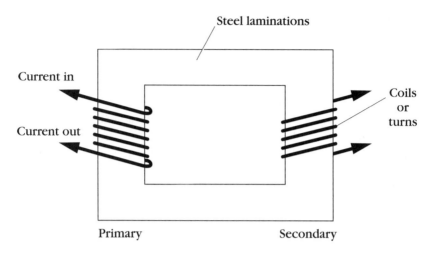

2-8 *A common transformer with primary and secondary windings.*

and, hence, are classified as inductors. For a transformer to work, there must be two coils positioned such that the flux change that occurs in one of the coils induces a voltage across each coil. Typically, the coil connected to the electrical source is the primary coil, and the coil applied to the load is the secondary one. Transformers are rated in volt-amps (VA) and kilovolt-amps (KVA). There are dry-type and oil-filled transformers, with the size sometime dictating which type to apply. There are also iron-core types, air-core types, and units called autotransformers.

Autotransformers are a type of transformer that has one winding common to both the input and the output. Instead of having isolation between the coils and employing the traditional two-circuit principle, one winding is used and higher KVAs are transformed. The cost of the autotransformer is attractive, especially when voltages that are to be matched are not that far apart.

Transformers can be used as isolation devices as well as step-up or step-down devices. It can be desirable to include a transformer with similar primary and secondary voltages to isolate one section of the electrical system from another. With no direct physical connection, conditions such as ground faults can be prevented from traveling throughout an electrical system and destroying components. The step-down or step-up transformers used to match voltages in a given system also offer isolation. Much of today's electronic equipment requires power of 460 V, 230 V, or 120 V to function properly. It is impractical to run 460-V power many miles in large diameter cables, so higher voltages are carried to substations containing transformers. The substations provide matching voltages to other equipment.

The rate of temperature rise is a constant attributed to transformers and other peripheral equipment subjected to higher electrical currents. The rate of temperature rise is defined as the allowable rise in temperature over the ambient temperature when the device is fully loaded. Ambient temperature is that of the surrounding air at the device itself, which acts as the heatsink for cooling.

When comparing two or more different transformers for the same application, a rule of thumb is to compare the weights of the units. Less iron and copper (or aluminum) in the winding content will show in the actual weight. Weight can be a factor when a transformer is used where harmonics might be present. The harmonics create eddy current losses within the transformer. The severity is hard to predict, but a transformer could overheat. To deal with this phenomenon, the K-factor value is attributed to transformers, along with the amount of harmonics it can handle.

The *K-factor*, sometimes inappropriately called the form factor, refers to the ratio of root-mean-square (RMS) current to average current. The form factor, more often is the effect of rectifiers on motors. This phenomenon is discussed more in the upcoming chapter on electric motors. Any time devices perform power conversions (ac to dc or dc to ac), some distortion to the waveform occurs somewhere. Transformers must therefore be analyzed with regard to their K-factor.

Power semiconductors

Electrical energy, or power, can be used for a multitude of applications. Many of these applications involve the rectification, or conversion, of incoming power into a different form to run a specific piece of equipment. Power semiconductors and similar devices are found in much of the rectifying equipment in plants today. Often called *thyristors*, these devices can be silicon-controlled rectifiers (SCRs), gate turn-off thyristors (GTOs), or their predecessor, the vacuum tube.

The vacuum tube, or electron tube, was the standard method for rectifying alternating current and dates back to the late 1800s. It can also be referred to as a *diode*, as it acts as a valve to control the flow of electrons. Its basic construction is that of a glass or metal enclosure from which the air has been evacuated (hence the name vacuum tube). Pins at the base of the unit provide connections to the cathode, which is made from tungsten, and supplies electrons via a filament that heats the cathode, thus emitting electrons. The anode, which is a plate, collects the electrons, and the grids control the overall emission. In industry, the ignition was used to rectify currents in higher ampacities.

Another version of the electron tube is the *thyratron*, commonly used to convert ac to dc. It is a tube filled with gas using three elements. With semiconductor technology and advanced solid-state electronics, however, electron tubes have been all but forgotten. Many old installations still depend on their rectification, but spare parts are not readily available anymore, and no one really wants to support this old technology. Tube technology is a thing of the past.

SCRs

Vacuum tubes evolved to thyristors for power rectification. Thyristors are a type of transistor in which there are several semiconducting layers with corresponding positive to negative (p-n) junctions. The thyristor is a solid-state version of the thyratron. The most common thyristor is the SCR. The SCR is still a fairly common device used in power rectification and is similar in operation to a diode. The SCR can block the flow of current in the reverse direction, much like the diode, but it can also block the flow of current in the forward direction.

The SCR has both a blocking condition and a conducting condition. In its blocking condition, or state, no current can flow through it. Likewise, in its conducting state, it acts similar to a closed switch. The SCR receives a small current signal called the *gate signal* when triggered into conduction. In the conducting state, the SCR continues conducting until the gate is removed and the current flow reduces to zero. This turn-on and turn-off function of the SCR allows for extremely good control and very small losses of current leakage. The SCR has a very good forward voltage drop value while conducting, which means that large amounts of current can flow through it with very little energy loss.

The "hockey-puck" design of an SCR is the most common. It looks like a hockey puck, usually white in color, with two leads for controlling its gate function. Very high values of current can be run through the SCR, thus making it the rectifier of choice in higher horsepower applications. Additional ratings are attributed to SCR design. The peak impulse voltage is typically in the 1400-V range. When designing the location of the heat-dissipating components of an SCR or any other rectifier system, surface area for heat dissipation, typically for the heat sink, is important.

GTOs

GTOs are power semiconductors that have self-commutating capability. Once commanded, the GTO turns on and off repeatedly without ongoing commands. These devices are available in high current ratings and have high overcurrent capabilities, thus making them very suitable for larger horsepower applications. Their turn-on and turn-off times are good, and the speed at which they switch is adequate for the typical higher horsepower applications in which they are used. A major drawback to using them is their cost, which, when compared to similar phase controlled devices, is five to six times higher than that of a conventional SCR device.

Transistors

The transistor has become the power semiconductor device of choice. These solid-state components have been around in some form since the early 1950s, but have become quite popular in the past decade. Basically, they consist of different semiconductor materials, sometimes with arsenic or boron, in conjunction with silicon. How the electricity moves through the silicon is directly related to the amount of arsenic or boron contained. Transistors provide fast switching capability for a relatively low cost. There are three general types of transistors: the bipolar transistor, the field-effect transistor (FET), and the insulated gate bipolar transistor (IGBT).

Over the years, the bipolar transistor has been used a lot in applications. It can be found in oscillators, high-speed integrated circuits, and in many other switching circuits, such as variable-speed electronic drives. It is available in rated currents six to seven times lower than thyristor-type devices. The transistors can then be paralleled in operation to achieve greater current-carrying capacity, but the drawback is higher cost. This self-commutating device also cannot withstand too much overcurrent. Bipolar transistor switching speeds, for their time, were very adequate. They were in the 2- to 4-kHz range, thus making them a very good switching device for their relative cost, even though they had marginal turn-on and turn-off capabilities. With the introduction of faster switching devices, however, bipolar transistors are being replaced.

One such replacement that is extremely fast switching is the MOSFET, or metal-oxide semiconductor field-effect transistor. It is a self-commutating device that is not too costly to incorporate into a system. Its drawback is that it is not available in current ratings much above 20 A, thus limiting its applications. It does have switching capabilities in the 100-kHz range, good overcurrent capability, and very good turn-on and turn-off conditions.

Many of today's electrical switching applications are well suited for the insulated gate bipolar transistor. This self-commutating device is available in 300-A current ratings with good turn-on and turn-off ability and switching speeds of 18 kHz. It is somewhat cost-effective to manufacture and can be implemented into an electrical circuit at a relatively low cost. With costs and performance driving the semiconductor industry, a newer, better, and faster transistor will emerge within the next couple of years. Competitive manufacturing and costs drive development.

Wiring and cabling

All the electrical components mentioned in this chapter would not be capable of performing any real function unless the electricity can get to and from each device. Wire and cable, or the conductors, is routed throughout the plant and from device to device. Most wire and cable is made from either copper or aluminum and has a specific sheath. Accepted standards exist for the coloring of the sheath for wire uses. Sizes for American wire gauge (AWG) for solid round copper wire are listed in the *National Electrical Code Handbook*. Every plant in this country with large quantities of electrical and electronic equipment should have a copy of this book. Included with the tables are resistive loss values of a select diameter of wire over distances.

Industrial cable can be used for many other purposes and can be made from other materials. Fiberoptic cables are constructed of glass fiber or plastic. Hydraulic and air-line cables can be padded, or bundled, into the same single cable to save space. There is also ribbon cable, coaxial cable, and shielded cables. Cable can be flat, round, or even come in a package made for flexure.

When selecting electrical cabling, make sure it is resistant to electromagnetic interference (EMI) and radio frequency interference (RFI), is flexible, has a relatively low capacitance, is lightweight, has a small bending radius, and can be resistant to chemicals and other nasty items found in a typical industrial environment.

Wire connections

Just as important to the size and type of cable is the actual connection, which can make or break the application. For instance, if the connection is loose, a dangerous condition exists. First, the wire can fall off and harm a person or another component. Second, heat can build up near the area of the loose connection, which can result in fuses being blown. The loose connection can also result in intermittent equipment faults and failures. It is best to ensure that the connection is made soundly and properly the first time a cable is installed.

Wiring and cabling

Specialized connectors aid in connecting multiple wires. Many are screw-down type, requiring a screwdriver be used, while others are clip-on. The multiple strands of wire can also be brought into the connector and soldered into place. In this case, only one physical connection is needed. Soldering saves time and hassle. One commonly found style of connector is the MS, or military-style connector. This connector is shown in Fig. 2-9. All the wires are connected into the connector head, in this case with a female fitting, and the head is then mated to the matching male portion with the outer housing screwed down around the actual connection.

2-9 *A military style connector on a servo motor.*

Two most common reasons given as to why an electrical component does not work are the infamous *short circuit* and the omnipresent *loose connection*. The short circuit is not always the problem, but it is worth defining. Figure 2-10 shows a typical circuit with a resistive value. The dotted lines show a path of lesser resistance where the current wants to flow freely. In these short-circuit cases, the current flow can increase to dangerous levels until a protective device is installed or another electrical component is destroyed. Wire and cabling can even melt from the heat. This problem is accompanied by a distinct smell of something electrical burning. Shorts are not as frequent as predicted and reported to be the problem. Out-of-the-ordinary occurrences must happen for a short circuit to appear. Foreign conducting debris, the presence of water (yes, water and the molecules carried in it make it a conductor), and breakdowns in motor windings are some common

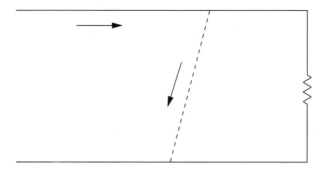

2-10 *The infamous short circuit.*

reasons for short circuits. The best solution is to place inexpensive, replaceable fusing at those pieces of equipment best protected.

A loose connection is a bit more forgiving in its malfunction properties. In the case of a loose connection, it's more difficult to locate the loose connection, but it doesn't typically create the havoc that a short circuit does.

Tools of the trade

When a machine is not working or a piece of electrical equipment is not responding, shorts and loose connections are the most frequently quoted solutions. Finding out if indeed these are the actual problems and where they are located, then figuring out how to correct them so it does not happen again, is the real solution. This solution requires a methodical troubleshooting approach and the proper equipment to help find the problem. Some basic tools that the electrician must possess in the factory include a multimeter, or volt-ohmmeter, to measure values of voltage, current, and resistance in components; a digital, stored-image oscilloscope, which can display and analyze the waveforms and then store them in memory; and a wire tracer.

These troubleshooting devices can locate energized wires, circuit breakers, and wires shorted to ground, otherwise commonly referred to as a *ground fault*. A wire tracer is shown in Fig. 2-11. It can trace wire in conduit to locate circuit breakers, fuses, and outlets in a breaker panel. Besides all the aforementioned, it can also locate buried conduit and metal pipe and find shorts, opens (where no current is flowing), and actual ground faults. It can do all these functions by transmitting and receiving various levels of electricity through conductors and, by adjusting the sensitivity setting on the wire tracer, find exactly what it sets out to find. Application notes are mentioned in

Fusing and circuit protection

2-11 *A wire tracing kit.* Amprobe Instrument, a Division of Core Industries, Inc.

this section that describe how each of the different electrical wires and problems can be located.

Other types of wiring and cabling schemes also exist. Shielding and grounding methods are extremely important, especially with low-voltage control wiring. These techniques are discussed in greater detail in Chapter 9, primarily for control wiring. Grounding is fully discussed in the *National Electric Code (NEC) Handbook* under Article 250. Methods for grounding, which equipment should and should not be grounded, and locating ground connections is addressed. For further standards and methodology regarding wire and cable of most types, consult the NEC handbook. For issues concerning sizes and ampacities, articles 300 to 400 address different applications and routing methods. As always, it is just as important to consult the local building codes for your locality.

Fusing and circuit protection

The fuse is a safety device that protects components from overcurrent conditions. The common construction of the fuse is a current-carrying strip of metal or wire that can melt in an excessive current situation. In industry, the cartridge fuse is the most common. The cartridge is placed ahead of the component to which it is designated to protect. Most often, fuses are placed at the point where power enters a given installation, component, or piece of equipment. For further clarification on where fuses should be located consult Article 240, Overcurrent Protection, Sections E and F, of the NEC handbook.

Circuit breakers are other means of protecting devices. A circuit breaker works using magnetic fields. In an overcurrent condition, an electromagnetic field draws a metallic portion of the circuit out of the circuit, opening the circuit and thus protecting downstream componentry. Unlike the fuse, the circuit breaker can be reset when the electromagnetic condition has passed and reused again. Like fusing, circuit breakers are covered in the NEC handbook under Article 240, Overcurrent Protection, Section G. Interrupting capacities, where to locate them, and using them in parallel or as switches are covered.

Electronics

Without electricity there can be no electronics. Without electricity, motion and power transmission would be extremely difficult. Ben Franklin and his kite can be credited with the present day industrial automation as we know it—the energy source! Electronics in its basic sense is the study of the motion of electrons and their control. In the world of industrial automation, it combines power, computer processing, conductors, and semiconductors, plus a host of other technologies to control the industrial automation world. From the video and audio equipment in the home to the 13,000 volts of energy generated by the power plant, electronics is making things happen. Much like the Industrial Revolution, which was a mechanical change, we are in the midst of an electrical-based change in the world today.

Controlling electrons is happening all around us in the home, office, and factory. As more equipment is based on a digital scheme rather than an analog scheme, more electrons can be controlled faster and better. As computer power increases to handle still greater amounts of digital data, so too will electronic capabilities. Imagine every appliance produced having its own microprocessor chip included! More to the point, however, the factory environment needs electronic capabilities to solidly automate. Without electronics and computer-aided devices, industrial machines and processes could not be as proficient as they are today.

Despite all the efforts to control electron flow, there are some unpleasant side effects, such as harmonic distortion. Any device that converts ac to dc power can disturb the supply line to which it is connected. This disturbance is in the form of choppy waveforms that interfere with the 60-cycle, three-phase feeder. While the jury is still out on what harmonics are and what they do, it is evident that harmonic distortion is responsible for excess noise on a particular electrical line, and it can always cause computers and peripheral-sensitive equipment to fail.

As new developments occur, there will be inevitable side effects, but these are problems we can deal with. Sometimes these side effects even create new jobs, such as electrical technicians, who are in demand and will be for years to come.

The age of electronics is upon us. The pace is 10 times faster now than it was even 20 years ago. The times are exciting and the technology is constantly changing. Industrial electronics is here to stay!

Bibliography and suggested further reading

Beiser, Arthur. *Physics*. 1973. Menlo Park, CA. Cummings.

Boylestad, Robert L. *Introductory Circuit Analysis*. 1982. Fifth Edition. Columbus, OH. Charles E. Merrill.

Matthews, John. *Solid-State Electronics Concepts*. 1972. New York. McGraw-Hill.

Matthews, John. *Experiments in Solid-State Electronics*. 1972. New York. McGraw-Hill.

Tomal, Daniel R. and Widmer, Neal S., *Electronic Troubleshooting*, First Edition, 1993, Blue Ridge Summit, PA. TAB Books.

3

Power transmission basics

The electrical energy used in a factory must be converted into work to produce anything. The conversion of electrical energy into mechanical power is called *power transmission*. Volts change to watts and then to torque as products are moved out the door. Many power transmission devices are common in the industrial automation sector. This chapter looks at many such devices, some old and some new, and how they play a key role in automation.

Although automation existed years ago, as electrical energy has been around for many years, today's factory is more dependent on digital electronics and control than in the past. Much of today's mechanical power transmission equipment has some electrical controls. Consequently, plant personnel must have a good understanding of what is occurring mechanically to assimilate electrical needs. Often, there is an electrical source to the power transmission system.

In addition, electronic controls in the factory have become more and more a part of power transmission products. Clutches, brakes, speed variators, variable speed fluid couplings, and many other devices have either a closed loop controller or another electronic means to adjust their speed. These controls are electrically fed and usually control the voltage or the current to the power transmission device. As we move forward into the twenty-first century, electronics will play a major role in power transmission. Controllers for power transmission components are discussed later in this chapter; first, I'll cover the basics of power transmission.

Power transmission

Production in the factory and work are well related. Work in its basic definition is a force acting through a distance. Work is equal to a force multiplied by distance (F × D). A force is a push or pull that causes motion of an object. The object moves in a straight line in the direction of the force applied to it. Power is the amount of work done in a period of time and is usually expressed in horsepower (hp) or watts (W) for electrical power. Figure 3-1 is an example of work and how it is expressed. It shows that if 25 lbs is lifted 10 ft, 250 ft-lbs of work have been expended.

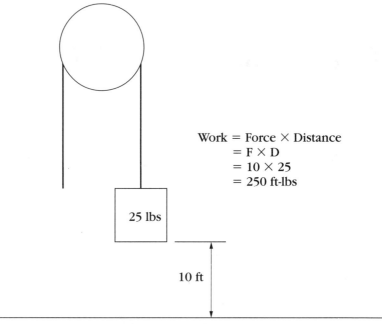

3-1 *Work equals a force times a distance.*

Power transmission can be electromechanical, which is the conversion of electrical energy into mechanical power. Power transmission can also imply that electrical or electronic devices are used in conjunction with mechanical components. Either way, as we look at work, forces, power, and torque, many of these terms seem to be interchangeable. The fact is, many times they are substituted in dialogue to get a point across. If everyone using the terminology had a good, basic understanding of each term, there might be less confusion and maybe a few less problems in the factory with respect to machinery and automation. For instance, as you'll see in more detail later in the chapter, torque and work are very similar. Work is a force multiplied by a distance, whereas torque is a force multiplied by a radius. Close, but yet different. It all depends on where the term is applied and what is trying to be accomplished.

Obviously, the most significant device used to convert electrical energy into torque is the electric motor. Electric motors are discussed in greater detail in a later chapter. This chapter is more concerned with how that torque is transmitted from the motor shaft to get the actual work done. Figure 3-2 illustrates a typical mechanical drivetrain and its associated components. Each of the individual pieces are further analyzed later, but a basic understanding of torque and horsepower is necessary first.

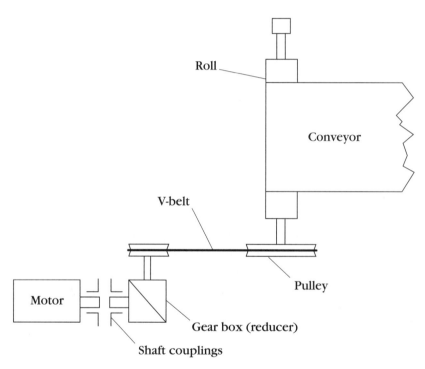

3-2 *The mechanical pieces from the motor to the conveyor.*

Horsepower

Terminology and usage has created some confusion regarding horsepower. In the late eighteenth century, James Watt of Scotland determined that one horsepower was equal to 33,000 ft-lbs of work in one minute, which is equivalent to the amount of power required to lift 33,000 lbs one ft in one min (see Fig. 3-3). A horse, being the work-

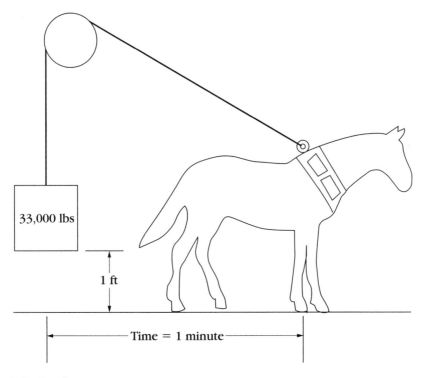

3-3 *One horsepower.*

ing beast in those days, was selected as the "prime mover," and the rest is history. (If ducks were the work animals in the eighteenth century, would we then size according to duckpower?) Horsepower can be derived from the following equation, where force is equal to F, D is distance, and t is time:

$$\text{Horsepower} = \frac{F \times D}{33,000 \times t} \quad \textbf{[Eq. 3-1]}$$

The term *horsepower* has probably caused more costly undersizing situations, when it comes to motors, than we can imagine. When sizing the pound-foot requirements of a given application, we are really concerned with torque. After torque, it is important to look at the electrical current requirements of the application and how they relate to torque. The electrical equivalent of one horsepower is 746 W. To find a value for horsepower when other values are known, the following formula is used:

$$\text{Horsepower (hp)} = \frac{\text{torque (ft-lbs)} \times \text{speed (rpm)}}{5250} \quad \textbf{[Eq. 3-2]}$$

Horsepower

Motors are rated in both horsepower and current, specifically, full-load current or full-load amps (FLA). When sizing the actual power requirements and selecting a motor, brake horsepower (BHP) is used, which pinpoints the actual power requirements at the motor shaft. Always check the torque requirements of the driven load to ensure the motor can produce the necessary output. Motor speed versus torque curves are available to help.

For the motor shaft and the appropriate load to be driven, many engineers and technicians actually turn the connecting shaft to a motor by hand. If turning the load by their own power is possible, most likely the electric motor can provide the torque necessary to turn the load. Turning the motor by hand also indicates if any binding or unwanted friction exists, which could be a problem later. Of course, when process speeds allow, a certain amount of gear reduction can go far in gaining the mechanical advantage to turn a given load. The gearbox can be a lifesaver in getting an application to work. These concepts are looked at closer later in the chapter.

The following is a common scenario in a typical factory: a piece of machinery has been installed in a plant location and has run for many years. It has produced thousands of parts. It sometimes quits running at higher speeds and loads. No one seems to have a good answer as to why, but everyone agrees that a larger motor must be needed. Joe, the current plant engineer, is assigned the task of increasing the motor size from 5 hp to 7.5 hp. His boss, Henry, says to Joe, "It had better work!" Joe then talks to Bob, the maintenance foreman. Bob suggests going to 10 hp for no other technical reason than to be safe. Joe thus determines that a 10-hp motor is his safest approach. This motor is installed and works fine (they never check to see that the motor is only running 60 percent loaded).

Joe and Henry both retire, and Bob leaves the company. The new plant engineer, Pete, is revamping the entire machine several years later. He figures he will double the horsepower from 10 to 20 hp because no records exist as to how the motor was selected originally. The project is reviewed and, by committee, all decide to select a 25-hp motor to run the same mechanical components on the machine. What we now have is overkill: a 25-hp motor running a machine that a 7.5-hp motor could do efficiently.

The moral of this story is that, while it is true that a larger motor can probably do the job, most likely, an optimum, smaller motor could do the same job. This margin of safety that engineers and plant personnel factor in can also work against them if the motor is required to make fast accelerations. A larger motor has more mass to accelerate, thus taking longer than a smaller motor. The more obvious reasons of wasted up-front cost on a larger motor, wasted energy, and poor power factor while running the larger motor could also become issues. That is why it is best to fully look at the machine, its components, speed, and so on to make a good selection. No one ever likes to look bad, especially in today's industrial climate where production and uptime are the only objectives, and individuals often take the safe route. Thus, motors are sized by the all-too-common motor sizing technique: "Well, a 10-hp motor runs a similar machine in our other plant so we'll use a 10-hp motor also." A look at the torque requirements is the proper method.

Torque

Torque is defined as a rotating force. Further definition is any twisting, turning action requiring force. As shown in Fig. 3-4, torque is the product of a force multiplied by a distance, or $T = F \times r$. If a force, F, is applied to the lever arm at a distance equal to the radius, r, a resulting torque is produced. Derivations of this formula can give us the amount of work as a torque acting through an angular displacement. There are mainly two types of torque, static and dynamic. Factory personnel are mostly concerned with dynamic torque. Rotating apparatus exhibit dynamic torque, a state of constant movement, correction, and change. Static torque is a more consistent value. Static torque might be an issue with a holding brake.

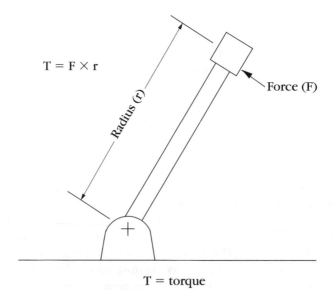

3-4 *Torque equals a force times a radius.*

Torque

Torque can take on different values depending on what is required. To start a machine, we might be interested in the *breakaway torque* rather than other types of torque in the operating cycle. In another instance, we might be concerned with *running* or *process torque*, if the application is sensitive to torque variations once at rated speed. Running torque can be expressed as the torque required to keep the machine running at a given speed. Process torque could factor in the requirements to compress, cut, or periodically act on a material, possibly increasing torque needs for an instant. *Accelerating torque* is also important, because all too often someone wants a machine or process to cycle faster, and this torque value plays a major role in sizing.

The universal formula for determining torque of a rotating piece of equipment is:

$$\text{Torque} = \frac{\text{horsepower} \times 5250}{\text{speed (rpm)}} \qquad \textbf{[Eq. 3-3]}$$

There are several derivations of this formula, as follows. These are merely the same formula, restated to find the proper unknown value when the other values are known.

$$\text{Horsepower} = \frac{\text{torque} \times \text{speed}}{5250}, \text{ where torque is in pound-feet}$$

$$\text{Horsepower} = \frac{\text{torque} \times \text{speed}}{63{,}000}, \text{ where torque is in inch-pounds}$$

$$\text{Speed (rpm)} = \frac{5250 \times \text{horsepower}}{\text{torque (pound-feet)}}$$

$$\text{Speed (rpm)} = \frac{63{,}000 \times \text{horsepower}}{\text{torque (inch-pounds)}}$$

Speed, which is sometimes referred to as rpm, or revolutions per minute, can also be shown in formula form as N. At 1750 rpm, 3 ft-lbs of torque is equal to 1 hp. Note that 5250 is a constant used with foot-pounds of torque. Likewise, the 5250 constant is changed to 63,000 when torque is in inch-pounds. These inch-pound calculations are more typical for servo and stepper motor applications. The sizing can even be in ounce-inches. The aforementioned formulas are for rotary motion. For linear motion, the formula to find horsepower is the following:

$$\text{Horsepower} = \frac{F \times V}{33{,}000} \qquad \textbf{[Eq. 3-4]}$$

where F is force in pounds and V is velocity in feet per minute. This formula is useful when trying to determine the horsepower required for a line moving a material a given speed (V), linearly, at a given tension (F).

Beyond the linear and rotary torque and horsepower basic formulas, other equations are useful for sizing motors and prime movers. One is finding the accelerating torque of a rotary device. Accelerating torque is the torque required above and beyond that required to drive the given load. It must be added to the usually larger value of load torque to adequately size an application with substantial acceleration rates. Acceleration torque is found by using the following formula:

$$\text{Acceleration Torque} = \frac{(WK) \times \text{change in speed}}{308 \times t} \quad \textbf{[Eq. 3-5]}$$

where acceleration torque, or T, is expressed in pound-feet, and WK is the total system inertia, which includes motor inertia (from the rotor) and the load's inertia and is expressed in pounds/feet^2. t is time.

If a gear reducer or other important inertia-containing component is part of the system, its inertia value must also be included. Any power transmission component that must be part of this acceleration must be factored into the equation.

Torque is a more appropriate value to determine what size prime mover is needed to move the load. Once the torque value is known, determining which motor to use is simpler. Motors are sized by horsepower, however, which can be potentially dangerous. The better scenario is to provide the motor supplier with speed and the torque requirements at that speed to size the motor for duty cycle and complete heat dissipation. Let the motor supplier select the motor and frame, who knows the motor constructions. For example, some frames can be more capable for overload conditions than others.

A torque wrench can be used to tighten nuts onto a seated surface; it measures the amount of force in inch-pounds. A similar device can be used to measure the torque requirements of a particular shaft. This evaluation is good for any rotating component but will not factor into the equation acceleration or peak torque requirements. A motor is a stupid device. If a motor cannot turn because its load is too great, the current to the motor increases to try to move the load until the supply is shut off. It is for this reason that electronic overload protection devices should always be implemented in a motor system. Both ac and dc motors should be protected. If a motor does not turn, the motor might be undersized, and the torque output from that motor inadequate to perform the application.

How to use speed, torque, and horsepower curves

Figure 3-5 shows a typical curve with the designations for speed on the x axis and torque on the y axis. Most speed/torque (or speed versus torque) curves have speed on the horizontal axis and torque on the vertical axis. This particular curve is for a variable torque application. It can also be equated to a centrifugal load or even a soft start device. The torque increases proportionately as the speed increases until full, 100 percent speed has been reached, which is also where full torque capability is met.

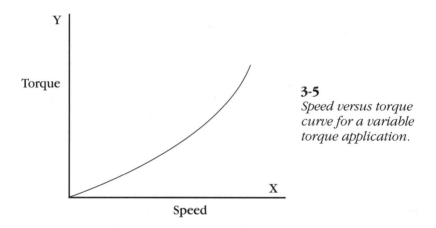

3-5 *Speed versus torque curve for a variable torque application.*

As discussed earlier, horsepower indicates how high in speed a motor can run and yet maintain usable torque. The peak on a horsepower curve indicates the maximum power available and at which speed that power is attained. Selecting a speed beyond the peak yields no additional usable power. It is said to be in a state of *constant horsepower* at that point. The power achieved at higher speeds is also available at lower speeds, so we can save undue wear on the motor by running it at lower speeds. In any given mechanical or power transmission system (gears, pulleys, reducers, etc.), the value for horsepower is constant throughout the system at any given time.

Figure 3-6 is a speed/torque/horsepower nomigram. A point can be selected on any of the graphs, such as speed, and that point can be one of the known values. Then, if either torque or horsepower is also known, a corresponding line can be drawn through the two points (the known components) to find the unknown value. A nomigram saves the time of calculating the values using the equations.

Power transmission basics

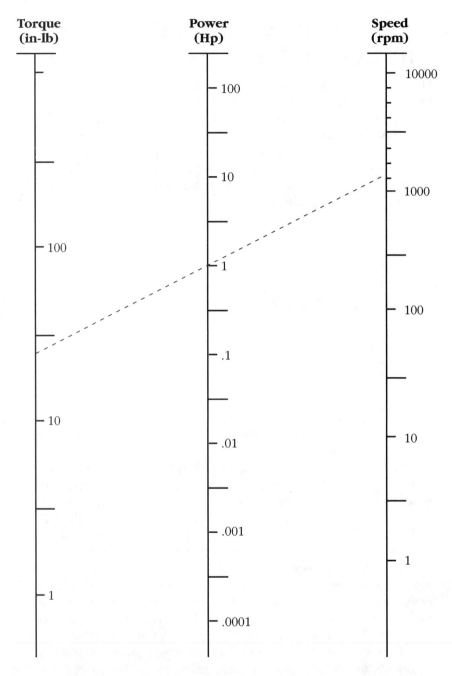

3-6 *A speed–torque–horsepower nomogram; connecting two points will yield the third unknown.*

Other valuable charts and graphs for torque and horsepower are shown in Figs. 3-7 and 3-8. Figure 3-7 indicates the expectations of torque when an application or motor is in a state of constant horsepower. Torque begins to diminish in a constant horsepower application after reaching 100 percent of rated speed. The converse is true in Fig. 3-8, which shows constant torque. If we increase speed while maintaining constant torque, horsepower continues to increase. Examples of constant torque applications are plentiful, such as conveyors, rotary tables, and countless other constant-torque applications that rely on an operating zone that does not go beyond 100 percent speed, which would diminish torque requirements. Constant horsepower application examples are center-driven winders and machine tool systems, including drills, milling machines, and so on. Remember that torque is that component doing the work, so we must be concerned that enough torque is available.

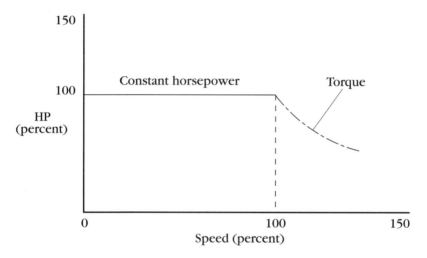

3-7 *Constant horsepower.*

Inertia

We often hear the term *inertia* used in motor sizing and system evaluation. Inertia is often designated by the letters *J* or *I* or even the term WK^2 used in conjunction with or in place of inertia. WK^2 is actually the weight of an object multiplied by its radius of gyration value, K, which is squared in the equation:

$$\text{Inertia} = WK^2 \qquad \textbf{[Eq. 3-6]}$$

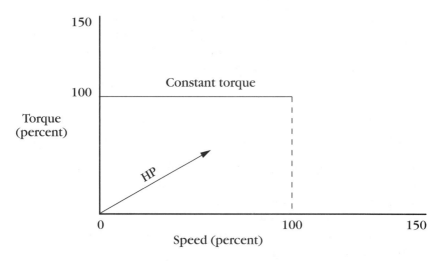

3-8 *Constant torque.*

The inertia formula is sometimes called WR^2. Inertia is that property by which an object in motion will stay in motion until acted upon by another force. Thus, a machine once in motion stays in motion until an external force acts upon it. If a motor running full speed is stopped, the rotating elements come to rest because friction and other forces act upon the moving parts. Clutches, brake systems, and regenerative methods can all stop objects that have substantial inertia values.

There are two distinct types of inertia. One is applicable to rotating elements, and the other is mainly relative to linear motion. Ninety percent of plant applications involve rotating elements; thus, plant people are usually referring to this type of inertia during a discussion. Inertia in a linear motion application is inherent in an elevator or conveyor type of power transmission system. Inertia (rotating or linear) is generally only a factor when the speed of a machine is changing. In practice, WK^2 is the total value of all rotating components of the system in which the main components are the inertia of the rotor (motor) plus the inertia of the load in lb-ft^2. The total values of all the inertia except that of the motor can be referred to as the *reflected inertia* back to the motor.

Reflected inertia is an important property. Its value can have a dramatic effect on a system's performance. When incorporating gear reduction into a motor and power transmission scheme, inertia values are affected by the gear ratio. With a change in speed, the reflected inertia changes inversely by the square of that gear ratio. For example, if speed increases by a factor of four (because of the ratio), the reflected inertia decreases by a factor of 16. This ratio can work in one's favor in a power transmission system where a high reflected inertia exists and acceleration rates are important. This ratio can also work against performance when an increase in speed, rather than a reduction, is needed at the output of the motor. This hindrance can be overcome using a higher torque output motor. Be aware of these types of instances; this is one of those "gotcha" surprises that if not planned for ahead of time can later cost extra time or money. Figure 3-9 shows a typical gear-reduction scheme compared to a speed-increasing scheme.

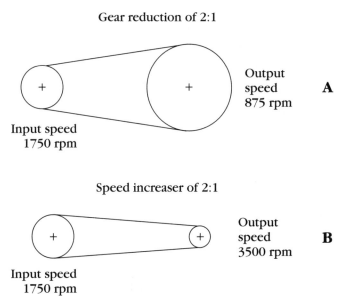

3-9 *Top: gear reduction; bottom: speed increasing.*

The *flywheel effect* is the phenomenon whereby half of an object's mass is on the outside and half is on the inside. A large flywheel is both hard to start and hard to stop. It can assist the transition through intermittent but dramatic load changes when the machine is running. A flywheel can work against the machine when acceleration and deceleration rates must be increased. A good way to illustrate the flywheel effect is to understand the relationship between the two objects in Fig. 3-10. Object A is solid and Object B is hollow, but both have the same mass. Object B is much harder to set in motion because of its hollowness. Thus, the shape of power transmission components can play an integral role in an application's performance.

 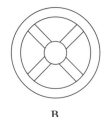

3-10
Both A and B have the same mass. Object B has more inertia even though its diameter is greater than A's.

A B

When considering inertia, ask the following questions: Must the motor change directions or speeds quickly? Must it start and stop frequently? What components in the drivetrain contain high inertia values? Will inertia of certain objects, such as rolls, accumulate more and more material on them as the machine runs? If inertia does accumulate, it must be accounted for in system sizing. The rule is to size a power transmission system for its worst case and factor in a margin of safety to ensure that when the operator hits the run pushbutton, every component starts and stops in synch.

All objects have some mass; thus, all have some inertia. Inertia is an important factor when finding the desired performance for a mechanical system, especially when using mechanical devices with electronic controllers. Electronic controls do not magically provide the escape route if a power transmission line does not respond as desired. Electronic controls can deliver electrical current, which can be converted into power, but we often run into an immovable object. We then need to find out why that object won't move. A troubleshooting tree later in this book examines these issues.

Friction

All mechanical systems exhibit some amount of friction. In sizing motors and torque requirements for a given application, friction should also be accounted for in the equation. A small amount of friction is actually desirable, as it can aid in a given load coming to rest sooner. Some machines even have enough friction that during an emergency stop condition they can stop instantly. Of course, mechanical and electronic braking should always be considered from a safety point of view.

Friction is actually welcome for many power transmitting products. V-belt drives need friction to transmit torque from one sheave to another. A variable-speed pulley relies on this friction. Clutches and brakes also operate on a medium of friction. Friction is not desirable in gears because it creates heat, which results in losses and inefficiencies.

The term *windage and friction* can be used when discussing rotating components. Windage and friction means that when a high-inertia device is at full speed, the only torque requirements at that time to maintain speed (with a steady load) are to overcome friction, compensate for the deflection losses from wind, and keep the device from running away. Another term relative to friction is *stiction*, which is used to imply that a mechanical component can exhibit the tendency to stick in a particular location. Stiction indicates that the coefficient of friction is currently very high. Friction exhibits different tendencies for rolling, or rotating, friction and static, or at rest, friction: an object at rest must overcome more friction than an object in motion.

Motors, or prime movers

Electric motors are discussed at great length in Chapter 6; however, there are many types of motors and many ways to power them. In the case of pure power transmission, the device that initiates the motion might be designated as the *prime mover*. Motion can be accomplished in several ways. Electric motors are the most common by far, but there are pneumatic and hydraulic motors and steam-powered drivetrains. The prime mover or motor has common traits relative to the operation and control of the power transmission. These include some of the aforementioned characteristics: inertia, torque production, shafting, coupling to the load, and overload and service factor values.

Because the electric motor is the most prevalent, and many important features are often relevant to both motors and other devices, I focus on the motor in the following discussion. An electric motor has many mechanical features and constants. Its rotor and shaft have a measurable value of inertia, which can help determine if the motor can be used in high acceleration and deceleration applications. A motor with a high rotor inertia takes longer to accelerate and decelerate than a low-inertia rotor. Longer acceleration/deceleration times might be desirable in certain instances, but high rotor inertia is often used to keep a load moving once the motor is at full speed. Thus, an evaluation of the motor and load versus the application is important.

The motor also has mechanical time constants and a torque constant. The mechanical time constant is basically the time that passes from when an electrical signal is introduced to the motor to when the motor actually produces new torque output. The torque constant is attributed to a motor per the actual torque output for every amp of current supplied.

A typical nameplate of a motor is shown in Chapter 6, Fig. 6-22. Additional motor data and test data can usually be furnished. For older motors in service for many years, however, the original paperwork is probably nonexistent. Motors and gearboxes are usually the only components in a drivetrain that carry a full nameplate. Couplings, gears, belts, rolls, and such do not give the plant personnel much data. Of course, data is usually available on the engineering drawings or specifications. Wouldn't it be nice if more pertinent data were available on machine parts? It would certainly help future maintenance and operating personnel.

Stress

In any mechanical power transmission scheme, there has to be a weak link. When severely overloaded, some component eventually fails. The hope is that all components have been sized with enough of a service factor that nothing breaks, but something usually has to give. Sometimes it is the coupling, other times the belt. There is always a maximum amount of stress that a material can take. This stress can be called the *strength of materials*, or *stress analysis*, or even *finite element analysis*. It is, however, necessary to check the strength of a material, even to its individual elements, to see if it can hold up in intense applications.

A motor's weakest component can be the shaft. If the coupling is sized adequately and the motor is also well sized, the shaft can twist or break under the load. Shaft material and stress factors should always be reviewed. Shaft stress can be calculated as follows:

$$\text{Stress} = \frac{(\text{hp} \times 321{,}000)}{[\text{rpm} \times (\text{shaft diameter})^3]} \quad \textbf{[Eq. 3-7]}$$

This equation yields shaft stress in pounds per square inch and should probably be calculated if the motor might be overloaded.

Gear reduction

Gear reduction is a very important facet of power transmission. Gearing can provide the motor, and the application, with a tremendous mechanical advantage. As long as the speed reduction can be tolerated, torque advantages can be achieved simply by implementing gear reduction. Instead of paying more for a larger motor, it might be more economical to install a gearbox. A good example of gear reduction as a mechanical advantage is that of a revolving restaurant high above the ground. The diners are seated and the restaurant turns very slowly, maybe one revolution per hour. The motor used as the prime mover for this application could be very small in torque output. It might equate to a 15-hp motor running at 1750 rpm. With so much gear reduction, a tremendous amount of torque is produced—enough to move a several-thousand-pound object, such as the restaurant. Because the net revolutions per minute are so minuscule, it is possible to use all of this gear reduction.

Couplings

Many types of couplings exist, and the use determines the type. A coupling that connects a motor shaft and its loadside shaft should be selected based on use, abuse, and environment. Types of couplings include jaw type, shaft couplings, elastomeric, flexible, rigid, and metal and flexible disc couplings. Duty cycles and service factors must also be factored into selecting any coupling. The common couplings used in direct drive applications include flexible and compliant types, which provide a degree of nonrigid, forgiving connection in the application.

A pair of jaw type couplings are shown in Fig. 3-11 with their spider and mating parts, one assembled and one not. Typical couplings

3-11 *Typical jaw couplings.*

consist of many common components. The mating halves of the couplings have an appropriate bolt configuration, along with a bore requirement (the coupling must fit over two different shafts with different diameters).

The perfect coupling should transmit full power efficiently. Enough places in the drivetrain allow electrical and mechanical losses. The coupling area should not be one of the loss areas. Ideally, the coupling can compensate for misaligned shafts. Some couplings can handle more misalignment than others. In addition, the coupling should provide a certain amount of cushion in shock-loading applications.

There is also the torque-limiting coupling, sometimes called a *torque limiter*. A torque limiter is a coupling design made to break at a certain level of torque, or stress. It ensures that no damage occurs to the more-expensive power transmission devices, such as motors and high-precision gearboxes. It is much more desirable to lose an inexpensive coupling and be out of service for a short period while the problem is identified than to lose an expensive machine and be out of service indefinitely.

Coupling selection should also be based on ease of installation, low maintenance, and simple repair. The unit selected should be relatively inexpensive, and spares should be readily available. Although the coupling is a smaller, somewhat less-significant component in the power transmission system, it is important for a number of reasons. When selecting the right coupling, many parameters must be considered, including torque, maximum speed, shaft sizes, availability, and cost. Other, not so obvious, questions include the following: Should a service factor be considered? How much misalignment do we anticipate? Is this a shock load situation, and do we require more endplay? Is the environment corrosive? What is the distance between shaft ends? Up-front analysis of these issues can save downtime and headaches later.

Gearboxes

Sometimes called a *gear reducer*, a *gearbox* is an assembled device that houses all the gear reduction needed for a given application. It has become the convenient device to house gears, input and output shafts, and lubricant. Before fully housed gearboxes, gears and pulleys, along with belts and chains, were installed to provide the necessary speed reduction. The gearbox evolved as a compact unit in which the same reduction was achieved in a smaller, safer package. The net result has been the gearbox; as long as the lubricant is maintained, the gearbox can remain in service for several years.

Gearboxes have rated outputs and appropriate efficiencies. If 20 hp is supplied at the input shaft, its output rating might only be 17.5 hp. Thus, the gearbox is 87.5 percent efficient. The lost horsepower is in the form of heat and noise. For drivetrain system analysis purposes, however, horsepower is assumed constant throughout a system. Losses and efficiencies are not initially considered.

Gears

Gears made of hardened steel are often found in lubricant within the housing of a gearbox. These components are crucial in the transmission of power in the factory and are often the source of much of the audible noise in the factory. The configuration of gears can change rotation and can provide a very high ratio of torque and speed. As a rule of thumb, the lower the ratio, the more efficient the geartrain system. Each introduction of another gear is called a *stage* in the geartrain, as is shown in Fig. 3-12. Each additional stage adds another 2 percent loss to the overall system efficiency and adds more noise. Those 2 percent losses consist mainly of friction, heat, and noise. An odd number of stages in a geartrain reverses the rotation, and an even number keeps the rotation the same, as seen in Fig. 3-12 by the rotational arrows on each gear.

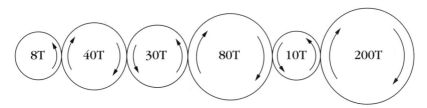

Total gear reduction = 40/8 : 30/40 : 80/30 : 10/80 : 200/10
= 25 to 1

3-12 *A six-stage gear system.*

There are different classifications of gears used in gear reducers. One type is called the *spur gear*, and its teeth are parallel to the shaft. It is typically noisy. Another is called the *miter gear*, which is used to change direction at a 90° angle and a 1:1 ratio. Another common gear is the *bevel gear*, which is very similar to the miter. Common to most of these gear arrangements is *backlash*. Backlash occurs in most mechanical systems where two pieces of metal are in physical contact with each other. Typical backlash is shown in the partial gearset of

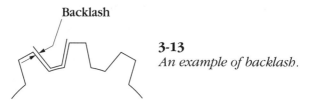

3-13
An example of backlash.

Fig. 3-13. Zero backlash implies that no play occurs between the components. While backlash can never be zero, because of wear from the initial engagement, it can be minimized. Minimizing backlash takes extra-precise machining of the gears, however, which can be expensive.

Different types of gears and configurations of gears within a gearbox have different names. One is called the *worm gear reducer*, and its main components are the worm and worm gear. One attractive feature of the worm reducer is that it is difficult to drive backwards, providing a built-in safety feature. A disadvantage is that the worm gear is not as efficient as the standard helical gear. The *helical gear* engages more teeth at any given time, thus transmitting more torque with less backlash and less noise. More teeth can engage at a time because of the angles on the teeth. Another gear arrangement is the *planetary gear*. As the name implies, this gear has a sun gear around which the others rotate. This package yields the highest reduction ratio in the least amount of space. The sun gear is the driven gear, and the ring gear provides the output. The planetary gear design has low backlash and is fairly efficient.

V-belts

As the name implies, V-belts are power transmission belts with a cross section in the shape of a *V*. Thus, the bottom portion of the *V* seats well into the pulley or sprocket into which it is placed. This concept is illustrated in Fig. 3-14. The power transmitted is a direct function of how much actual surface area is common to the belt and the pulley. When slippage occurs, surface-to-surface contact is lessened. An industrial V-belt is made of high-grade rubber product with tensile and other cords interlaced within it. Multiple belts are often used for transmitting larger amounts of power. Advantages of using V-belts include lower installation and replacement costs, practically no noise, no lubrication (oil on a V-belt actually causes slippage problems). V-belts are fairly forgiving and can absorb shock loads. They are easy to install, require very little maintenance, and can be used for high-speed applications.

Chains and sprockets

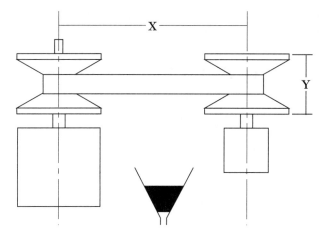

3-14 *A V-belt drive and V cross section.*

Some disadvantages of V-belts are that they are not recommended for oily, hot, and harsh environments because they can slip, and this slip possibility makes them impractical when trying to synchronize components. Low-speed operation is typically not recommended, either. A common practice in industry is to use a minimum of two belts, even if one will do, for redundancy. If one belt breaks, the other can keep the machine running.

Often, a belt must connect a critical device to another without any slippage. For example, a pulsing feedback device cannot be directly mounted to a motor shaft and thus must be mounted in parallel to the shaft. This mounting is often accomplished using a timing belt. Timing belts are flat belts with teeth. They are highly efficient, lightweight, and eliminate slippage. They do have substantially less torque transmitting capability than do their V-belt counterparts, however, and should not be used for extreme power transmission conditions.

Chains and sprockets

Often referred to as *roller chains*, chains and sprockets are the choice to transmit power in nasty environments. A roller chain suffers no slip (maybe a little backlash), runs in high ambient temperature and oily atmospheres, and is very efficient. It requires accompanying sprockets with teeth to be a complete package. The sprockets and the chain are more expensive than belt drive systems. The chain system must be lubricated, is obviously noisy, and is not as forgiving as V-belts. Shock loads can be amplified or even can cause the chain to snap.

Over time and because of use and wear, the chains actually become elongated. At some point they must be replaced.

Clutches and brakes

A clutch is defined as a device that engages and disengages motion. It must be used in conjunction with a prime mover, usually an electric motor. Clutches can also be actuated by air or fluids. The electromagnetic clutch uses a permanent magnet and electromagnet. By providing dc voltage via a voltage controller, the amount of magnetism is controlled, thus providing as much clutch engagement as necessary. Engagement material, which transfers the torque and engagement times, is crucial to the performance of a clutch. Current engagement material should contain no asbestos, but some is still in use. Other friction media can include ceramics, leather, or other nonasbestos material. These newer materials are wear items, however, and must be replaced.

Clutch actuation is manual, electrical, or mechanical. Air, electric, hydraulic, centrifugal, or magnetic means are all common. *Engagement time* is defined as the amount of time required from when a signal is sent to a clutch until enough contact has been established with the friction material to cause the desired motion. Engagement time is especially important in high-cycling applications where the load must be picked up and released often. Heat can become a major issue with the clutch. An air-actuated clutch operates under the same principle as an electric clutch, but air is the medium by which pad pressure is achieved. Hydraulic clutches use oils and synthetic fluids to provide engagement. The control of the flow of these fluids is proportionate to the engagement of the clutch.

Other types of clutches include the *magnetic particle type*, which uses magnetized particles to provide certain degrees of clutch strength. These clutches are used often to set a tension or drag on a rotating component. They usually are not seen in high-horsepower applications. Two other types of clutches are the *over-running clutch* and the *centrifugal clutch*. The over-running clutch is used as a backstop, such as on a conveyor system, to ensure that backdriving cannot occur. These clutches also perform simple indexing applications and will free wheel when not engaged. The centrifugal clutch works on the principle of centrifugal force. At zero and low speeds, the clutch is not engaged to any other component. When speeds are such that the outer housing can centrifugally fly out to engage with another rotating device, however, the clutch is fully engaged and working. These clutches are inexpensive and used quite a bit with two-cycle machines. Another clutch type worth mentioning is the *workhorse eddy-current clutch*, discussed in further detail later in the book.

A brake is very similar to a clutch in operation, but its sole purpose is to stop all motion, most often rotary. The friction materials and actuation methods are common to both brakes and clutches. There are many types of brakes. Some are air-actuated, but most are electrical. The strength of the electrical signal is directly related to the amount of braking, or engagement, applied. Similar to the clutch, heat dissipation is a major issue with the brake. When engaging and disengaging in high-cyclical applications, it is necessary to size a brake unit not only for torque requirements but also for heat buildup. If the heat is not dissipated, the brake can lose its braking capability.

There are different uses for brakes in industry. Some brakes perform a holdback function, providing less than full braking for tension, pressure, or drag on a rotating component. Other brakes are called *fail-safe brakes*. Their operation is exactly opposite of typical brake usage. Normally, a brake is commanded by using a signal, which indicates that we want to stop our machine or process. The brake is in an off, or nonengaged, state when electrical or supply power is present. Sometimes, however, it is desirable to have the brake disengaged when electrical power is present for safety reasons and, in the event of a power failure, the brake becomes fully engaged. Fail-safe brakes ensure no motion when a loss of electrical power occurs. All brake applications generate heat.

Variable-speed pulleys

The working components of a variable-pitch, or variable-speed, pulley system is shown in Chapter 7, Fig. 7-3. This device has been used extensively in industry for many years. With so many other more efficient and modern ways to obtain variable speed, the pulley system has lost popularity. Its disadvantages, besides being inefficient, include frequent adjustments and a high frequency of belt wear. A pulley requires a constant-speed prime mover, which is usually an ac motor. Variable-speed pulley systems are typically not seen above 50 hp.

Fluidized speed variators

Fluidized speed variators use a fixed-speed electric motor to move a fluid, usually oil through holes, or jets, to reduce speed output. These devices are typically expensive because of the intricate machining required to manufacture them. The fluid must also be changed frequently to prevent clogging. The jets can leak over time and require frequent maintenance to keep them in good running condition. Their advantages, however, include the ability to handle shock loads and perform instant reversing without damage to internal parts. The speed ranges are very good, from high to low. Most variators are not available above 50 hp.

Proportionately infinitely variable (PN) transmission

A PIV is supposed to be proportionately infinitely variable (PIV) in its speed; however, it is really nothing more than a transmission box and a high-maintenance item. Its actual speed range is dependent on how many gears are in the box. The more gears there are, the more expensive the box. The PIV needs a prime mover running at a fixed speed. PIVs are found in many older plants. Today, it has been replaced with electronic drives and controls.

Ball screws

The ball, or lead screw, changes rotary motion into linear motion. It is a very common power transmission device found in industry, especially in machine tooling. It gets its name from the many steel balls located in a sleeve-like housing, called the nut. This nut moves on the grooves, or leads, of a screw. The balls actually roll in the grooves, thus moving the whole nut assembly. Thus, when an electric motor turns the screw, the ball bearings move the whole assembly as seen in Fig. 3-15. Ball-screw assemblies are most commonly found in positioning applications much like the requirement in machine tool systems. The motors used are often stepper- or servomotors with more accurate positioning controls.

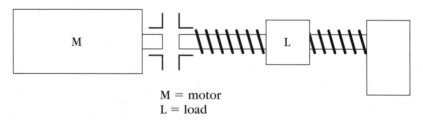

M = motor
L = load

3-15 *A ball-screw driving a load (L).*

When sizing the torque requirements of a ball-screw application, these three main criteria are needed:
- What is the load, W?
- What is the lead screw pitch, P, in revolutions per inch?
- What is the efficiency of the lead screw? (Efficiencies range from a high of 95 percent for a ball nut to a low of 50 percent for an acme brand type of screw with a plastic nut. Consult the lead screw manufacturer for efficiencies.)

The information is then factored into the following equation:

$$T \text{ (total)} = T \text{ (friction)} + T \text{ (acceleration)} \quad \textbf{[Eq. 3-8]}$$

To find the torque requirement of friction, use the following formula:

$$T \text{ (friction)} = \frac{F}{2 \times \pi \times \text{pitch} \times \text{efficiency}} \quad \textbf{[Eq. 3-9]}$$

where T is torque and F is frictional force. Inertia also plays a role and must be considered.

The pitch of the lead screw is expressed in revolutions per inch and is usually shown in symbol form as p. Pitch is not to be confused with the lead of the screw which is actually $1/p$. The efficiencies of lead screws are fairly close, in the 80 to 90 percent range. The variations depend on whether the ball nut is metal or plastic. One type of lead screw, the acme, is very common and inefficient, with efficiencies as low as 30 percent. The acme screw frequently must be adjusted for wear to maintain performance. It is sometimes necessary to hold to very tight tolerances when positioning with a lead-screw system, even to zero backlash. To meet this requirement, one ball nut can be preloaded against another, minimizing looseness and end play.

Drives

Drives can take on different meanings. Many of the previously mentioned devices that transmit power are often referred to as drives. They can be electrical or mechanical. They can be called drivers or drivetrains or drive systems. Whatever the nomenclature, the purpose of the device is to set or keep an operation, machine, or process in motion. Most have an electrical requirement because they usually need a fixed-speed prime mover, which is often an electric ac motor. Some drives have mechanical needs while others do not. Without drives, industrial automation would be impractical.

Direct drive

It can be desirable and possible to drive the load directly from the motor shaft, eliminating gears, pulleys, or other mechanical linkages. Fewer components mean less initial cost and less problems. The motor, or prime mover, must be sized to move the desired load adequately. In many rotary motion applications, inertia is relatively low, and mechanical simplicity is desired. These applications are good candidates for direct drive. The coupling should provide a degree of damping and help compensate for any misalignment.

Direct driving at a constant speed allows for an ac motor to be powered directly from the factory-supplied electricity and is called *line starting* or *line operation*. The ac motor accelerates extremely fast to its full-rated speed. Full voltage is applied to the motor. This type of operation usually allows 5 to 6 times in-rush current to the motor, which reduces the insulation and winding life if it is started and stopped frequently. Both electrical and mechanical devices can softly start a motor and system. Of course, a dc motor cannot be line-started nor provided with a bypass system.

Variable-speed drives

Electronic variable-speed drives are being used in more and newer applications every day. As the cost of componentry drops, they can be justified for many more applications that require variable-speed control of either an ac or dc motor. Chapter 7 is dedicated to motion control and details the different ac and dc drives available.

Many times, ac drives have been misapplied because users felt the speed could be reduced, thus eliminating any gear reduction. This fact is simply untrue. The application's torque requirements do not change. If gear reduction was in place, a variable-frequency drive (VFD) cannot produce the extra torque needed just because the speed is reduced. A drive can provide a certain amount of overload and even deliver continuous current for constant torque, but gear reduction can provide an application with a tremendous mechanical advantage. It might not be possible to install a variable speed drive and take out a gearbox. These misapplications have given electronic drives a bad reputation. It is important to understand torque, speed, and horsepower and how they all relate to understand how gearing helps achieve higher torque output at lower speeds.

Power transmission controls

Power transmission componentry often has an electrical content. For instance, electromagnetic brakes need to convert ac control power into dc and then vary that dc voltage to the electromagnet. These devices are sometimes called *power supplies* but are actually electronic controllers. As microprocessors become more prevalent, so too will simple mechanical component controls become more complex. The microprocessor allows for more precise control of a given signal and also provides valuable information about failures. Enhanced control allows logging of faults and diagnostics. Electronic control also requires more troubleshooting and more spares inventory.

Fluidized speed variators, clutches, and variable-speed pulleys all incorporate an electric motor, thus making the overall system electromechanical. Electrical power, in watts, is eventually converted to foot-pounds of work. Likewise, the controls for many of these power transmission devices are digital and electronic. Some controls are current-source type, while others are voltage-source type. A current-source controller monitors the current in the controller-power-transmission device circuit, whereas a voltage source unit monitors and controls using the circuit voltage.

Pumping applications

A common application that uses many power transmission components is that of pumping. Motors, couplings, VFDs, and other components are used for a given pump system. It thus becomes necessary to size a given motor for a certain flow rate (gallons per minute, or GPM) and a certain pressure (or head, in feet). The equation is as follows:

$$\text{Horsepower} = \frac{(\text{GPM} \times \text{head (ft)} \times \text{specific gravity of the liquid})}{(3960 \times \text{efficiency of pump})} \quad \textbf{[Eq. 3-10]}$$

where 3960 is a horsepower constant, and head pressure in feet is equal to 2.31 pounds per square-inch gauge.

The specific gravity of water is usually referred to as 1; however, this value can change based on temperature. Many other liquids have different specific gravities; common liquids and their specific gravities are shown in Table 3-1.

Table 3-1. Various specific gravities for common liquids (English system of units at 14.7 psia and 77°F)

Liquid	Specific gravity
Acetone	0.787
Alcohol, ethyl	0.787
Alcohol, methyl	0.789
Alcohol, propyl	0.802
Ammonia	0.826
Benzene	0.876
Carbon tetrachloride	1.590
Castor oil	0.960
Ethylene glycol	1.100
Fuel oil, heavy	0.906
Fuel oil, medium	0.852
Gasoline	0.721
Glycerine	1.263
Kerosene	0.823
Linseed oil	0.930
Mercury	13.60
Propane	0.495
Sea water	1.030
Turpentine	0.870
Water	1.000

Fan and blower applications

Another common power transmission application is for fans and blowers. Cooling towers, dryers, and other fan systems all move air or other gases. Air has different characteristics when cool compared to when it is hot. Likewise, gas also must be treated differently when under pressure. In addition, horsepowers and speeds have specific relationships, or laws, often referred to as the *affinity laws*. A simple equation for calculating the horsepower of a simple fan or blower is as follows:

$$\text{Horsepower} = \frac{\text{CFM} \times \text{Pressure (psi)}}{33{,}000 \times \text{efficiency}} \quad \textbf{[Eq. 3-11]}$$

where Cubic feet per minute, or CFM.

This calculation works for a system with normal temperature and

pressure characteristics. For centrifugal, or variable, torque situations, the affinity laws described as follows apply:
1 The flow in a system is proportional to speed, whether increasing or decreasing.
2 The speed in a given system increases by a square function as the pressure increases.
3 The speed increases in a given system by a cube function relative to the horsepower increases, and vice versa.

What the affinity laws mean is simply this: a fan's characteristics follow these curves for flow, pressure, and horsepower. If a change is made to one or the other, one must be willing to make the necessary changes to the rest of the system to get the performance desired, which can mean installing larger motors, changing base speeds, or changing gear ratios. These system characteristics are also common to centrifugal pump applications.

Hydraulics

Hydraulics is an often misunderstood term. It is not a water-based aerobic exercise. Rather, *Hydraulics* pertains to fluids in motion. These fluids, usually water or petroleum, can exert pressure in the piping in which they are contained. Hydraulics is also concerned with the pressures and resistances to flows. In the power transmission world, we are concerned with the means of transmitting power via fluid. Accurately controlling flows and pressures also helps control a process or machine.

There are hydraulic cylinders, hydraulic brakes, and transmissions. Like air-driven devices, pressurized fluid can be compressed or decompressed to attain motion, as in cylinders, and to gain a mechanical power advantage. To perform the pressurizing functions, an electric motor typically runs the compressor. Larger systems entail more electrical and electronic content. Pressure and flow controls are becoming more powerful, thus requiring the mechanic to troubleshoot the electronics.

A hydraulic system consists of five main components:
- An **electric motor**, or prime mover such as an engine, which drives a pump.
- The **pump** increases the pressure of a fluid, which is controlled by valves.
- **Valves** are the medium for transferring fluid.
- A **piping system**.
- The **load** itself.

The fluid needs to be moved through the piping, but the one electric motor can often be the drive source for both pump systems. As can be seen, electronic and electrical control can be used in many of these components.

The motor is often an ac motor requiring an electronic variable speed controller, or drive. The control valves can be electrically actuated based on a transducer's feedback to a valve controller. These valve controllers can monitor pressure, flow, and temperature. In fact, flowmeter control has spawned some fairly sophisticated devices. The overall process for which this hydraulic system is part of can be controlled by a more complicated controller, such as a programmable logic controller. Thus, understanding the electronic hardware is just as important as understanding the mechanics of the process components and the process itself.

Process control is a big part of industrial automation. Good control of the system is dependent on the sensors, limits, and transducers. Specific devices have been developed for the controlling fluids. One such device is the *magnetic flowmeter*. Sometimes called *magmeters*, these magnetic flowmeters measure the flow of liquids in an enclosed pipe. The fluid must be able to conduct electricity for the magmeter to function properly. These devices are specialized process control products for industry and are covered in greater detail in Chapter 5 on process control.

Pneumatics

Air-driven systems in the factory have a definite niche. As long as compressed air is available and the distance from the compressor to the device is not too great, a pneumatic solution is viable. Pneumatics is that branch of physics concerned with the mechanical properties of gases. Most often, pneumatics is concerned with the properties of air, especially compressed air. Air must be pressurized to be used in factory processes, which is done via a machine commonly known as the *air compressor*.

There are basically two types of air compressors, static and dynamic. The dynamic types include centrifugal and fluid jet. The static type is the most common and acts as a volume changing or displacement device. It is also known as a positive displacement compressor. Its action is much like the hand tire pump, where a moving piston within a cylinder decreases the actual volume of space in which the air resides. This action thus compresses the air into a smaller space, thus increasing the pressure of that air. The air is typically rated at 90 to 105 pounds per square inch gauge (psig), which is usable air pressure in a plant where air-powered tools, air motors, and actuators are driven by air.

Air-powered systems have several advantages. Air devices do not create shock hazards and thus can be applied in wet applications. They also do not create sparks and are fairly safe in explosive environments. A small compressor can provide a lot of available air when used in conjunction with a storage tank. Air systems provide flexibility and economy.

The control of air systems is simple. Tubing, piping, valves, pistons, and cylinders are the main components; there are few moving parts. Reliability is high and, from an electrical standpoint, the motor used at the compressor is one of the wear, or maintenance, items. The flexibility of an air system is in the fact that all one needs to do is add another valve and some piping to introduce more capability. Relief valves can be incorporated to guarantee the protection of a particular system. As for performance, an air-actuated piston within a cylinder can change its motion quickly in discrete steps and with very little shock.

Air systems have been replaced by electrical systems wherever practical because of convenience. The cost of electrical systems has decreased, and the performance of an electrical system has increased with the advent of digital control. Disadvantages of air systems include noise, mostly from the air relief, and maintenance. The air lines in a compressed air system can get water in them, ultimately contributing to loss of pressure. All in all, there is always a place for air-powered devices in the factory, but as electronic control continues to be developed, fewer air systems will be specified.

Noise

Walk into any production facility and you will probably be given a set of earplugs or wish you had a set. Electromechanical operations are loud. Motors spinning at high speeds produce one band of audible frequencies; metal banging against metal produces yet another. Some factories are simply too noisy. Maybe this condition can be blamed for misinterpreting torque and horsepower terms on the plant floor—nobody can hear too well when shouting above noisy machinery! In the chapter on industrial safety, noise, allowable limits, and how to deal with them are discussed. Noise also indicates that the process could be more efficient. The presence of noise means that energy is being lost.

Efficiency

Efficiency is defined as the amount of work output divided by the work input. This value can be attributed to a single piece of equipment or calculated for several components that compose a full system. It is sometimes more important to know the entire machine's efficiency to get a clearer picture. One component, by virtue of its design, might be somewhat inefficient, while all other components in the system are high in efficiency. That one inefficient component does not lower the entire machine's efficiency much, but it is always worth investigating alternate methods and designs to further increase efficiency. Power costs are constantly rising, and if machines run continuously, substantial savings can be attained.

Efficiencies can be broken into mechanical and electrical segments. Some components in a machine might even exhibit properties of both. Energy losses are usually in the form of heat, but noise and surges of power from starting and stopping should also be factored into the overall equation. As we strive for perfect efficiency, we must accept that nothing is 100 percent efficient, even though manufacturers claim to be approaching this value. Will superconductive components cause 100 percent efficiency? Can we be over 100 percent efficient? If theoretically possible, competitive industry will drive it to fruition!

Service factor

When selecting power transmission components, it is customary to size the component with a factor of safety, or service factor, to ensure that a catastrophic failure does not happen. Service factoring is theoretical and does not account for the fact that there are many components in a power transmission line, and one might have been overlooked. Also not considered are flaws, defects of material, and external forces, which all might cause failure. Perhaps, if money and size are not an issue, a component can be designed and selected with a service factor 10 times its required value. Then it will finally be safe.

Realistically, experience and manufacturers of power transmission equipment have given guidelines for choosing service factors. A service factor of 1 is the base and the absolute minimum. Service factors are used in selecting couplings, brakes, gears, gearboxes, drives, motors, and almost every single component in a power transmission scheme. With 1 as the base, it is practical to establish values for service factor. For example, centrifugal fans, liquid mixers, and variable torque pumps are given a value of 1. Conveyors and feeder equipment applications are 25 percent higher, or 1.25. Machine tools, heavy material mixers, mills, cranes, and elevators require service factors of 2. The safety of personnel and users is also considered, but so is the severity of the application and its duty cycle.

Service factors differ with the type of equipment and component being applied. A variable speed drive has different service factor needs than a V-belt. The type of prime mover is also a consideration. Smooth prime movers such as electric motors, steam, and gas turbines are predictable. Diesel and gasoline engines are not always smooth and thus require additional service factor attention. All in all, the service factor is the designer's "ace in the hole." Certain manufacturers, upon fielding a request for additional service factor, often select a frame or model based on their experience with the specific application and use, sometimes making it difficult for the designer to discern between similar equipment offerings. Double checking with the manufacturer, comparing weights, and doing physical comparisons usually can provide the answers.

Maintenance

Most mechanical systems need constant attention. Parts are physically wearing on or against other parts. Lubrication is necessary, but it must be monitored. Lubricants get dirty with metallic particles and dirt and act as coolants for machinery. Today's synthetic lubricants can hold up longer than before, but machines still need the periodic oil change. Scheduled maintenance for equipment is often hard to adhere to, but it keeps machines, which are normally huge capital investments, running longer. A machine can run 24 hours per day, seven days per week, never stopping except for tool changes and product changes. Thus, in management's eyes, there is no convenient time to shut down. When a coupling breaks or a motor overheats, however, there is usually time to panic! Mechanical and electrical systems all need periodic maintenance. Suppliers of machinery and equipment provide a schedule for maintenance and which components need more attention than others.

One movement is to minimize components and get maintenance-free equipment in the plant. This plan is becoming more practical with electronic equipment available to perform work that once took several mechanical components. Electrical equipment needs scheduled maintenance, too. Heat, dirt, and moisture buildup will shut the plant down just as fast as a broken coupling!

A new breed of service person is also emerging. The electrician must be cross-trained on mechanics, and the mechanic must be trained on electrical systems. Even operators of machines must know a little about every part of their machine. They are usually not looked upon favorably if they constantly call service every time a problem occurs (even if they are operating the worst machine in the facility). The person well versed in multiple disciplines will get compensated accordingly and be better prepared for change and new technology.

Many electronic controls are equipped with onboard diagnostics, and some can even pinpoint a problem within themselves. Common diagnostic features are also appearing between pieces of equipment.

Power transmission authorities and standards

A large base of experienced personnel and documentation exists for use by technicians and engineers in mechanics and power transmission, such as the Power Transmission Distributors Association (PTDA), SAE, and ASME. Some organizations even overlap, such as the AIME. Because so much technology has worked its way into other disciplines, the standards and organizations overlap. Just look at the term *electromechanical*. It says it all. Electrical, process, manufacturing, and industrial engineers must all wear many hats in today's factory. The term *electromechanical* might even be outdated one day. The more appropriate term should be *digital-electromechanical* (or DEM)! The new breed of factory engineer is on its way; automation engineers are today's plant experts.

Conclusion

Power transmission has been the brawn of industrial production while the controls have been the brains. They must exist together because they cannot exist independently. A computer is not going to force water through a pipe, just as a coupling cannot count rpm (at least not with today's technology). One common thread runs between transmission components: the circular, rotating object. The cave dwellers discovered the wheel, and its principle is still in effect today, helping produce thousands of products daily. The motor shaft is round, the coupling is round. So too, are pulleys, gears, brakes and clutches. Technology has evolved around this principle and will probably continue to do so.

This chapter's intent is not to be the standard for power transmission but rather a quick reference for those necessary formulas and definitions that the plant people use every day. Maybe this chapter can settle an argument or two. Various facts are presented, but it is always a sound idea to get another's opinion on a subject, especially when safety is an issue. As technology becomes increasingly complex, one should rely on the vendor's expertise and competence. After all, the vendors and manufacturers of high-tech products are constantly upgrading and trying to outdo their competition. Put the onus on them to provide good, sound facts and judgment on critical issues.

Bibliography and suggested further reading

Beiser, Andrew. 1973. *Physics*. Menlo Park, CA.
Cummings, Electrocraft Corporation. 1980. *DC Motors, Speed Controls, Servo Systems*. Hopkins, MN.
Mott, Robert L. 1979. *Applied Fluid Mechanics*. Second Edition. Columbus, OH. Charles E. Merrill.
Shigley, Joseph E., *Mechanical Engineering Design*, Third Edition, 1977, New York: McGraw-Hill.
Electrical Apparatus Service Association, *Electrical Engineering Pocket Handbook*. 1988. St. Louis, MO.

4

Industrial computers

The microprocessor created a new multibillion dollar industry, the computer industry. It has influenced almost every other industry on the face of the earth, including the automotive, oil, food, medical, pharmaceutical, and housing industries. Computers are all around us. Sometimes we do not even know they are there. A microprocessor can be in an appliance or even a toy. Just as computers have become a part of our everyday life, so too has computerization become a crucial part of industrial automation.

In the factories, computers and their microprocessors are required for all types of control. Motion and process control, communications, machine vision, engineering, and even the front office, all use computerized equipment every day. It is sometimes scary to think about all the daily events now linked to a computer. With every passing year, the computer content to our work and home life rises. Get those backup power systems, now, before they run out!

So what is this computer phenomenon anyway? Who started it? Why? People like Pascal, Boole, and Babbage had a major influence on the beginnings of computerization and, because their names are used in the computer industry, are widely known. Charles Babbage built an analytical engine in the 1830s. He built it to aid machines in making parts more precise. Unfortunately, the industrial world had yet to discover the need for tight tolerancing, and his computer concept fell dormant for about a hundred years. Prior to Babbage, Blaise Pascal invented a mechanical adding machine. Of course, the real origins of the computer are in the abacus, which dates back many centuries. Many individuals have made major contributions to computers. All these early inventions were created to help people perform tedious calculations more consistently and faster.

Industrial computers

Tedious tasks drive computer development today. The "striving" forces are faster, smaller, and more powerful. Whatever the clock speed is this instant, another faster, more powerful device is being designed. Miniaturization, or putting the same capability into a smaller space, is also rampant. And it has to. If computers were still the size of those early units in the 1940s and 1950s, assembly and production lines would be dwarfed by the computer's hardware (not to mention the extra heat loads). The electronic numerical integrator and calculator (ENIAC), built in 1946, was huge. Computers have been getting smaller ever since, as evidenced by the development of the integrated circuit (IC). The IC is a solid-state device consisting of hundreds of capacitors, diodes, resistors, and transistors on one small silicon chip. The IC has allowed more powerful computers at much lower costs. And as usual, there is the demand that the next-generation microprocessor be more powerful. Throughput is the name of the computer game. Newer and faster methods of switching continue to evolve. Figure 4-1 shows the evolution of switches and where it exists today. Computer hardware and switches continue to develop faster than ever.

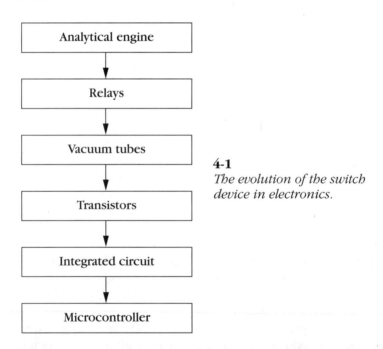

4-1 *The evolution of the switch device in electronics.*

Digital electronics have evolved over the past two decades from the binary system. Digital logic is the basis of most modern electronic equipment. Whether or not a condition is true or false, on or off, high

Industrial computers

or low, its state determines how and when an event should be controlled. Various forms of logic exist because they are easy to implement. One is called transistor-transistor logic (TTL) and is used in a wide variety of electronic controllers. By using transistors, input and output circuitry can be used to collect and send information in a system. In any TTL scheme, a network of resistors, diodes, and a power supply can be found.

From the TTL scheme, complementary metal-oxide semiconductor, or CMOS, circuitry design came about. The CMOS digital circuit used metal-oxide semiconductor field-effect transistors (MOSFETs) to achieve better input and output collection, emission, and gating from the transistor. These digital logic hardware schemes preceded the current trend of accomplishing everything via software. Not all inputs and outputs can be furnished routinely as very low voltage signals, which are readily used by a microprocessor.

Computer architectures, or the techniques and components that are the building blocks of any computer system, have changed quite a bit over the past few years. Binary systems abound, but many digital systems employ logic functions consisting of AND, OR, and NOT logic. These architectures can be used in conjunction with the microprocessor's capabilities. For example, flash memory, discussed later in this chapter, is based on NOT logic. AND, OR, and NOT logic can be used in programmable controllers, cellular phones, and even microwave ovens. The term *logic* is key: inputs and outputs must be accounted for somehow and are generally on or off. The AND and the OR operators handle the I/O. Sometimes these operators are referred to as open and closed *gates*. As the system determines the status of inputs and outputs, it classifies high-voltage readings as *on* and low-voltage as *off*. The NOT operator indicates when an event is *not* happening. Combining the functions of AND gates and NOT gates yields the NAND gate. Combining OR and NOT gives us the NOR gate. While a multitude of methods exist for programming digital devices, it usually gets down to a binary level at some point.

Today's industrial computer actually consists of many elements, including hardware, firmware, and software. There is a microprocessor and sometimes a math coprocessor if enormous amounts of numerical data must be processed. The microprocessors have grown from 8-bit and 16-bit to 32-bit and beyond. The microprocessor is dependent on the machine's crystal, which can be likened to the heartbeat of the computer. The crystal oscillates at a high frequency when electricity is applied, giving the crystal, or computer clock, speeds of 33 megahertz (MHz), 66 MHz, 100 MHz, and beyond. This crystal establishes the speed by which the microprocessor can step through the software. The microprocessor's bit configuration and the system's crystal are the two main elements of the computer for power and throughput.

Besides the actual microprocessor and the crystal, there are memory, input, and output devices, along with a communications protocol and standards for these external devices. This chapter assumes the reader is somewhat familiar with computers and focuses mainly on newer techniques, industrial applications, and the basics.

Microprocessing basics

As we look at industrial processes and how to automate them, we find that many events are simply *on* or *off* states. Those states simulate an open or closed valve, for example. Starting and stopping motors is also on-and-off logic. The computer takes this basic principle to its extreme. The processor just switches on and off states. Therefore, our control scheme is not as complicated as we think. It is just vastly immense. This switching is done millions of times per second. Binary digits, or bits (1s and 0s), change their on and off status constantly to control the on and off equipment in the plant. The computer also multiplies and divides by adding and subtracting at amazing speeds, allowing the machine to make complicated process decisions.

The *bit* is the building block. Bits are made into bytes, and bytes into characters. One byte is equal to a given number of bits. Eight bits is one byte in an 8-bit machine, and 16 in a 16-bit machine. This proliferation will continue. Table 4-1 is a helpful guide for those instances when the bit terminology is flowing and the calculator is not handy. As bytes constitute characters in a program, a standard must exist for identifying these characters. This fact is especially true when transferring characters from one computer to another. The current standard is the American standard code for information interchange, or ASCII. This standard allows for seven bits per character, making 128 unique characters and a further 256 characters with 8 bits per character. The ASCII character code equivalents are shown in Table 4-2.

The computer system is simply a coordinated system of functional components, as shown in Fig. 4-2. The blocks shown designate the rudimentary elements of any computer system: input devices, output devices, arithmetic and logic processing, and data storage. These are the focal points of development. For industrial applications, special attention is given to the input and output devices. The

Table 4-1. Binary number decimal equivalents

Power of two	Bit number	Decimal equivalent
2^1	1	2
2^2	2	4
2^3	3	8
2^4	4	16
2^5	5	32
2^6	6	64
2^7	7	128
2^8	8	256
2^9	9	512
2^{10}	10	1024
2^{11}	11	2048
2^{12}	12	4096
2^{13}	13	8192
2^{14}	14	16,384
2^{15}	15	32,768
2^{16}	16	65,536

For example, when referring to a 12-bit number, the total units are 4096.

Table 4-2. ASCII character code equivalents

Binary	Octal	Decimal	Hexadecimal	ASCII character
0000000	000	000	00	NUL
0000001	001	001	01	SOH
0000010	002	002	02	STX
0000011	003	003	03	EOA ETX EOM
0000100	004	004	04	EOT
0000101	005	005	05	ENQ
0000110	006	006	06	ACK
0000111	007	007	07	BEL
0001000	010	008	08	BS
0001001	011	009	09	HT
0001010	012	010	0A	LF
0001011	013	011	0B	VT
0001100	014	012	0C	FF
0001101	015	013	0D	CR

Table 4-2. Continued

Binary	Octal	Decimal	Hexadecimal	ASCII character
0001110	016	014	0E	SO
0001111	017	015	0F	SI
0010000	020	016	10	DLE
0010001	021	017	11	DC1
0010010	022	018	12	DC2
0010011	023	019	13	DC3
0010100	024	020	14	DC4
0010101	025	021	15	NAK
0010110	026	022	16	SYN
0010111	027	023	17	ETB
0011000	030	024	18	CAN
0011001	031	025	19	EM
0011010	032	026	1A	SUB
0011011	033	027	1B	ESC
0011100	034	028	1C	FS
0011101	035	029	1D	GS
0011110	036	030	1E	RS
0011111	037	031	1F	US
0100000	040	032	20	SP
0100001	041	033	21	!
0100010	042	034	22	"
0100011	043	035	23	#
0100100	044	036	24	$
0100101	045	037	25	%
0100110	046	038	26	&
0100111	047	039	27	'
0101000	050	040	28	(
0101001	051	041	29	)
0101010	052	042	2A	*
0101011	053	043	2B	+
0101100	054	044	2C	,
0101101	055	045	2D	-
0101110	056	046	2E	.
0101111	057	047	2F	/
0110000	060	048	30	0
0110001	061	049	31	1
0110010	062	050	32	2
0110011	063	051	33	3
0110100	064	052	34	4

Binary	Octal	Decimal	Hexadecimal	ASCII character
0110101	065	053	35	5
0110110	066	054	36	6
0110111	067	055	37	7
0111000	070	056	38	8
0111001	071	057	39	9
0111010	072	058	3A	:
0111011	073	059	3B	;
0111100	074	060	3C	<
0111101	075	061	3D	=
0111110	076	062	3E	>
0111111	077	063	3F	?
1000000	100	064	40	@
1000001	101	065	41	A
1000010	102	066	42	B
1000011	103	067	43	C
1000100	104	068	44	D
1000101	105	069	45	E
1000110	106	070	46	F
1000111	107	071	47	G
1001000	110	072	48	H
1001001	111	073	49	I
1001010	112	074	4A	J
1001011	113	075	4B	K
1001100	114	076	4C	L
1001101	115	077	4D	M
1001110	116	078	4E	N
1001111	117	079	4F	O
1010000	120	080	50	P
1010001	121	081	51	Q
1010010	122	082	52	R
1010011	123	083	53	S
1010100	124	084	54	T
1010101	125	085	55	U
1010110	126	086	56	V
1010111	127	087	57	W
1011000	130	088	58	X
1011001	131	089	59	Y
1011010	132	090	5A	Z
1011011	133	091	5B	[
1011100	134	092	5C	\

Table 4-2. Continued

Binary	Octal	Decimal	Hexadecimal	ASCII character
1011101	135	093	5D	]
1011110	136	094	5E	^
1011111	137	095	5F	_
1100000	140	096	60	`
1100001	141	097	61	a
1100010	142	098	62	b
1100011	143	099	63	c
1100100	144	100	64	d
1100101	145	101	65	e
1100110	146	102	66	f
1100111	147	103	67	g
1101000	150	104	68	h
1101001	151	105	69	i
1101010	152	106	6A	j
1101011	153	107	6B	k
1101100	154	108	6C	l
1101101	155	109	6D	m
1101110	156	110	6E	n
1101111	157	111	6F	o
1110000	160	112	70	p
1110001	161	113	71	q
1110010	162	114	72	r
1110011	163	115	73	s
1110100	164	116	74	t
1110101	165	117	75	u
1110110	166	118	76	v
1110111	167	119	77	w
1111000	170	120	78	x
1111001	171	121	79	y
1111010	172	122	7A	z
1111011	173	123	7B	{
1111100	174	124	7C	|
1111101	175	125	7D	}
1111110	176	126	7E	~
1111111	177	127	7F	DEL

Microprocessing basics

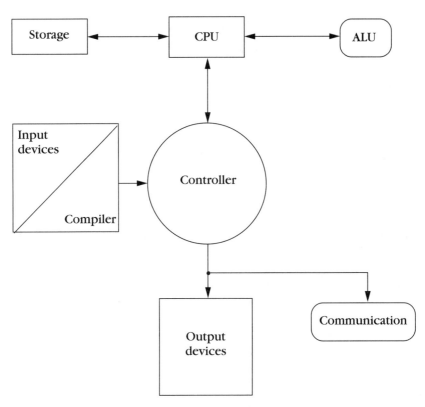

4-2 *The basic elements of a computer system.*

environment, use, and user dictate how all these elements are applied. The computer system is a vital cog in the industrial wheel.

Many features of the traditional computer appear in most facets of factory automation. Motion controls are computers. So too, are programmable controllers. Looking at the main elements individually, we find the computer to be a fast acting, repetitive machine. Input can come from different areas. Direct serial links, magnetic tapes, floppy disks, or operator interfaces can all provide the necessary data to be processed. The central processing unit, or CPU, handles the bulk of the work. It uses an arithmetic logic unit, or ALU, to crunch numbers. Bytes are broken down into bits for processing, and all this work occurs at the speeds at which electricity can functionally flow. Resistive and conductor constraints currently inhibit processing and transfer speed, which is one of the challenges confronting the computer industry.

Software

Some argue that computer hardware is more important than software, but thousands of applications programs have been developed, and most are powerful and effective. Everything from data acquisition to process simulation to machine control. Without them, hardware is useless. These packages have evolved to icon-based and graphical high-level programs. They are high-level because they need to be broken down before the computer's processor can run them, which could involve compiling the program, especially if the program is user-produced or a changed application. (Most of the standard purchased application programs are already compiled and ready to run.)

Prior to data being processed, a program (or set of instructions) must be resident to perform the task. This set of instructions is software and in the computer's storage. Storage can be random access memory (RAM), with additional storage called read-only memory, or ROM, available on most computer systems. ROM exists as hard-coded programs not readily changed. Other forms of storage are also available, and the next generation of computers will even have different methods of storage.

Hardware

If you can touch it, and it's somehow related to a computer, it's hardware. Microprocessor chips, memory chips, printed circuit cards, ribbon cable, chip sockets, batteries, LEDs, switches, and integrated circuits all compose the computer part of the computer system. The input devices are also hardware and consist of the keyboard, disk drives, touch screens, mouse, scanners, card readers, joysticks, and digitizers. Output devices include printers, monitors, displays, or other devices. Virtually thousands of different types of input and output devices exist. Many are used in the office, while many others are used on the factory floor. Different housings and enclosures need to be considered when installing on the plant floor.

Another advancement in printed circuit board design for computers is surface-mount technology. It is a newer technology that enables chips, resistors, capacitors, and other small components to be mounted faster and without much degradation to the actual board. These surface-mounted devices, or SMDs, are not the traditional, larger, through-the-board components that had to be soldered, hand-wired, or made into hybrid circuits (sometimes referred to as *breadboard design*). Rather, SMDs are much smaller block units soldered to a foil seating.

They are then heavily coated with a protective coating, which further solidifies their position on the board. By using SMDs, more components can now be placed on a similar-sized board. In addition, fewer holes are in the board, thus making it stronger. The SMD process is much faster, and many more boards can be produced in a shorter amount of time. These SMDs are, however, small and fragile; they are basically throwaway items if damaged. They are also very sensitive to heat, so board repair and soldering is limited.

Memory

Most microprocessor-based systems use some type of memory. Having enough, being able to quickly access it, and assuring that the resident data remains are the real issues concerning computer memory. Memory is often mistaken for storage and storage mistaken for memory. Both are similar but actually two different entities. *Storage* is where large amounts of data, other application programs, and other computer files reside while not in immediate use by the microprocessor. Storage can be in the form of CD-ROM discs, floppy disks, and magnetic tape (the best storage for the plant floor and for use with industrial computers is open for debate). *Memory* is a storage vehicle, but it is dynamically involved with the computer's present application. Memory could be classified as hardware, as it is resident on a printed circuit board to be readily accessed by the microprocessor.

Several different schemes of computer memory exist, each with its advantages and disadvantages. A computer stores data in its memory using an array of cells and binary on/off states. These cells are electrically charged and are called *volatile* because, if power is removed, the charge is lost and so is the stored data. Many computers rely on volatile random access memory, or RAM. Some provide battery backup in case of a power interruption. Both static random access memory (SRAM) and dynamic random access memory (DRAM) are volatile. DRAM is typically used by the microprocessor as its main memory and is the more common form of memory in use today. SRAM is more complex. It is often associated with cache, or memory buffers, between the main memory and the microprocessor and moves data more efficiently and quickly.

Because memory is mostly volatile, the computer industry has embarked on a trek to find the best, most flexible, and least expensive approach to storing data and not losing it when power is absent. This type of memory is nonvolatile and has evolved many times. The evolution is shown in Fig. 4-3. Volatile memory is also shown for a

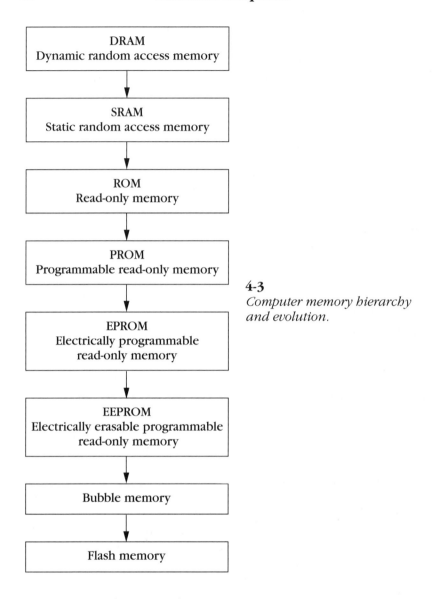

4-3 *Computer memory hierarchy and evolution.*

complete progression. Nonvolatile memory can permanently store data without constantly updating the electrical charge.

Nonvolatile memory devices store data until the devices are erased or reprogrammed. The most basic type of nonvolatile memory is the ROM, or read-only memory. Like the acronym suggests, this memory can only be read from. Data can be taken from this memory location but the data cannot be changed or written over by the active computer and program. ROM is not reprogrammable. Any change to the data or instructions in ROM must be fully redone. The advantage of ROM is that it is inexpensive and can be mass produced for products such as video games. The ROM chip is also sometimes called a PROM, programmable read-only memory, chip.

A step beyond ROM is the EPROM, or electrically programmable read-only memory. These devices can be erased and reprogrammed. They have a quartz window on top of the chip and are removed from their socket to be erased. An ultraviolet light is directed through the quartz window for several minutes to completely erase the chip. To reprogram an EPROM, approximately 12 V is required to "burn" the new program into the memory device. EPROMs get quite warm when loading software into them. They are used extensively throughout industrial computerized equipment.

The EEPROM is the electrically erasable programmable read-only memory. It requires no quartz window, which helps keep costs down, and only a low voltage to erase it. The EEPROM also does not need to be removed from its socket for erasure and can be reprogrammed fully or partially. Extra capacity for addresses is required for partial recoding and to aid in decoding. The main advantage of EEPROMs is that, for applications requiring numerous erase/write cycles, it can accommodate upwards of one million erase/write cycles per package. EEPROMs are typically more expensive than PROMs and EPROMs because many more transistors are required for error correction. Their convenience and practicality usually outweigh the added cost.

Flash memory, or flash ROMs, has helped create microcontrollers. Microcontrollers are self-contained units that can perform their own analog-to-digital (A/D) conversion, interfacing to other components, and other instructions necessary to perform a subfunction of the overall process. All this on one chip. As more and more data must be processed and more attention must be given to external events and communications, microcontrollers and flash memory are gaining momentum. Flash memory uses AND, NOR, and NAND logic circuits; each offers some advantage over the other. AND types are mainly low-power and high-speed, with the capability of writing small blocks of code. NOR flash memory devices are similar to AND types but offer random access to memory locations. NAND types are easily erased and consume small amounts of power.

The benefits of using flash memory are many. First, they offer all the erasing and writing advantages of the best EEPROM memory device. Once in their socket they do not need to be removed. They can be erased and reprogrammed with little effort and power. No EPROM eraser or EPROM burner. They also offer upload and download capability, even from remote locations—just dump the compiled program into the chip. You no longer need to erase and burn the EPROMs and then use express courier for overnight delivery for even the most minor of changes. Software testing and development can thus occur faster because the erase/compile/burn process is virtually eliminated. Even the need for software test fixtures is reduced. More immediate testing, on actual equipment and hardware, is at hand. Flash memory on various machines via microcontrollers makes self-diagnosis and calibration more practical. The evolution of memory devices has reached the point where battery backup is not required, and we can practically load and reload programming into memory at will. By the time this chapter is read, there will probably be a further advancement in memory devices, which is a good segue into firmware.

Firmware can be described in many ways and can be included as software or hardware. Because it is stored instructions and kept in ROM, it could be classified as software. But, because it is software stored on a chip, it could be hardware. These reasons are probably why it received its own name—firmware. A manufacturer's specialized, self-developed code is usually the firmware in a system. The user cannot access this code.

Peripherals

After all the inputted information is processed, the computer provides the output, or finished product. This output is the formatted data. It can be sent to any output device: a printer, a monitor, or uploaded or downloaded to another device for further use. Once output, the cycle is complete, and the computer can perform another function. The computer never tires and processes flawlessly and extremely fast.

In the industrial workplace, redundancy and battery backup are required so that data and programs are not lost because of power outages or brownouts. It is important to keep the process going in the factory. In some instances, if a process were to stop midstream, a dangerous or explosive condition could occur. Factory computers cannot fault. Newer technologies involving bubble memory and flash memory chips enable a machine to retain information when power is turned off even without using batteries or uninterruptible power supplies (UPS).

Online in the factory

Typically, the computer control scheme is a top-down, distributed arrangement. A main computer, often a mainframe plant computer, controls the scheduling and planning of the entire plant's operations.

Pertinent data is then downloaded to the appropriate smaller minicomputers at the cell or supervisory levels of the plant. From here, another download of data happens to the individual microprocessors controlling the machines, programmable logic circuits (PLCs), and industrial computers on the plant floor, and production can begin. Once production has begun, various exchanges of data occur between devices, and information is uploaded to the host or plant system for planning and more scheduling.

Personal computers (PCs) sit on almost everyone's desk in the workplace today. If one does not use a computer directly, his or her job most assuredly is affected by its use somehow. From the secretary's word processor to the plant engineer's industrial computer, the PC is a common component.

The plant engineer of today arrives at the office in the morning, grabs a cup of coffee, goes online, and finds out what is happening in the plant. How is the machinery running? What speed? How many products have we made? Are we making a good product? Did we have any problems through the night? These are all questions that can be answered at the computer. Thus, when the boss wants information, the engineer does not need to go out on the plant floor to interrogate operators or read gauges. The information has been brought to his or her PC.

The microprocessors allow the processed data to be moved to the proper location. We can monitor our work as we go and adjust the process accordingly. In essence, the throughput of data is proportional to the amount of product manufactured in industry today. Because the machines make the product, the faster the data is processed, the faster a product is made!

The computer lends itself well to the flexibility needed in the factory of today and the future. Industrial automation would lose its thunder if plants could not easily change their process. If expensive and time-consuming hardware changes were required every time a new or improved product was needed, automation would never have progressed as far as it has!

The software configurability of any computer allows for the existing hardware to be reused. Retooling is becoming a thing of the past, as retooling times become greatly reduced. Much of the creativity and growth in the computer field is in the software packages that run the machines and computers. These software packages usually are customized to the machine and the task to be performed. The instructions, or code, must be written in higher-level languages rather than in binary form to allow plant engineers the opportunity to maintain the software. Opportunity is the key. Often, several computer systems exist in a particular factory, each one with similar hardware but vastly different software. Plant support people must first have a basic understanding of microprocessors and then be willing to continually learn new programming languages. The instructions must be written so that operators can easily understand the instructions. After all, their job is to run the machine and make the product, not program the computer. Hence, the software packages must be very user-friendly!

Experience has shown, however, that once an electrician has mastered a particular machine or language, that knowledge is often put to use by leaving the company for another for a hefty salary increase. The first company is then left without a knowledgeable person on that machine. High hourly rate dollars then start flying to the contract software programmers. The programmers can be from the original manufacturer of the machine, a freelance programmer, or ironically, the person who left the company but is now moonlighting for more money.

Software must be as user-friendly as practical, but it must also be self-diagnosing. If the machine shuts the production line down, then managers want to know why they are losing money while the line sits idle. Today's machine must be able to quickly, if not instantaneously, tell a maintenance engineer what is wrong, or at least indicate the location of the problem so troubleshooting can begin. The complexity and training involved on new products is a big reason companies standardize on a particular brand of control.

Uploading and downloading

Some manufacturers of control equipment can communicate with their hardware via a modulator/demodulator (modem) link from machine to service center. Many controllers also provide a history of what happened prior to a fault in the equipment, using data accumulated as the machine ran. This history can be accessed and printed and sent via fax to the service center for analysis. Some systems can even be accessed via modem to interrogate the computer system to discover the problem. When a vital part of the production line is not working, its controller is most often blamed for the downtime. With newer memory types, changes to software can be almost instantly implemented by uploading (or downloading) to the machine. Uploading can be referred to as the sending of data to a supervisory, or host, computer. Downloading is the opposite: sending data from the host to a slave or less-powerful machine.

Although newer and faster methods for computer communications are discussed at length in a later chapter, it is worth mentioning that a large number of computer systems still rely on the old mainstays: serial and parallel. Many times a computer's only output device is a plotter or printer near the machine (and this output device usually gets pretty dirty). This scheme only requires a short RS-232 signal cable from the computer's serial port to the output device, which is adequate for certain applications. Some instances require data to be sent to a central point, which is still done with serial cabling. If the data does not need to travel quickly and only travels periodically (once or twice a day), high-speed communications is overkill.

Industrial computers and workstations

Industrial computers come in many sizes and shapes. The traditional office or home computer is common but not usually practical for use on a dirty factory floor. A standard computer placed on the plant floor often fails in a short time. These machines sometimes are placed within an enclosed casing to help protect them from the environment. Other times, the computer has the luxury of a clean control room, but steps still need to be taken to lengthen its service, including a clear plastic cover over the keyboard to protect it from dirty hands.

Many computers and workstations must be placed on the plant floor. Rack-mounted, hardened computers complete with sealed keypads are common. Oil and dirt resistance measures ensure uptime. Good, clean ventilation to the computer also helps keep it running, especially when dust and dirt can collect in an area. Dust and dirt can smother a heat-generating device quickly. A regular maintenance schedule of cleaning all computerized systems is always important. Many of the color monitors on the plant floor are actually smart terminals, or workstations, with built-in microprocessors allowing them to work independently of a master computer system. Most times, however, they do not; they often are sending and receiving process and machine data to and from their host. The onboard computers usually control the enhanced graphics and screens so prevalent with these types of systems.

Computers can come in various packages. From the elaborate console and workstation shown in Fig. 4-4 to the hand-held computer in Fig. 4-5, it is evident that whatever the needs of the application, the

86   Industrial computers

4-4 *A typical control console.* Micon, a subsidiary of Powell Industries, Houston, Texas

4-5
A hand-held computer.
Two Technologies, Inc.

computer industry is apt to provide. A small, compact design in a computer has limitations. The computer shown in Fig. 4-5 allows a technician to perform basic computer functions at a machine or on the shop floor, but that work must then be taken to a laptop or desktop machine for data transfer and reuse. Pendant-type displays with keypads are very popular for remote use, easily handled, and can be found at various machines, usually within that machine's enclosure. They use a simple connection to allow access into a computer system. The hand-held computers carry that process a step further by giving the user a microcontroller.

Industrial computers and workstations 87

Because computers can carry a complete list of specifications and features, it seems appropriate to furnish as part of this chapter a sample set of specifications for a computer. In this case, the specifications are for the hand-held computer being discussed. Figure 4-6 carries the sample specifications for that computer. The irony is that although the unit is physically small, its specifications are just as detailed as for a desktop computer. Standard specifications are for physical size, battery and memory information, display/keypad, interface capability, power requirements, and options available. Of course, it also discusses the CPU, its clock, and certifications. Environmental data

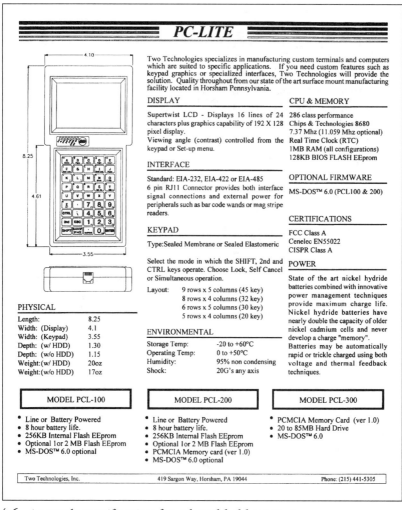

4-6 *A sample specification for a hand-held computer.* Two Technologies, Inc.

88   Industrial computers

shown is very similar to most other electronic pieces of equipment used in the factory. These are fundamental.

One feature the hand-held computer offers is a function called *virtual screen*. Figure 4-7 is a good example of what is meant by virtual screens. Virtual screens allow for standard computer application programs to be run on the hand-held unit. The liquid crystal display (LCD) on the hand-held device is physically capable of a 16 × 24 character display. By moving the cursors, one can navigate throughout a normal screen. This requirement is common in computer-aided drafting systems, as the drawings often exceed the available space on a workstation screen.

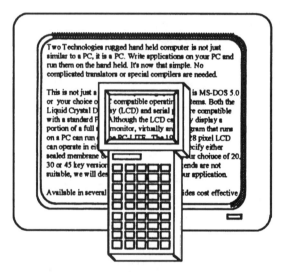

To take advantage of PC-compatible software, the PCL works with an entire full-size monitor screen in its memory but, as illustrated, only a portion of that screen will fit in the PCL's display.

With the PCL's "Virtual Screen" feature, the PCL display acts as a "window" through which you can view a portion of the "full" display screen. Keys are provided on the keypad to allow you to easily move this window around within the virtual screen.

4-7 *A virtual screen.* Two Technologies, Inc.

CAD, CAM, and CAE

Another phase of the manufacturing process affected by computers is design. Engineering and drafting are now computer-aided processes. Computer-aided drafting (CAD) systems are mature and efficient, evidenced by the fact that many powerful CAD software packages can run on desktop computers. Drawings require a huge amount of memory and overhead, however, so it is more practical to use hardware more specific to CAD use. Larger color monitors with high resolution allow for better viewing of a drawing. A clock speed of 66 MHz or higher is often necessary to retrieve, redraw, and save drawings faster.

CAD operators are plentiful and talented. Not only are they proficient in a particular discipline, but they can "click and drag" that mouse with the best of them. Technical schooling now includes hands-on use of a CAD system for drawing courses. Employers want these skills in a candidate. Employers are willing to teach the new employee about the product and the field, but not about the basics of computers and CAD systems. That experience must come with the employee.

More and more, CAD design files can be downloaded to another facet of the manufacturing process, such as estimating the cost of a product. Individual components and subcomponents of the product can be given attribute data, including prices and other pertinent information, to generate a bill of materials. In other instances, the product bill of materials can be downloaded to the shop floor to a computer-controlled machine ready to produce a part (CAM, or computer-aided manufacturing). A part can thus be designed such that when an order is entered into the interactive materials resource planning (MRP) system, the computers take over: a shortage list of subcomponents used in the manufacture of the product is run; those parts not in stock are automatically ordered, and a shop order generated. Engineering and shop drawings are already on file ready to be printed. Our product is produced with virtually little human or manual intervention. This approach does not yet apply to the majority of manufacturers—but it will!

Customers once requested their record drawings in blueprint and sepia form. Now they request a disk with the CAD software files to load onto their plant system. They can even use a plotter to produce their own drawings. The ammonia smell of the blueprint room will probably not be missed. A sample CAD drawing is shown in Fig. 4-8. Besides drawings, other technical data is available via the computer system. Technical publications, specifications, and manuals are then merged with the drawing files to complete the process. Much of this information needs mass quantities of memory and storage, thus requiring multimedia methods such as CD-ROM.

The design process of new products has also seen high-tech change, commonly referred to as computer-aided engineering, or CAE. Instead of producing a physical model of a conceptual design, the designer can portray a three-dimensional model on the computer screen,

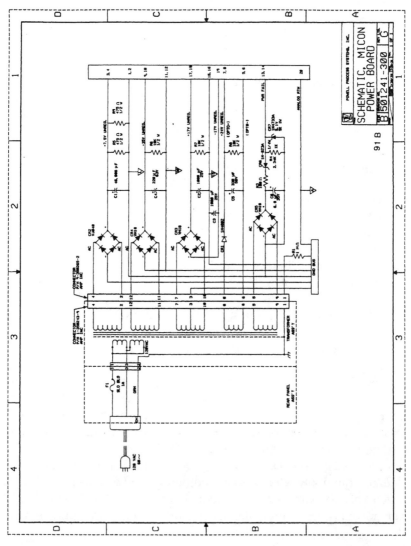

4-8 *A sample CAD drawing.*

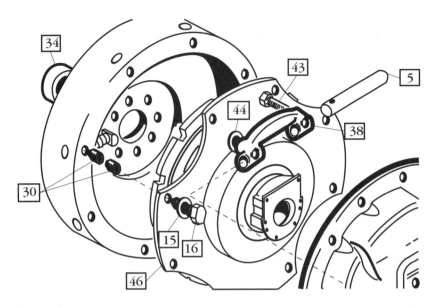

4-9 *A hotspotting screen.* Fisher Communications, Inc.

thus saving weeks of time. Simultaneously, a finite element analysis of that design can be done with software. This analysis checks for structural deficiencies and flaws before a prototype is even built!

Another tool of the CAE community is that of *hotspotting*. This technique allows for the graphic highlighting of individual components of an image. The graphic highlight is then stored as a transparent layer of data, to be used later as required. Figure 4-9 shows a typical drawing with certain parts labeled and numbered, allowing individual parts to be identified and linked or traced to other product parts families. By setting up the parts libraries this way, the designer can link similar part data with other product files. This method of linking throughout a part family is called *hyperlinking*. Design and processing time are greatly reduced, and the designer has much more flexibility. The growing field of CAD, CAM, and CAE has created numerous opportunities. Technical individuals are needed not only for the computer tools but also to write and support the software, troubleshoot the hardware, and continue to expand the horizons of this industry.

Industrial computer applications

The applications where computers can be used are endless. Many applications are very similar, while many others are completely different. Data acquisition systems are similar to one another; on the other hand, the data collected and analyzed by one manufacturer might be completely useless to another. Data acquisition seems to be the perfect application for a computer. Repetitive tasks, many numbers, quick calculations, and recording—the computer package addresses all these functions extremely well. Whether the data is used to control a process, as in statistical process control (SPC) or to ensure quality (statistical quality control, or SQC), or just simply to provide evidence to a customer that what was ordered was built, industry is now locked into computerized schemes. Faster and better products, both hardware and software, are being developed to ease increasing burdens, such as the mandatory requirements to computerize most manufacturing processes. If you do not computerize, your business is almost viewed as second-class.

Data acquisition systems actually gather large amounts of data from a process or a machine via automated measurement systems, process controllers, and even manual entry. The data, usually numerical, is processed by a specific program with custom instructions for the given process. Once processed, the data is either kept on file for future use or sent to a plotter, printer, or other output device in a form requested by the user (plots, graphs, or lists). This methodology can detect trending, predict future problems, and verify that the process is in good operating condition.

It is often necessary to share data and disk systems between computers. Sharing information is known as the *client-server scheme* and is graphically shown in Fig. 4-10. In this scheme, the computers are connected and one computer runs programs resident in the other computer without copying the programs. The computer used to key in the commands is the *client*. The other computer is the *server*, and it typically displays status information. The client computer can use the disk drives and output devices of the server, achieving a pseudo local area network (LAN). These schemes save time and disk space when properly implemented.

The LAN on the plant floor is created out of necessity. Often, a computer system is installed at a particular machine or section of the production facility. In time, it is decided that it is convenient to link that computer with another, such as to the plant's main computer. Networking can be accomplished with specialized software. Often, however, the hardware must be upgraded or peripheral equipment added to achieve the network. Networks are discussed further in a later chapter on plant communications.

Prior to networking the computers, standalone systems were the norm for many applications. For instance, at a milling or grinding station, the performance of a particular machining process might need to be plotted. The serial number of the part being machined, how

Industrial computer applications 93

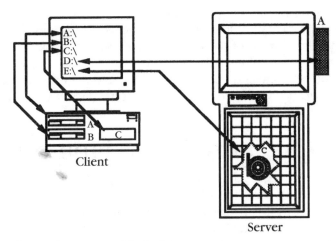

4-10 *Client-server block diagram.* Two Technologies, Inc.

long it took to finish, and other information is useful in tracking, planning, and scheduling this type of work in the factory. Many machine operators must learn how to run a simple computer program. Management now expects everyone in the plant to understand the computer and use it when available. Adequate training can minimize computer phobia with operators and technicians.

Another use of computers is that of parts inventory. There can be thousands of parts in the storeroom. Unless these parts are logged in and kept track of, they might never be used or can even be lost. Computerized inventory can be used by maintenance, engineering, and other plant people who need to find a spare part. The computer is, again, the perfect device for this work. Thousands of part numbers, classified by machine, type of part, or vendor can be placed into a common database. Instead of looking up a part number from an outdated book, the technician can enter a name or number and, within seconds, get the answer from an up-to-date computerized list.

In conjunction with spares inventory, periodic maintenance on machines and equipment needs to be performed. This data needs to be logged into a computer application for proper administration. When a machine needs maintenance, the details, such as what needs done, what was done last time, and by whom, can all be recorded.

Data acquisition is very common. Individual computers at various machine locations for use by operators is common. So too are the spares inventory and maintenance scheduling applications. All are frequent applications of computers. Perhaps the most important role of the industrial computer in the factory is that of process monitoring. The definition of the term *computer* is important here. If we define the computer as the microprocessor and its associated components, do we really have a distinct case for applications? If so, then computers can exist in most of the controllers on the plant floor. They can play many roles besides being just an industrial tool. Process monitoring and machine and motion control fall into the category of computer applications, thus demonstrating that the microprocessor is one of the most dramatic inventions ever.

Computers: today and tomorrow

Computers have played a key role in automating machines and processes everywhere. The traditional primary installations were in the defense and scientific environments. Today, high-speed computers are a way of life in airline, retail, telephone, and just about every other industry. In supermarkets, the checkout lines have been automated with the laser scanners of the universal product codes. Throughput of customers is enhanced and inventories are kept current through online monitoring.

Schools now offer several computer-related courses as minimum requirements rather than optional studies. Computer-assisted learning is in existence even at the elementary school level. We do not so much train children on specific application software packages, but rather on how to make computers part of their everyday life.

Automation can take many forms. A machine can be defined as a device that performs work. The work done is irrelevant to a computing device. It can be a computerized welding machine or the computerized reservation system of a hotel. Both are forms of automation. Both depend on a microprocessor for processed output data to operate. One system is more mechanized than the other. Both are electronic systems and both automate. Look at the new equipment around, especially in the factory. It will amaze you as to the computer and microprocessor content.

As higher-speed processors become available and as memory becomes more cost-effective, the power and flexibility of the controls and design products will continue to amaze! Industrial applications are dictating the ground rules: make a faster microprocessor, make it smaller, and make it less expensive. The limits seem boundless. Who knows where the technology will end? Maybe it won't. A sign of the times is shown in Fig. 4-11. Here, the typical computer system has CD-ROM disks lying about. It used to be magnetic tapes, then 5.25 floppy disks (the bendable kind), and then the high-density 3.5 microdisks. More data residing somewhere!

Artificial intelligence, neural networks, and fuzzy logic are making more progress than we think. Soon, many machines on the plant

Computers: today and tomorrow

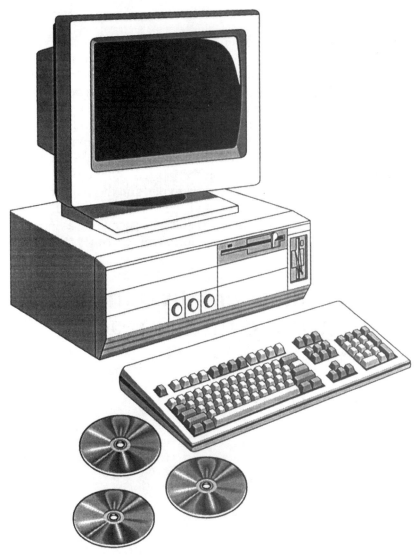

4-11 *CD-ROM disks around a computer workstation.* Fisher Communications, Inc.

floor will have adaptive capabilities that will render them almost self-sufficient. The good news is that we have the ultimate control—we can just pull the plug! (Although with battery backup, we might have to disconnect those systems, too). Maybe another major discovery will revolutionize the computer industry, but who can predict? Will superconductive products open the door? How miniature can chips be made? We'll find out soon enough. One prediction that is fairly safe: computer technology will continue to change and grow at a rapid pace.

As for development, moving too fast can be a detriment. Products are often introduced to the market before completely tested and debugged just to stay competitive. The technology is changing at a rapid pace. Can we, as a society, keep up with the pace? The generation of humans currently involved must remain computer literate. They must be continually trained on newer software packages and systems. A few basic computer courses cannot keep the person up-to-date! The computer learning process must be a never-ending process for as long as that person contributes on the job. The only other option is to pay expensive computer specialists to perform these functions.

Computers are here to stay, and they have changed the world dramatically. Without them there would be no such thing as industrial automation. Our cave dwellers did not know what they were going to miss!

Bibliography and suggested further reading

Tomal and Widmer, *Electronic Troubleshooting*, 1993. TAB/McGraw-Hill, Blue Ridge Summit, PA.

Mandell, Steven L., *Computers and Data Processing Today*, 1983. West Publishing, St. Paul, MN.

5

Process controllers and PLCs

Automating the factory really gained momentum when programmable logic controllers (PLCs) were introduced in the 1970s. Controlling processes was an important step in automating industry. Microprocessors made it possible to accept, analyze, and change what occurred in a process. As operations became repeatable, it was obvious that electronic monitoring of that operation could allow an operator to do another task or simply not worry about the monitored operation. Prior to the 1960s, no magic box or electrical device consolidated these efforts. Control relays and relay logic were the standard.

The process control of post-war industry was using many cumbersome control relays to perform a task. A simplified conventional relay is shown in Fig. 5-1, representative of the style used to build a process control system in industry. As an electromagnetic device, it connects a circuit when its other contact portion was electrically activated. When power was removed, the relay would return to its original state. Another type, the latching relay, operates similarly, but when power was removed, it remains in its latest position. Both types were used extensively to build control systems in factories. The word *build* is key because numerous relays had to be used to perform even the simplest function, with much interlocking between relays, such as shown in the simplified control sequence in Fig. 5-2. To run a motor takes several interlocking relays. Many wires needed to run from terminal to terminal to perform the simple function of Fig. 5-2, which is called the *hardwire approach* to control and relays.

These devices were bulky, taking up valuable factory floor space and giving off a lot of heat. Because so many were needed to perform the simplest of tasks, one could find an entire enclosure filled with relays. They were very hard to troubleshoot, also; it was often difficult

Process controllers and PLCs

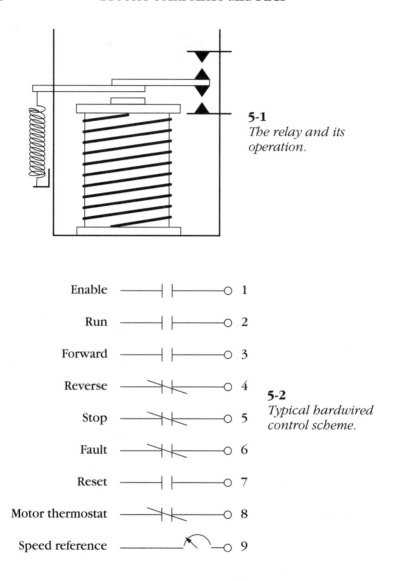

5-1 *The relay and its operation.*

5-2 *Typical hardwired control scheme.*

to discern which one had a problem and, if one had to be replaced, new wiring had to be run. Luckily, the control industry has marched on and we now have better electrical products. Today's relay is much more sophisticated. It comes in a solid-state package and can provide isolation if necessary. In many older factories across the country, however, you can still find an enclosure filled with relays. The good news is that everyone realizes these old devices have got to go!

Industry demanded an electronic means of sequencing these bulky relays into a sleeker, more compact, and more efficient form, which was the beginning of the programmable controller. The name itself is confusing, as many types of controls exist in the industrial environment. The PLC is sometimes referred to as a PC, or programmable controller. This acronym obviously shares initials with another PC, that of the personal computer (and this device also has a niche in the automation sector, as seen in the previous chapter). PC also can stand for process controller. Are these controllers the same or different? The answer is, sometimes! Therefore, when talking process and industrial control, it is simpler to refer to the controller as a PLC, or programmable logic controller. In addition, motor starters, ac and dc drives, motor protectors, timers, counters, simple pushbutton controls, and even limit switches can often be construed as controls. The importance of the relay still lives on. Newer versions are very much employed in industry every day. The relay and latching relay mode of operation is still used in control sequencing, ladder diagrams, and controller programming.

The desktop computer could be used as a PLC. It is a programmable device that, when set up to accept input data, acts on that data and performs the programmed function, ultimately providing an output. This input/output activity is a primary function of today's PLC. Using a computer for this type of task is a terrible waste of its power, so it is not common to see computers used as PLCs. Industrial computers are used quite extensively in factories, however, for other computing and processing functions, including performing minor I/O monitoring.

Some of today's PLCs can be simple and straightforward, while others can be very complex. Some types are of the building-block variety, whereby add-on modules can be incorporated as extra functionality is required. Others are simply standalone units. All have a main central processing unit (CPU). All are programmed in some way. And all basically replace that old style relay logic approach of yesteryear. Key requirements of any PLC include how fast the scan rates are, how much I/O can be incorporated, and how easy the programming is. What's more, the programmable controller is virtually synonymous with process controllers. These machines are alike in both function and look.

PLC hardware

Like most of modern day electronic equipment, a PLC comprises many circuits and components. So many manufacturers of these products have very similar offerings, it is difficult to tell one from another. Thus, product specifications are very important. Not only do these specs give a detailed look at the hardware and physical information, they also list software, firmware, and functional data. Figures 5-3 and 5-4 show a typical process controller specification and a product introduction, respectively. The introduction is an overview of the product and summarizes the specifications. Most electronic equipment manuals have both of these sections included.

Typically, the industrial-grade programmable controller consists of a rack-type superstructure, as shown in the photograph of Fig. 5-5. The rack is actually nothing more than a metal enclosure with corrosion-resistant paint, protective coating materials, and a framework to support several modules. The rack houses the common bus and slot configuration into which the add-on modules are seated. These modules have printed circuit boards attached and, once seated

---

1.2 CONTROLLER SPECIFICATIONS

INPUT-OUTPUT CONFIGURATION

Normally, the common terminals of all I/O circuits are grounded. The analog inputs have filtering networks that provide transient and spike suppression. Reverse voltage applied to the analog inputs or discrete outputs will not result in component damage.

ANALOG-TO-DIGITAL CONVERSION

a. A-to-D Conversion Accuracy +/-0.025%.

b. Long Term A-to-D Stability +/-0.05% of span shift/year max.

c. Resolution 12 bits.

SIGNAL INPUTS

a. Number Available - Six (6)

Signal Inputs 01 to 04 are high level analog inputs. The S-32 is available with one to four high level analog inputs converted to low level inputs.

Signal Inputs 05 and 06 are frequency/pulse inputs.

b. Signal Type

1. High Level -

0 -5, 0.25-1.25, 1-5 vdc, 4-20, or 10-50 mAdc, (with appropriate precision resistors installed at input terminals).

2. Low Level - (Isolated from ground)

Thermocouple
RTD
Millivolts
See page II-29 of the Installation & Maintenance Manual for details.

3. Common Mode Rejection -

80 db min. at 60 Hz Peak Input +/-10 vdc max.

4. Maximum Common Mode Voltage -

+/-10 vdc operating - non-destructive.
+/-150 vdc continuous.

5. Input Resistance -

>100 megohms.

6. Transmitter Type -

Two (2) or four (4) wire. 27.5 vdc XMTR excitation supplied from the controller or an external source.

7. Input Filtering -

12 Hz RC network for each channel.

8. Frequency/Pulse -

0 to 10 KHz square wave only or pulses converted to a resistance value of 400 Ohm or less (pulse on or positive half cycle) or 1 (pulse off or negative half cycle). The signal is connected to the S-32 via a discrete input (see Discrete Inputs below).

DISCRETE INPUTS

a. Number Available: Four (4)

DI-21 to DI-24. DI-23 and DI-24 may be used to interface to frequency (pulse) signal source.

b. Signal Type

ON/OFF Status Monitoring (dry contact).

c. Power Supply

27.5 vdc internally supplied or dc power supply voltage when the dc power supply option is used.

5-3 *Typical process controller specifications.* Micon

PLC hardware

d. Input Isolation	d. Power Source
Opto-isolated from controller electronics. Shares MICON• PC communications on return side.	27.5 vdc supplied either internally by the controller or from an external source (the dc power supply voltage when that option is used).
e. Open/Closed Definition	e. "Keep-Alive" Relay Contact:
The discrete input is OPEN when the external circuit resistance is 1 KOhm or greater (less than 11 mA flow through the contact or device) and is CLOSED when the external circuit resistance is 430 Ohms or less (more than 14 mA flow through the contact or device).	Rated at 27.5 vdc, 0.50 A.
	SERIAL PORTS
	a. Number of Ports: Four (4)
ANALOG OUTPUTS	1. Supervisory Communication Port (to/from workstation)
a. Number of Outputs: Two (2)	RS-485 @ 9600 19200 or 38400 baud. 1200 or 2400 baud are available and can be used when peer-to-peer communication is not used.
b. Signal Type	
4-20, or 0-20 mAdc current source.	
c. Output Load Range	2. Peer-to-Peer Communication Network Port
0-700 Ohms.	RS-485 @ 286 kbaud.
d. Signal Ripple	3. Local Configuration Port
+/-0.025% of 12-bit scale.	RS-485 @ 286 kbaud for hand-held configuration entry panel or @ 9600 baud for configuration by means of computer.
e. Power Source	
27.5 vdc internally supplied or the dc power supply voltage when that option is used.	4. Extended Discrete I/O Communication or MODBUS Protocol Communication Port.
DISCRETE OUTPUTS	RS-485 @ 38400 baud for Extended Discrete operations.
a. Number of Outputs: Two (2)	
b. Signal Type	**POWER SUPPLY**
ON/OFF logic. Open collector transistor conducts to DC common when discrete output is ON.	a. Line Voltage
	90 to 130 vac or 180 to 260 vac 45 to 400 Hz.
c. Rating	b. Optional 24 vdc (nominal) power supply (23-28 vdc).
125 mAdc @ 50 vdc max.	

5-3 *Continued.*

into the backplane, motherboard, or bus, can be screwed in so the module cannot be unseated. These boards and modules fit neatly into the rack assembly. This rack and module package in Fig. 5-6 is shown as an exploded view demonstrating how the pieces fit together. The face of each module might have nomenclature that identifies it as a CPU, I/O module, and so on. The face also has status lights, usually LEDs, to indicate fault and nonfault conditions. There might also be ports for external connections, depending on the module type.

The rack is of little use unless the CPU module is included. As a matter of fact, the PLC is not a PLC without the CPU! The CPU cannot operate without a suitable power supply. Thus, a true programmable controller has a rack, power supply module, at least one I/O module, and the CPU, at a minimum. The CPU module is where I/O is manipulated, information is processed, memory resides, communica-

c. Power Requirement	RF INTERFERENCE
35 watts max. 20 watts avg. (depending on I/O configuration).	Normal operation is not disrupted when the MICON• operates in the following RF fields:

MICROPROCESSOR (CPU)

a. Control Processor

Motorola 68B09, 8-bit I/O, 16-bit registers.

b. Instruction Cycle Time

500 nanoseconds.

c. Arithmetic Precision

15-bit signed fraction, 7-bit signed 2's complement exponent.

+/-0.003% accuracy, or greater, of all arithmetic operations.

d. Second Processor

Intel 8031

PROM (control section)
CMOS EPROM 32k x 8 (27C256).

PROM (comm. section)
CMOS EPROM 8k x 8 (27C64).

RAM (control section)
CMOS RAM 8k x 8 (6264) battery backed by on board NICad battery.

EEPROM (control section)
CMOS 2k x 8 (2817) used as non-volatile back up memory, suitable for transporting configuration.

a. 20 volts/meter @ 14 KHz.
b. 10 volts/meter above 14 KHz.
c. 5 volts/meter from 30 MHz to 10 GHz.

The above are equal or exceed the MID-STD-461B qualifications.

ENVIRONMENTAL CONDITIONS

The S-32 controllers meet the following environmental conditions without loss of accuracy

a. Operating Temperatures

Condition	Ambient Temp.	Ambient Rel. Humidity Non-Condensing
Normal Limits	0-50° C	5-96%
Operative Limits	0-55° C	5-96%
Transport Storage	-40-70° C	5-96%
Nominal Reference	.25° C	35%

b. Vibration

5 to 60 Hz @ 0.030 inch total excursion along the X, Y and Z axes.

c. Chemical Corrosion Resistance

Conformal coating is applied to boards for protection against H_2S, SO, and traces of sand, salt and dust.

CONTROLLER BLOCK DIAGRAM

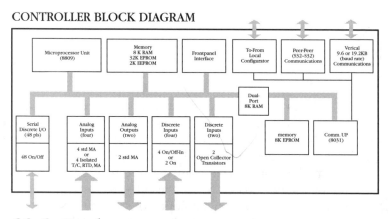

5-3 *Continued.*

tions are controlled, and diagnostics are monitored. The CPU module is the most pronounced module in the rack, usually having the widest profile, shown in Fig. 5-5, where the CPU modules are in the upper portion of the rack, positioned below the analog meters and above the I/O modules.

MICON• S-32 OPERATOR'S MANUAL

SECTION I
INTRODUCTION

1.1 PRODUCT INTRODUCTION

The S-32 is a "single/dual loop" microprocessor-based industrial process controller of the MICON• product line.

It is a compact self-contained instrument that takes less than 15 square inches of panel area and less than 0.18 cubic feet of space behind the panel and yet contains two microprocessors, ROM, RAM and associated electronic components, front panel, display components, control push buttons and power supply. All components, with the exception of the Front Panel and Power Supply components, are mounted on a single multi-layer board.

Besides compactness, the design and manufacturing of the S-32 has built in power, flexibility and reliability that characterizes the MICON• controllers.

The S-32 has four (4) high level analog inputs, two (2) analog control outputs, four (4) discrete inputs and two (2) discrete outputs, eighty (80) steps for loop configuration (4 loops, 20 steps per loop) and thirty-eight (38) steps for analog and discrete input configuration.

Two of the discrete inputs can be used to interface to frequency (pulse) signal source. See Section IV, FNC #06. As a hardware option any number of the analog inputs can be converted to accept low level signals from thermocouples and RTDs. (Hardware, Rev. B).

The design of the front panel hardware and software provides easy to read displays and simple "one finger" operation.

A library of pre-programmed functions (algorithms) resides in the S-32 memory. Most of these pre-programmed functions may be combined freely in loop configurations without restrictions or considerations of available memory, sequence or linking each function to the next. See Section's II and V for details.

The function library approach, used in the MICON• controllers, relieves the user from the task of designing control algorithms. Long and complicated loop configurations are broken into a number of small and easy to understand and remember building blocks.

Configuration entered into the controller is stored in the CMOS Random Access Memory of the S-32. In the event of power loss or temporary removal of the controller from active duty, data integrity is assured for a period of up to two months, by a NI-Cad battery mounted on the controller main board. For long term storage the S-32 configuration can also be uploaded into the non-volatile EEPROM memory (back-up memory) and is retained in storage until erased or another configuration is uploaded. The configuration can be easily downloaded from the EEPROM into the main controller memory at will. See Section V.

Self diagnostic routines continuously monitor the controller's "health" and interrupt train of pulses generated by the controller, the so-called "keep alive" signal, in case a malfunction is detected.

The self diagnostic routines also generate alarms and error codes very useful in pinpointing hardware problems and rapidly restoring full normal operation.

Uninterrupted control of the user's process can be assured by one-to-one automatic back-up. A MICON• Integrity Detector (MID) that monitors the controller "keep alive" signal, is included in the S-32 and will cause control of the process to be transferred to a "hot" back-up unit in case of primary controller failure. The transfer of control is instantaneous, bumpless and complete, i.e., all I/O analogs and discretes are transferred to the back-up controller.

Two very useful loop tuning features included in the S-32 are: The SELF-TUNING feature that uses the Ziegler-Nichols method to determine the optimum loop proportional gain, integral (repeats/minute) and derivative (rate) for the particular process and, the **ADAPTIVE TUNING feature that is used to optimize control of a given process by adjusting the loop proportional gain. The gain adjustment is based on the process dynamics and is continuous.**

Conservative design parameters, high quality components, strict quality control and extensive pre-conditioning are combined to make the S-32 an extraordinarily reliable electronic instrument. Each and every controller is burned-in for 100 hours at a temperature of 50° Celsius, and a minimum of 100 hours at ambient temperature.

Reliability and versatility, in conjunction with the powerful functions and the number of loops available, make the S-32 an instrument that can be successfully used in almost all process control applications regardless of how complex and sophisticated they may be.

5-4 Process controller introduction. Micon

Within the CPU module exist many subcomponents, shown in Fig. 5-7, the block diagram for the controller. This diagram shows two controllers, the primary and the backup. (Redundancy and "hot" backup control schemes are discussed later in this chapter.) You can see the microprocessors, the math coprocessors, and the communication processors in this block diagram. Memory includes main and

Process controllers and PLCs

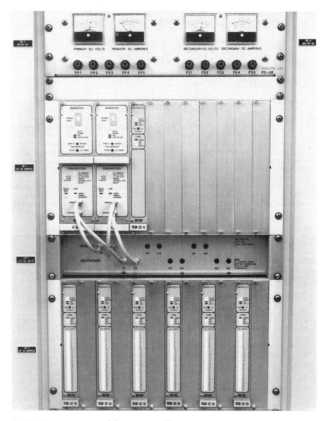

5-5 *Programmable controller.* Micon

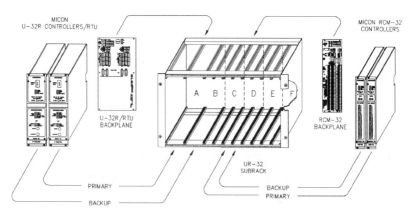

5-6 *Controller rack and backplane.* Micon

PLC hardware

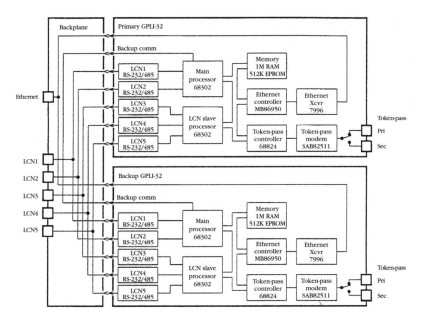

5-7 *A CPU module.* Micon

shared memory devices using both RAM and EPROM. The back portion of the module, which is incorporated into the backplane, has seated connections for the fieldbus, primary and secondary, and peer-to-peer communications. On the front, or face, of the module is the coaxial connections for the token-passing network, which can be seen on Fig. 5-5.

The CPU module has a different set of lights from the I/O modules on its face, including message codes for system diagnostics, several communications LEDs, low-voltage and fail-alarm LEDs, along with the coax connections. The I/O modules, on the other hand, have many of the same indicator lights, but also contain a status LED for every I/O in use within that module. These I/O modules, as shown in Fig. 5-5, are mounted below the main controller modules in the rack. The I/O modules tend to be more slender than the CPU modules. In Fig. 5-8, the entire floor-standing assembly is shown. The door of the rack assembly is removed for the photo, but provides the controller hardware protection from the factory environment. There is even a filtered ventilation scheme to cool the cabinet inside with clean air. The photo also shows a dual-terminal, dual-keyboard workstation. An operator interface must always be present in a process control system.

The CPU is the heart and soul of the programmable controller, and its processing capability is paramount. It must contain the latest

106   Process controllers and PLCs

5-8 *Complete lineup of controller enclosure and operator console.* Micon

and fastest microprocessor to achieve fast execution times. Typical execution, or scan times, are expressed differently. Many manufacturers of PLCs specify scan times per one thousand or 500-word increments. These scan times can be 1 to 2 milliseconds for a 500-word program, further broken down to microseconds per basic instruction. Table 5-1 better illustrates how fast a microsecond is compared to a millisecond. The CPU usually has a 120-V built-in power supply; some units have a 24-V power supply. The power supply furnishes the appropriate voltages where required. Power is required to run the internal clock, supply voltage for the LEDs, and be available for local I/O.

In the factory, it is generally not acceptable to lose any data, control, or programming at any time, regardless of power outages, electrical noise, etc. within the process controller. Once this electronic device has control of the entire process, it is difficult to tolerate any downtime. Downtime can cost thousands of dollars in actual production lost, and the process might require a lengthy startup and setup once interrupted. Elaborate steps are taken to ensure downtime does not happen. One method is to provide battery backup power in case of electrical supply interruption. Battery backup is mainly for the memory. Similarly, read-only memory is used for much of the soft-

Table 5-1.
Various powers of 10 and their alpha prefix designations

Power of 10	Alpha prefix
10^{18}	exa-
10^{15}	peta-
10^{12}	tera-
10^{9}	giga-
10^{6}	mega-
10^{3}	kilo-
10^{2}	hecto-
10^{1}	deka-
10^{-1}	deci-
10^{-2}	centi-
10^{-3}	milli-
10^{-6}	micro-
10^{-9}	nano-
10^{-12}	pico-
10^{-15}	femto-
10^{-18}	atto-

ware and firmware. Another precaution is to incorporate a constant-voltage transformer between the main power supply and the PLC to keep electrical disturbances, or dips, to a minimum. The most elaborate, most expensive means of ensuring PLC uptime is a redundant PLC system, often called a hot backup. A hot backup is a CPU, fully programmed and ready to go online should the main CPU fail. The expense is worth the price when a production line is producing thousands of dollars in products a day.

Digital, discrete, and analog I/O handling

The CPU typically is supplied with a complement of input and output modules. The rack's physical size depends on the number of slots required by the application, including the insertion of all I/O modules. Each module has a dedicated, specialized function, including temperature sensing, motion and positioning modules for motors and servo systems, high-speed counting, ASCII or BASIC modules for two-way serial communication, and proportional-integral-derivative (PID) loop control. The most prevalent add-on modules are for discrete I/O and analog I/O.

The PLC, as a computerized input/output device, must be capable of handling hundreds, if not thousands, of individual, or discrete, inputs and outputs. Discrete inputs and outputs can also be called on and off states. Inputs in a typical factory situation could be low-voltage signals of a designated level. For example, an input can be a contact closure or opening of that circuit to indicate that a limit switch has been triggered. Triggering could be the result of a relay sending a 24-V or 120-V signal to the PLC. The PLC, in turn, scans all the input addresses and recognizes when a change has been made, which is recognized as a change in the on and off status of the particular address.

Another method by which the PLC recognizes a change at a particular address is by a change in actual voltage coming into that address. If a particular input is said to "go high," its voltage level has increased to a predetermined value. If it is said to "go low," the voltage has approached a value of zero. High can equate to on and low usually equates to off.

Another input scheme is that of analog signals. These signals are usually in the form of 1 to 5 V, 0 to 10 V, or 4 to 20 mA. Their signal is in the form of dc electricity. These analog inputs can be furnished from potentiometers dialed by an operator to an appropriate value that the operator perceives as a setting, maybe for speed or time or another pertinent function. This signal, upon arrival at the PLC, is wired into the analog module. The PLC program then recognizes that it has received a 5-V signal.

First, the PLC must determine which scale that signal must be equal to. For instance, if a 0 to 10 Vdc potentiometer signal comes into an analog module, what does it mean? The PLC program must be set up to know that zero volts are a message to do nothing while a full 10 V means to panic, or do something rapidly. Likewise, analog outputs can be furnished by the PLC in the same form: 1 to 5 Vdc, 0 to 10 Vdc, and 4 to 20 mA. These signals can be used to drive meters with the appropriate values to display accurate and up-to-date readings.

An important issue with analog inputs and outputs is that of resolution. Typical resolutions are 12-bit and higher, which determine how accurately the analog signal can be scaled. Analog-to-digital devices are used to convert the signal into usable, processing form. A high-speed, super-powerful signal is not of value to the process if the A/D resolution is low. For comparison, assigned arithmetic values for each binary designation are shown in Table 5-2.

Table 5-2. Binary number decimal equivalents

Power of two	Bit number	Decimal equivalent
2^1	1	2
2^2	2	4
2^3	3	8
2^4	4	16
2^5	5	32
2^6	6	64
2^7	7	128
2^8	8	256
2^9	9	512
2^{10}	10	1024
2^{11}	11	2048
2^{12}	12	4096
2^{13}	13	8192
2^{14}	14	16,384
2^{15}	15	32,768
2^{16}	16	65,536

Other modules

It is not accurate to say that the PLC is just an input/output device. The inputs can have many forms, values, and meanings. The outputs can have the same effect and are usually planned beforehand by the PLC programmer. They all must be defined, one by one, as to function, timing, and interrelation to other I/O. A useful document to log and keep track of I/O is shown in Fig. 5-9. These summary sheets can be and should be used before, during, and after the controller's program is written. The same needs to be done for both discrete and analog I/O.

The common mode of communication with most PLCs is via a high-speed network, or bus. The bus is usually a built-in component of the main CPU control module. Different data transmission speeds are usually set up within the controller. Typical network speeds are in the 1 to 2 Mbps region, with more elaborate systems capable of speeds well over 10 Mbps. Different scan rates are assigned for certain crucial pieces of information, and those scans are prioritized.

Prioritization is important because if every I/O had to have the fastest scan, and an enormous amount of I/O existed, a single scan could take too long and possibly get "hung up" waiting for data. Some

I/O just is not as important to the process at every instant in time. For instance, when performing a high-speed counting routine, bits must be scanned quickly, whereas the monitoring of a motor thermostat can be looked at less frequently. Remember, we are talking about milliseconds, or even microseconds. A motor in a temperature-overload condition takes minutes to overheat. A half-second delay in faulting the motor and cutting off power will not destroy the motor.

		MICON• P-200 ANALOG I/O SUMMARY			
			MICON Co.		
			Powell Process Systems, Inc.		
			Houston, Texas		
MICON NO.	PASSWORD	JOB NO.	FIRMWARE	DATE	SHEET NO.
INPUT NO.		AI-01	AI-02	AI-03	AI-04
FIELD XMTR TAG NO.					
DESCRIPTION					
SIGNAL RANGE					
INPUT NO.		AI-05	AI-06	AI-07	AI-08
FIELD XMTR TAG NO.					
DESCRIPTION					
SIGNAL RANGE					
INPUT NO.		AI-09	AI-10	AI-11	AI-12
FIELD XMTR TAG NO.					
DESCRIPTION					
SIGNAL RANGE					
INPUT NO.		AI-13	AI-14	AI-15	AI-16
FIELD XMTR TAG NO.					
DESCRIPTION					
SIGNAL RANGE					
OUTPUT NO.		AO-01	AO-02	AO-03	AO-04
FIELD DEVICE TAG NO.					
DESCRIPTION					
SIGNAL RANGE					
OUTPUT NO.		AO-05	AO-06	AO-07	AO-08
FIELD DEVICE TAG NO.					
DESCRIPTION					
SIGNAL RANGE					

5-9 *A sample form for keeping track of analog and discrete I/O.* Micon

Other modules

MICON• P-200 DISCRETE INPUT SUMMARY

			MICON Co.	
			Powell Process Systems, Inc.	
			Houston, Texas	
MICON NO.	PASSWORD	JOB NO.	FIRMWARE	DATE	SHEET NO.
INPUT NO.		DI-21	DI-22	DI-23	DI-24
FIELD DEVICE TAG NO.					
DESCRIPTION					
NORMAL					
INPUT NO.		DI-25	DI-26	DI-27	DI-28
FIELD DEVICE TAG NO.					
DESCRIPTION					
NORMAL					
INPUT NO.		DI-29	DI-30	DI-31	DI-32
FIELD DEVICE TAG NO.					
DESCRIPTION					
NORMAL					
INPUT NO.		DI-33	DI-34	DI-35	DI-36
FIELD DEVICE TAG NO.					
DESCRIPTION					
NORMAL					

MICON• P-200 DISCRETE OUTPUT SUMMARY

			MICON Co.	
			Powell Process Systems, Inc.	
			Houston, Texas	
MICON NO.	PASSWORD	JOB NO.	FIRMWARE	DATE	SHEET NO.
OUTPUT NO.		DO-21	DO-22	DO-23	DO-24
FIELD DEVICE TAG NO.					
DESCRIPTION					
OUTPUT NO.		DO-25	DO-26	DO-27	DO-28
FIELD DEVICE TAG NO.					
DESCRIPTION					

5-9 *Continued.*

ASCII and BASIC modules allow the PLC to communicate with other devices. Printers, other PLCs, and computers can receive information from the PLC via a serial communications port using these modules. Drivers that set up the communications ports to send and receive data must be in place or otherwise programmed. This module allows blocks of instructions to be transferred to another device, such as a drive controller. The BASIC module must have a certain amount of memory (mostly battery-backed), an adequate amount of capacity for a transfer, multiple baud or transmission rates up to 19,200 bits per second, and good diagnostic abilities. The cross-referencing table for ASCII characters was shown in Table 4-2. The table showed binary, hexadecimal, and decimal equivalents for all the available ASCII characters and designations.

Another module that can be added to a PLC system is a temperature-sensing module, which is a common requirement for temperature watching in many processes. The adage "a watched pot never boils" is an appropriate analogy here. Variations of these boards can be used to input thermistor signals from motor resistive temperature detectors (RTDs) and most type-J or type-K thermocouples. A thermistor measures temperature via electrical resistance using semiconductor properties, while a thermocouple is a temperature-sensing device that can provide a value for temperature by using a metallic two-wire junction scheme. A simple thermostat just provides a contact closure when two metals within the unit come into contact with one another. It is important to match the temperature-sensing device with the correct PLC module. Alarms are common with this add-on module, and most are capable of working in a Celsius or Fahrenheit environment.

The position control module has become a means of closing the loop around a speed control device for a motor. Feedback is provided into the position control module using pulses from an encoder or other similar device. This feedback is then used in the PLC program, and a quick decision is made by the program as to the output to the motor being controlled. Many dedicated motion controllers are in existence that are more appropriate to controlling motors; however, the module inserted into the rack of a PLC can serve well in motor control.

High-speed counting modules allow for the accounting of parts and production pieces. They accept input in the form of pulses from encoders and higher-frequency devices. This type of module is required because, under the normal I/O processing and scanning scenario of a PLC, the input speed from such devices would be much greater than that of the PLC's microprocessor. This dedicated high-speed counting module is thus required for those applications. The counter data can be incorporated into the PLC program at the appropriate time for action. It can also be accumulated for later use in reporting and documentation. Bar code devices, vision systems, and other types of scanners can be used in conjunction with a high-speed counting module.

Other modules available to the factory are even more specialized. They include voice modules to send, record, and receive messages, and the radio frequency module, which recognizes wireless transmissions and provides the appropriate interface. Even beyond these, it is possible to mix and match high-speed, high-density I/O modules at the PLC for whatever the application's needs.

Man-machine interfaces

As previously mentioned, data must be exchanged between the programmable controller and operator, usually called the *man-machine interface*. What this term really means is that there must be a central area for displays and data entry. Many times, this interface is accomplished using an industrial terminal. Some are mounted and equipped with a keyboard to enter alphanumeric data, which is transmitted to the PLC over a fiberoptic, coaxial, or serial communications wire. A module is at the PLC to accept the data and appropriately route it to the right place within the CPU. This module can be an LAN controller on the CPU module, the BASIC module, or another communication module. Many factories use flat, sealed screens and keypads as these operator interfaces. Some even use touch-screen devices that virtually eliminate the keypad, allowing data to be selected and entered simply by touching the screen at the correct location with one's finger.

Remote I/O, I/O drops, peer-to-peer, and master/slaving

Connectivity is a term relative to the communications of many smart electronic devices in the factory. With so many controllers scattered about the plant, it is necessary to connect as many as possible. The local area network got its start and has gained popularity ever since because of this need to link devices. Wherever more than one computing, processing device exists, so does the need to communicate among them. This need might be to share information on an as-needed basis, to issue instructions, or to monitor. Different schemes

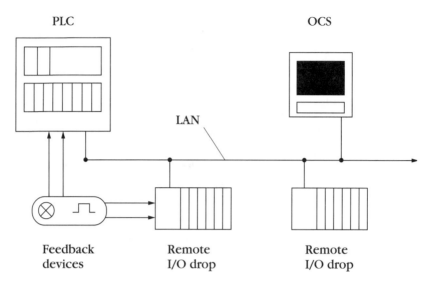

5-10 *PLC with remote I/O drops.*

are possible to tie together controllers and peripherals. One basic scheme is the PLC and remote I/O drop.

Figure 5-10 shows a simple block diagram with a common PLC, various modules, peripherals, and their interfaces. Multiple locations appear to have PLC racks. These are called *remote racks* or *I/O drops* and usually have a shorter profile. In some instances, depending on the application and use, extra processing power might be local to the remote rack. Most often, these are racks of I/O located closer to the machine or process, which eliminate long runs of wire. Once these remote racks are wired, the I/O status data is sent at high speed to the host PLC via a consolidated communication link, which is often a single strand of cable.

A variation of the remote I/O drop is the *master-slave* format. Here, a master controller, or PLC, is in charge of many other slave devices. The key difference from an I/O drop is the capability of the slave device. The slave unit has some processing power and does not rely on the master for all control and processing. The slave does rely on the master controller for instructions, and all communications are routed through this master. This approach has connotations relative to distributed control. These master-slave arrangements almost always require a redundant master in case the original master fails. If the master is out of service, the slaves are also down. Figure 5-11 shows the master-slave arrangement in simple block form. This approach is generally less expensive than a complete peer-to-peer system.

Networks

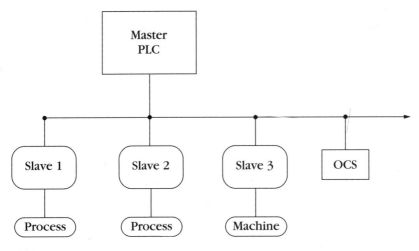

5-11 *PLC master/slave arrangement.*

Figure 5-12 shows a typical peer-to-peer, or PLC-to-PLC, lineup. In this scheme, several PLCs (each able to process on its own) are connected via a high-speed network. In this configuration, each controller is in charge of its own process and its own communications. Each device must send, receive, and decipher its own transmission data and then use that data to control the machine or process to which it is dedicated. In this scheme, it is customary to rotate communication control from device to device, known as token passing. The peer-to-peer system can keep other sections (peers) running when one of the other peer units has faulted or is down.

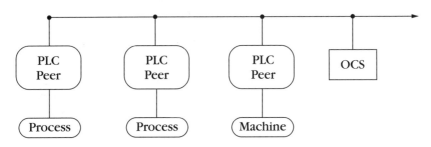

5-12 *Peer-to-peer controller arrangement.*

Networks

The common theme throughout the aforementioned control schemes has been communications. The method of communicating between factory controls must be extremely fast. It must also transmit over thousands of feet of cable without signal degradation. This communication is known as the *data highway*, or the local area network. It is local to the plant or factory in which the communication control is being administered. These data highways must connect hundreds of computers and controllers throughout the plant. Data transmission rates are in the 1 to 10 million bits per second (megabaud), depending on the sophistication of the system.

Just as important to the rate of data transmission is the method of cabling the network. A broadband coaxial cable can transmit the data in the 10 megabaud range by simultaneously transmitting more than one signal at a time. This transmission is in contrast to the baseband cable setup, where one signal is transmitted at a time. Obviously, a premium in price exists in using the broadband cable over the baseband, but, it might be worth it to move "tons of data." Telephone lines and fiberoptic cabling are often used in data transmission. Their limited capacities need to be dealt with as communicating schemes become more complex. One way to maximize communication schemes is to initially configure the system in the most efficient arrangement.

To get the most out of a communications system, one must use controllers and peripherals to their maximum potential, and the physical arrangement of the equipment must be laid out in the most streamlined, effective manner. Arrangement of the system equipment is called the *topology*. Planning ahead and incorporating capacity is a challenge for the LAN designer, because LANs often must connect office computers, the engineering department, and other parts of the factory together.

Cell control

Another control concept is that of the cell controller, which incorporates a lot of power in a single hardware package. Rather than distribute many slaves, I/O drops, or peer controllers throughout a given area of the plant, a cell controller can be installed to handle these applications. The plant is often divided up into individual work cells or manufacturing centers. These cells often generate products independent of the rest of the plant. Because they stand on their own, they can be controlled in a similar manner. The cell controller is not required to report to a supervisory system, nor is it dependent on external communication. It has enough processing power to handle what is required for its manufacturing cell. It could be likened to a PLC with multiple CPUs extensively programmed to handle many tasks simultaneously. Or it could use one high-capacity CPU to handle several tasks, called *multitasking*. While impressive, multitasking exploits the CPU and can get a process in trouble when least expected.

For example, assume a single CPU system is solving the position loops for many motors on a high-speed machine, scanning I/O left and right, and handling communications with a host computer while running the machine. The quality control manager wants you to collect data periodically for statistical process control charting. Collecting data will time-out the microprocessor at some point, and the machine will physically stop for no apparent reason. This stoppage of the machine will most likely be attributed to an intermittent glitch because no electrical nor mechanical source for the problem was found. Your program is solid and you feel pretty sure that you can make the machine and controller perform. The CPU, however, is being asked to take on too much. All it takes is an instant of waiting or indecision on the part of the CPU, and the machine stops. The solution is to use more than one CPU, each with its own dedicated function, so one will not time-out.

Distributed control systems

Much has been argued over which control scheme is best. Should the control and communication "traffic cop" be housed in a master, single unit or local to the process? The application and the available moneys often dictate the direction the control scheme takes. Distributed control is another one of those misunderstood terms and is certainly much broader than this book's intent. The prime aspect of a distributed control system (DCS) is that one large and powerful PLC shares, or distributes, its controller functions with lesser subsystem control peripherals.

Typically, a PLC-based system is less expensive than a true DCS system but does not have the full power and flexibility of a DCS. One important question that must be asked by factory personnel when designing the control scheme is costs versus downtime. If the control scheme is such that a master device must control many lesser components (processes, machines, etc.), what are the consequences when (not if, because it *will* happen) the master controller faults, completely fails, or is simply out of service? Can the plant tolerate having all subsequently controlled processes down because the master is down? If not, the answer is DCS, in which the control and communication capability is installed at each machine and process, and each controller is in charge of its own local area. If one controller goes down, the rest of the plant is still running.

DCSs are sometimes masked by other high-tech facility equipment. Valves, sensors, electronic drives, and other devices have microprocessors built right into their architecture, and a duplication of processing effort can occur. Duplicate processing is not uncommon with today's equipment. Ironically, costs are often not an issue when considering an analog versus digital device. Consequently, many products use microprocessors as part of their standard package so that plenty of functionality is built into the device. The control loop desired might actually be part of a manufacturer's standard software, which can answer the question of DCS versus PLC control.

PLC software and programming

PLCs are really nothing unless somebody has defined an application or need and then installed the instructions to fulfill that need. This is the PLC program. It is, as the name implies, a method of programming the controller to handle the inputs and outputs, or the PLC's logic. Just as the computer must be given instructions to perform its tasks, so too must the PLC. As there are many manufacturers of PLCs, there are many different languages and protocols. Let's look at the basis for these languages and instruction sets.

A mathematician from 1847 England, George Boole, is credited with the development of a system that uses symbols to perform logical relationships between entities. This system is called *Boolean algebra* and has become the standard in computer and digital circuit design. Because this algebra system incorporates truth values and binary numbers, it lends itself well to the on and off states of electrical circuits, which is why digital devices and computers can be given instructions based on what can be predicted by on/off states. Usually, a true condition is an on state and represented by a 1. A false condition is represented by a 0 and is an off state.

In Boolean algebra, further postulates extend beyond ordinary algebra. There are identity elements, such as the following:

Addition: $x + 0 = x$

Multiplication: $x \times 1 = x$

Certain mathematical laws in Boolean algebra also allow for high-speed calculations within digital circuits. The more common and certainly the most familiar (dating back to high-school algebra) are the following:

The associative law:

Addition: $x + (y + z) = (x + y) + z$

Multiplication: $x(yz) = (xy)z$

PLC software and programming

The commutative law:

Addition: $x + y = y + x$

Multiplication: $xy = yx$

The distributive law (combining addition and multiplication):

$x(y + z) = xy + xz$, and $x + yz = (x + y)(x + z)$

Other laws of Boolean algebra are DeMorgan's theory and the absorptive law:

DeMorgan's theory:

Addition: $\overline{(x + y + z)} = \overline{x}\overline{y}\overline{z}$

Multiplication: $\overline{(xyz)} = \overline{x} + \overline{y} + \overline{z}$

Absorptive law:

$x(x + y) = x$

It is important to have a basic understanding of Boolean algebra as Boolean logic is often employed in programming a PLC. It and ladder logic are found in many PLCs as the fundamental, low-level program that controls a given process. Of course, a good understanding of binary (base-2), octal (base-8), decimal (base-10), and the hexadecimal (base-16) numbering systems is essential to programming and working with computers and controllers. In working with numbers, especially the decimal and binary systems, the farthest digit to the right is the least-significant digit. Likewise, the farthest digit to the left is the most-significant digit. This concept comes into play when manipulating data and communicating between devices. These functions electronically process binary equivalents for everything from numbers to letters. You will hear values expressed as the least-significant and most-significant bits (binary digits).

As process requirements become more complex and more extensive calculation is required of the PLC, higher-level languages have also emerged. Some are in the form of blocks, which can be incorporated into the ladder scheme. Each block contains several instructions and is more or less self-contained. Thus, the programmer simply applies the block to the application program and does not need to list, step-by-step, every equation. This shortcut is most important in performing PID calculations. PID (the acronym for proportional-integral-derivative) loops are discussed later in this chapter.

PIDs are closed loops that aid in the correction of system and process errors. They could not be performed using common Boolean and ladder logic; it would be too cumbersome and time-consuming. Thus, higher-level languages have emerged. Many even go beyond block diagrams. Today's PLC programming incorporates symbols and coding much like BASIC, C, and other high-level languages. Many programs are now written with macros and function blocks. These functions, although having different names and acronyms, have the same end result when the program is implemented. The functions differ with each PLC manufacturer, and each has customized its own programming to suit the microprocessors and types of memory used in the PLC hardware.

Boolean algebra is sometimes called a *mnemonic*, which means it uses symbols to represent operators. Once the truth value is known of a particular entity, it can be further used in more comparisons. For example, there are conditional operators AND and OR. The AND function merely implies that a condition, or state, can only be on or off, whereas the OR function needs to have one or another condition true to be on or off. They are sometimes called gates, or open or closed states. Gating is a term used mainly in electrical and control dialogue whereby something is turned on and off.

These expressions and operators become the building blocks of a PLC program. Inverse values also exist for these operators. The NOT versions of these operators are NAND (not AND) and NOR (not OR). These and other operators handle the logic in a given control scheme. This logic is graphically depicted by common symbols and ladder diagram abbreviations not only used in ladder diagrams but also used frequently in basic electrical schematics and wiring diagrams. This graphical depiction of the application is called *ladder logic*, or a *ladder diagram*. It can also be called the *relay ladder diagram*. The relay ladder diagram's shortened version is the contact ladder diagram, which condenses the diagram into contacts and outputs. This diagram is actually the hierarchy of events and control for a given PLC application.

As suspected, the ladder diagram gets its name from its resemblance to a ladder, as shown in Fig. 5-13. Lines of the diagram are called rungs and define where in the diagram a particular event or sequence of events occurs. The rungs should be numbered to be able to navigate through the ladder diagram and reference them when necessary. Comments usually go to the right of the rung, and the rung numbers are on the left. The emergency stop and fault logic go at the top of a lengthy ladder diagram. Figure 5-14 shows an example of this rung. Various zones can exist within the ladder diagram. A zone is a certain routine or sequence of events, with each zone having designations for beginning and end.

A complete ladder diagram would have a control transformer, the electrical positive and negative parts of the ladder, and the earth

PLC software and programming

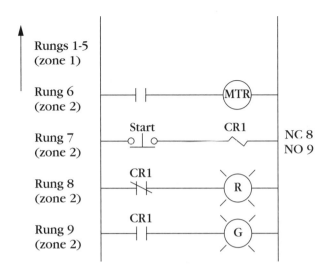

5-13 *Relay ladder.*

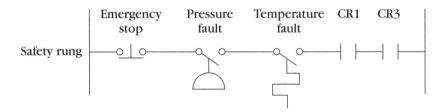

5-14 *The safety or emergency stop rung of a ladder diagram.*

ground point. Each vertical leg of the ladder is powered such that when a rung completes a circuit, current flows through it. Ladder logic is actually the relay logic of the machine or process and details the events and actions for the relays. How and when do we want electricity to flow through that rung? For instance, in Fig. 5-13 when the start pushbutton is selected, power goes to the coil of CR1. By reading the comments to the right of that rung, we find that rung 8 has the normally closed contact of CR1, and rung 9 has the normally open contact, showing how the green and red lights will be energized.

Of course, this diagram is very basic and there would most likely be many other conditions and actions present up and down this ladder. Most ladder diagrams for PLC programming use are shown in the contact ladder diagram scheme illustrated in Fig. 5-15. Figure 5-15 is the contact ladder diagram equivalent of the relay ladder diagram in

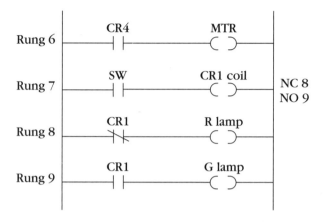

5-15 *Contact ladder diagram.*

Fig. 5-13. The relay ladder diagrams are very useful in the field when starting and troubleshooting equipment; however, contact ladder diagrams are essential when developing the program for the PLC. The root of PLC usage is programming, after all, and a PLC is not really justified for the application unless several inputs and outputs exist, along with many other subsequent actions.

The ladder logic contains different components common to most PLC and computerized systems. Addition, subtraction, multiplication, and division are necessary for processing and decision making. Get (for input) and put (for output) functions must be available to manipulate I/O and for comparisons. As a matter of fact, many times the ladder is the actual program and can be used to run a machine or process. This type of programming contains many necessary functions, or permissives, and the safety interlocks that those bulky relays provided 20 years ago. The logic is now handled electronically. Completely defining all functions of a process or machine's operation are just as important, but if a change is made, we don't have to pull and run new wires!

Techniques used to program a PLC vary from programmer to programmer. Some programmers like to use flags to help them locate where they are in the program, especially during troubleshooting. Likewise, counters are often used as a programming tool. Many times, after the PLC's CPU has counted a predefined quantity of events, an output is triggered. The output could be turned on or off to alert another device or operator that the desired count has been reached. As discussed earlier, if the actual countable signal is too frequent, another dedicated counter module must be used. In typical operation, however, the PLC's microprocessor is adequate for counting and logging repeated events.

Several instructions allow the programmer to move about within the ladder program. One is the jump command. Once activated, the program "jumps," or goes to another location in the program and executes from that point. It is much like a GOTO statement and is even similar to its jump-to-subroutine counterpart. Every PLC manufacturer has its own specific set of instructions, many of which are similar from manufacturer to manufacturer. Each has its own language and means of programming, making it imperative to completely understand the PLC system involved and obtain all the pertinent documentation. You can see why facilities attempt to standardize on control equipment. Relearning and retraining when using new technology or another manufacturer's control equipment can be very time-consuming. Production facilities do not have the luxury of much free time to use relearning or reteaching. Often, technicians are hired because of their in-depth knowledge of a particular manufacturer's control equipment, both hardware and software.

Keeping track of all that data

Part of the ladder program must monitor locations and where information, whether discrete I/O or strings of data, is actually coming or going. Looking at any complete ladder diagram, we can see numerical designations given to certain operators, designated in different ways by different manufacturers. This means tracking is very important and can virtually shut down a machine because the program is looking for information in the wrong location. Depending on the function and information needed, different designations are given to PLC terms. For example, a *register* is a location in the CPU's memory that can be compared to a simple variable in a computer system's array of data. Its value can change. Likewise, the register must have a memory location address. Addressing and registers go hand-in-hand. Addressing is sometimes called *polling*. In a master/slave PLC arrangement, information can be requested from the slave device's memory. In effect, the address of that register information must be known to retrieve the data.

To keep track of all the different memory locations on paper, a sample summary sheet is included in Fig. 5-16. This information makes the programming and troubleshooting much easier for the current PLC programmer, as well as the electricians, technicians, and eventually the new programmer who inherits the project or system!

MICON• P-200 LOC-MEMORY LOCATION SUMMARY							
					MICON Co.		
					Powell Process Systems, Inc.		
					Houston, Texas		
MICON NO.	PASSWORD		JOB NO.	FIRMWARE	DATE		SHEET NO.
LOCATION NUMBER		DESCRIPTION					SRC WHERE DEFINED
LOC-01							
LOC-02							
LOC-03							
LOC-04							
LOC-05							
LOC-06							
LOC-07							
LOC-08							
LOC-09							
LOC-10							
LOC-11							
LOC-12							
LOC-13							
LOC-14							
LOC-15							
LOC-16							
LOC-17							
LOC-18							
LOC-19							
LOC-20							
LOC-21							
LOC-22							

5-16 *Sample form for keeping track of memory locations.* Micon

As seen previously, multiple CPUs or even PLCs are often connected via a local area network. Multiple racks must be cabled together, so it is essential to have all memory locations identified. Obviously, a random storage method would not work at all. Information must have a designated address to be manipulated later. One further description on location is called a *node*. A node is one of the components on the local area network. When looking for data, the search begins at the node or device that has the memory holding the data needed. Subsequent channels and subchannels within the particular node are then used to further define where to find the desired data.

It is clear to see that the more devices linked together in the plant, the more complex and likely there are to be problems in data transmission, which can make locating the problem very difficult. One method of assuring that data can be passed from device to device is called *collision detection*. This function checks the activity on the bus or network before a stream of data is passed, helping minimize transmission errors.

Programming summary

Programming a PLC is not a simple task. It takes years of experience. It also helps if that experience is with one particular type of PLC and its specific language. Another piece of the equation, just as important to the programming, is that of completely understanding the machine or process. If it can be fully defined as to what must occur when other events happen, the programmer can get that information into the PLC. The biggest problem that exists today for programmers is that many functions of the process or machine are not known ahead of time. The programmer learns of important facts later, after the machine is started with the new PLC program or even when the machine is running and at 3:00 AM on a Sunday simply stops. This is frustration at its best.

PID loop control

The PLC is often used to perform more extensive calculations to control the process. Monitoring, manipulating, and turning I/O on and off is the primary job; however, as processes and machines have become more complicated and captive to many microprocessor-based systems, it is now expected that the PLC be more of a process controller. If the PLC can handle the extra functionality by adding a module or two, other devices (another motor, bigger tank, longer dryer, etc.) do not need to be added. If a sensor can be added and the feedback directed into the PLC for evaluation, the process might be more accurately controlled, thus averting more elaborate and expensive modifications to the process. It is still true that the fewer the components in a system the more reliable that system is. The more interconnecting cable and wire between multiple devices means more chance of data transmission errors. Use the capability of the equipment you've got first!

The PLC is often used as a setpoint, or PID loop controller. Today's PLCs are expected to have PID capability, the software and predefined function blocks to perform it, and the flexibility to perform variations of the algorithm. Figure 5-17 shows a process controller dedicated to PID setpoint and process control. It differs from the traditional rack-mount controller in that it is flexible, modular, and all the add-on boards are contained within the long, narrow housing. In addition, the operator interface and display is built right in. This unit can perform single or multiple PID loop control.

5-17
Modular Process Controller. Bailey-Fischer and Porter

Plenty of variations of PID exist, such as PID control loops and interactive and noninteractive PID loops. Some are adaptive, which means the controller can make additional corrections above and beyond the set PID loop changes as the process warrants. These changes can involve gain changes to the overall system. The gain of any system is defined as the ratio of the system's output signal to its input signal, as seen in Fig. 5-18. Full discussion on PID loop control is, however, beyond the scope of this book. Refer to the bibliography at the end of this chapter for additional sources.

A PID loop is just that: a closed-loop system that receives feedback from a device and compares that feedback to a reference, or set-

PID loop control

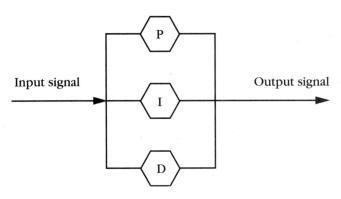

5-18 *Proportional, integral, and derivative (PID) control loops.*

point, for the process it is controlling. Often, all three components of the PID loop are not incorporated in the actual control scheme. For example, a control signal output to a device proportional to the sum of the error signal and the setpoint is the P portion of PID. The control signal is directly correcting for any deviation seen in the process. This control is often called PID, but in actuality it is simply P, or proportional control. For any given action, there is a proportional reaction by the system's controller.

Likewise, the I portion of the loop control is the integral of the error signal with respect to the setpoint. This step is where a good understanding of calculus comes into play. Integrals consider time. Resetting values through a corrective control loop during a period of time occurs in this type of control. The looping is dependent on the severity of the deviation to the setpoint and the processor's capability. Some systems employ both the P and I portions of the PID control scheme.

Last, the D portion of the PID loop is the derivative. A derivative is a change to, or derivation, of the resultant. This loop is concerned with rate of change in the deviation. Similar to the integral loop, controller scanning and how often the deviation is seen become major components of the equation. The D portion of the PID loop can be coupled with the P portion when the application dictates and is then referred to as a PD control loop.

PID control can be as simple as shutting a valve off when a desired temperature is met or as complicated as instantaneously correcting a motor's speed, inertia, and torque as a material builds on a roll at high speeds. An important facet of PID control is how fast, in milliseconds, or even microseconds, the entire loop takes, which is a function of the microprocessor speed, where data is coming from (and how fast the transmission is), and how well the program and calculation methodology are laid out. High-frequency filters can be used to keep PID signals true. Any deviations to these signals can be perceived as major fluctuations in a sensitive, high-response, and high-resolution control scheme.

The process control involved with hydraulics is a big part of industrial automation. Good control of the system is dependent on the sensors, limits, and transducers. Specific devices have been developed to control fluids. One such device is the *magnetic flowmeter*. Sometimes called *magmeters*, these magnetic flowmeters measure the flow of liquids in an enclosed pipe. The fluid must be able to conduct electricity for the magmeter to function properly. These devices have actually been around for over four decades.

Today, magnetic flowmeter devices can handle flows from as little as 0.003 gallon per minute up to 750,000 gallons per minute. This range covers very small and very large diameters of pipe. In a simplified sense, the magnetic flowmeter is a transducer integral to controlling a hydraulic system. When selecting a flowmeter, there are a few important issues to consider:

- The flowmeter must provide an obstructionless design. The object is to control the flow as smoothly as possible in a piping scheme. Any obstruction defeats the purpose. Any moving parts in a flowmeter design can pose an obstruction (and a bigger problem area over time).
- The output from the flowmeter should be as linear as possible. The signal to a supervisory computer system must be clean, stable, and fast. It should be directly proportional to the velocity of the flow of liquid through the pipe.
- The flowmeter itself, because it is installed right into the piping system in line with the flowing liquid, must be made of materials resistant to any corrosive chemicals that could be present in the flow. The parts of the flowmeter that come into contact with the fluid are called *wetted parts*.
- The flow measuring device must be accurate. Not only does the device and its particular design need a wide range in which to be fully accurate, it must also be repeatable, consistent, and able to adjust output quickly.

A magnetic flowmeter is shown in Fig. 5-19. The basic operation of a magnetic flowmeter is similar to many other controllers and devices discussed throughout this book. The basic operating principle is induction. Figure 5-20 illustrates how and what subcomponents are necessary to achieve the desired result. The fluid, being conductive, flows

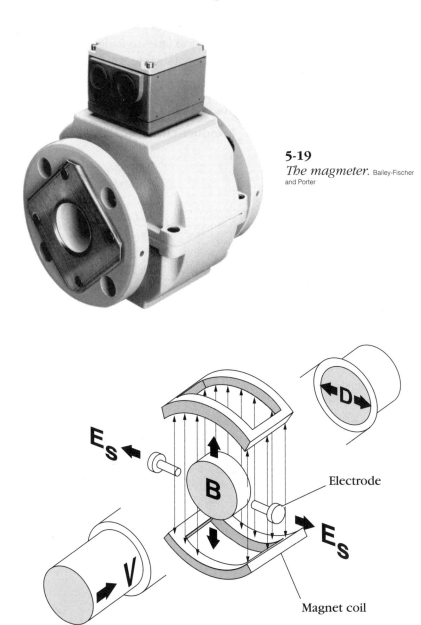

5-19 *The magmeter.* Bailey-Fischer and Porter

5-20 *Operating principle of the magnetic flowmeter.* Bailey-Fischer and Porter

through a magnetic field. Table 5-3 lists the various conductivities of water. A voltage directly related to the flow is produced, regardless of pressure, temperature, densities, or viscosities. In this manner, a signal is produced and then supplied to a supervisory system, which determines whether the flow is too much or too little and adjusts the device that initiates the flow or changes the flow (such as a valve).

Table 5-3. Conductivities of water

Water type	Conductivity (siemens/cm $\times 10^{-6}$)
Boiler feed (condensate)	1.2–20
Circulated coolant	800–8000
Demineralized (industrial)	0.1–0.9
Distilled (in glass)	1.0–7.5
Marine evaporator distillate	5.0–60
Mineral waters (soda water)	800–2500
Raw waters	80–900
Rinses (electroplating)	20–70; 750–3000
River and stream waters	80–1000
Ultrapure water	0.05–0.1

Although it might seem that a magnetic flowmeter is a fairly simple device in concept, its implementation has many behind-the-scenes functions occurring, including the ac excitation, conductivity adjustments, transmitter scaling and configurations, displays and readouts, calibrations, electrical noise reduction, self-diagnostics, and full communications. Besides these issues, material and liner types must be considered, along with orienting and properly installing such a device. All must be done correctly for proper operation. Some liquids have lower conductivity rates than others, and special attention might be needed when selecting the flowmeter. A technical bulletin exists that fully describes all facets of magnetic flowmeters from design to installation and is noted in the bibliography at the end of this chapter.

PLC applications

Figures 5-21 and 5-22 show in block form common yet different uses for programmable controllers: a data acquisition system and a chemical plant process. All applications cannot be shown graphically, but an extensive list of applications does follow. As PLCs become more powerful and software becomes more plentiful, challenging applications can be pursued. Some are now done routinely where before,

PLC applications

Typical SCADA architecture

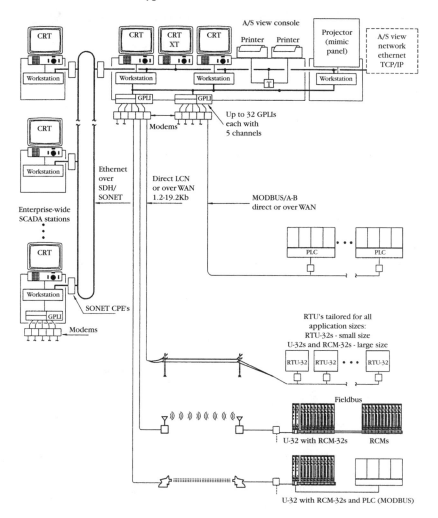

5-21 *A SCADA application.* Micon

even the thought of a PLC on that type of a machine was unheard of. PLC general applications include the following:
- *temperature control systems* to monitor and control a process's temperature; mixing, cooling, and heating of liquids
- *machine tools*: grinders, saws, milling, drills, presses, welders
- *pressure control systems*: pumping stations, air-handling systems

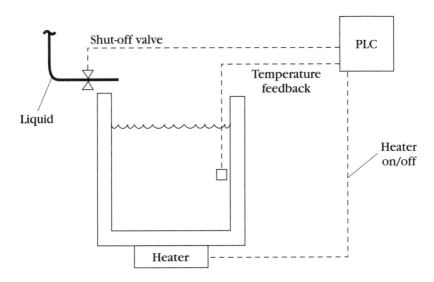

5-22 *A chemical facility mixer-agitator control application.*

- *conveyors/run-out tables*: starting/stopping, counting, coordinating
- *robots*: servomotor loop control, pass/fail, parts insertion, gantry
- *SCADA systems*: many data acquisition schemes, SQC, SPC, ISO-9000
- *material handling*: counting, sorting, loading, unloading
- *assembly machines*: automatic insertion, pick and place, fastening
- *energy management*: HVAC, air balancing, cooling towers, chillers
- *work cell control*: robots, storage, retrieval, staging
- *molding/casting*: temperature/pressure control, mixing, drying, painting
- *furnaces/boilers*: blower/burner control, induced draft, forced draft fans
- *mining*: conveyors, pumps, safety monitors, weighing
- *packaging*: incoming/outgoing materials, weighing, sorting, bar coding
- *power plants*: brownout avoidance (switching), cooling, temperature
- *food processing*: batch processing, weighing, packaging
- *petrochemical*: valve/piping control, instrumentation, safety monitors

- *transportation*: tracking, switching, dispatching, item handling
- *telecommunications*: switching, logging, line checking, polling
- *aerospace*: airplane manufacturing, air traffic control, parts tracing
- *metals*: hot/cold mill control, gauging, raw material received

The list goes on. It goes on because PLCs and process controllers are becoming extremely diverse. Computers and microprocessors are so prevalent that everyone and every business is getting into the act, using the high-tech machines to do or help do the work for them. Add-on modules are capable of doing local processing, thus freeing the master CPU for other functions. A big overlap exists in controller function, architecture, and appearance. A traditional PLC manufacturer might have many types of computing and process control devices available. This diversification creates solutions and new advances. Competition makes the control product's vendor base more reliability conscious and hastens the product's maturity, which all adds to bettering industry and the people who keep it going!

Troubleshooting, fault handling, and diagnostics

When PLCs burst onto the factory scene, they were a novel, "where-has-this-useful-device-been" product. There was so much excitement and eagerness to implement them that it was hard for supply to keep up with demand. It was evident, however, that PLCs were a young science. For many early models, there never were any initial hours of service because they did not work at all. New problems, never before experienced with relay logic, were showing up at various installations. Electrical, memory, microprocessor, programming errors, and wiring problems had not been encountered in real-time factory environments. As time and application experience marched on, there was a movement to make the controller self-diagnosing and better able to detect faults. The controller had the circuitry and the ability to check itself and other devices attached to it. All that needed to be done was to instruct the unit to do these things. And so diagnostics and fault-handling evolved.

It has evolved to the point where each product manual (sent with each unit shipped to a plant) has a major portion dedicated to the startup, proper configuration, fault and error message descriptions, self-tests, lights, indicators, and a host of other wellness checks. Besides being many times more reliable, if a problem with the controller exists, it can quickly be identified and corrected. Moreover, the family of products today uses mature subcomponents, technicians have worked through the early years with the products, and many manufacturers have plenty of real application experience. Good documentation, however, is still essential.

One excellent way of troubleshooting a controller system is via the *troubleshooting trees*, or flowcharts. By listing the problem or symptom at the top of the tree, we then answer yes or no to questions about the problem as the tree branches out. Figure 5-23 shows several corresponding troubleshooting trees for a typical process controller. These trees cover the basic walkthrough from power-up to checking for blown fuses, bad circuit boards, to checking the I/O.

If the controller finds a problem, it might light an indicating light or display a fault message. The maintenance manual still must define what these mean and pinpoint where to look for correction. These products are complex and have not yet become self-diagnosing enough to find exactly where all the problems are and then alert the outside world how to go about repairing it. Much self-monitor and self-test circuitry occurs in controller products, and the industry is continually striving for better diagnostics. Having no problems by virtue of good burn-in and factory testing are also preferable scenarios to the technician.

Bibliography and suggested further reading

Batten, George L. 1994. *Programmable Controllers, Hardware, Software, and Applications.* Second Edition, Blue Ridge Summit, PA: TAB Books.

Mills, Raymond C., 1991. *Introduction to Magnetic Flowmeters*, Bulletin no. 10D-14a. Warminster, PA: Fischer and Porter Company.

Bibliography and suggested further reading

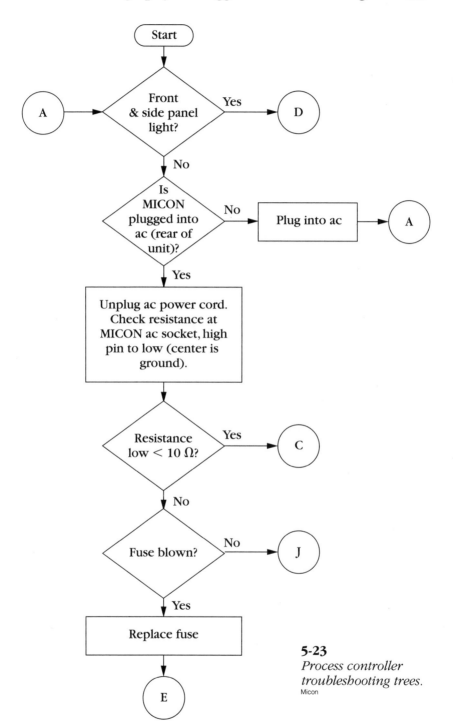

5-23
Process controller troubleshooting trees.
Micon

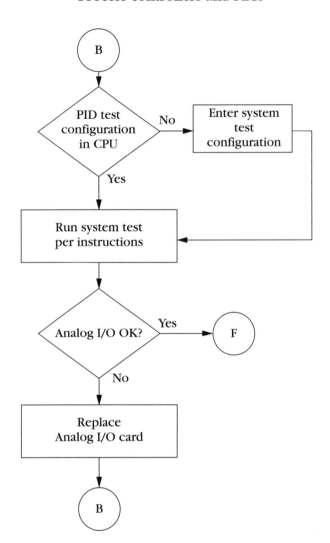

5-23 *Continued.*

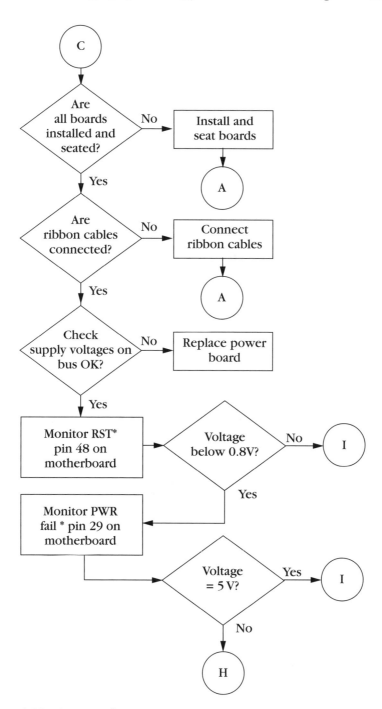

5-23 *Continued.*

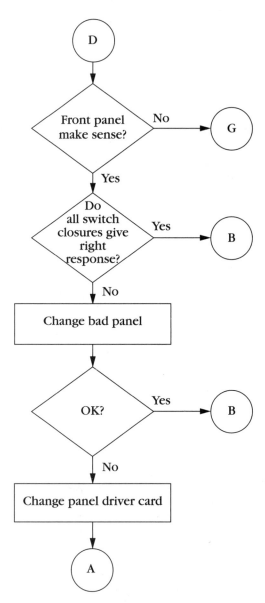

5-23 *Continued.*

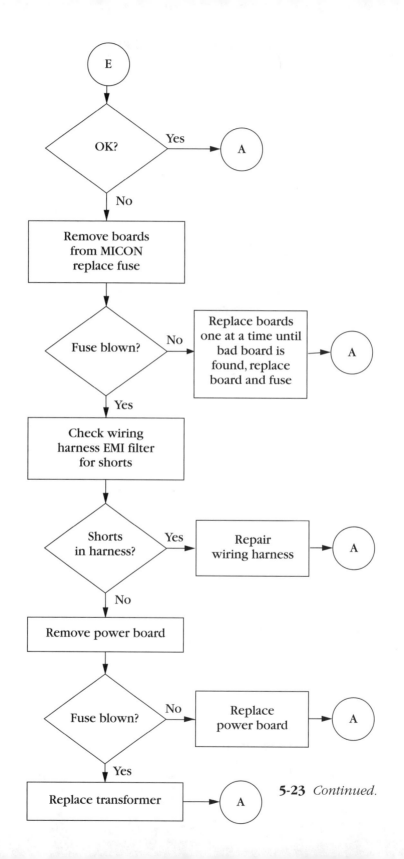

5-23 Continued.

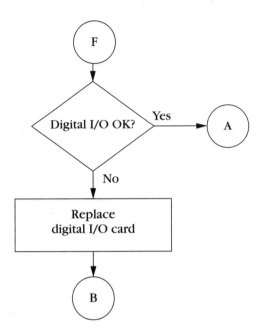

5-23 *Continued.*

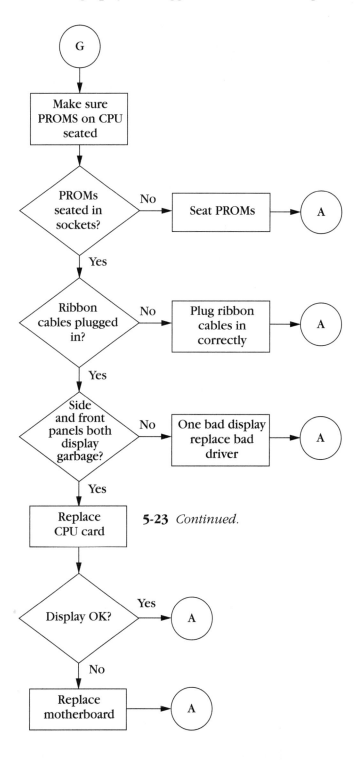

5-23 Continued.

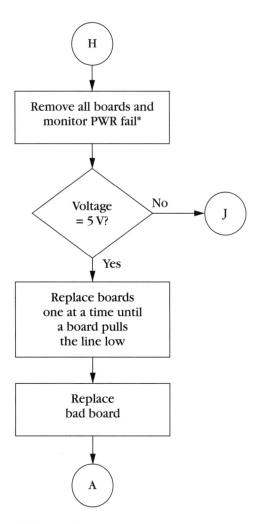

5-23 *Continued.*

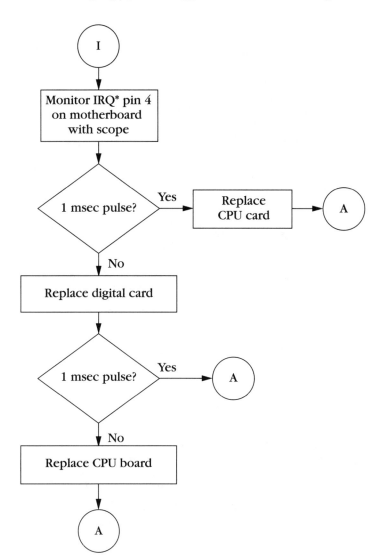

5-23 *Continued.*

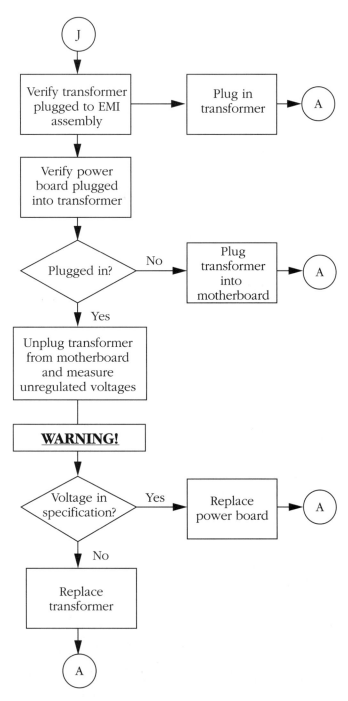

5-23 *Continued.*

6

Electric motors

Walk into any factory and observe all the processes involved in manufacturing. There is a lot of movement going on. You'll also note that behind each movement is a prime mover. In today's factory, that prime mover is probably an electric motor, powered by ac or dc energy. The electric motor is converting electrical energy into mechanical energy, or work. The electric motor is the force behind the product being manufactured. Without it, quite a few steam lines, diesel generators, turbines, and so on would be needed to act as the prime movers, which is not practical.

All motion takes on a rolling or rotating theme. The impact of the invention of the wheel is still being felt today and will be for many years to come. Some form of the wheel is an element in most power transmission systems and especially in the construction and operation of an electric motor. For example, most electric motors incorporate a rotating, round element called the rotor. It is called the rotor because it rotates and is the inside of the motor. One can be found in both dc and ac machines. The complementary component to the rotor in the electric motor is the *stator*. The stator is the stationary part of the motor and can be called the outside of the motor. Figure 6-1 shows the rotor and the stator. Regardless of how powered, the electric motor must have both a rotating element and a stationary element to produce torque. Nomenclature and the actual means of applying power to obtain speed and torque differentiate electric motors from each other.

Figures 6-2 and 6-3 show simplified views and cross sections of an ac induction motor and a dc brush-type motor, respectively. Besides the rotating and stationary components, other common pieces are the front and back bearings, the motor base, the terminal or conduit box, end plates (sometimes called bell housings), and the motor frame. Depending on the type of motor and its desired operation, construction varies as does material type. Motor stators can be made

146   Electric motors

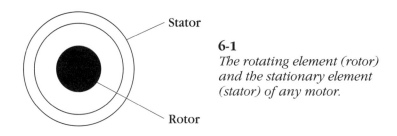

6-1 *The rotating element (rotor) and the stationary element (stator) of any motor.*

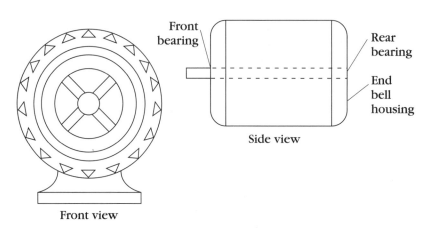

6-2 *Simplified front and side views of a typical ac motor.*

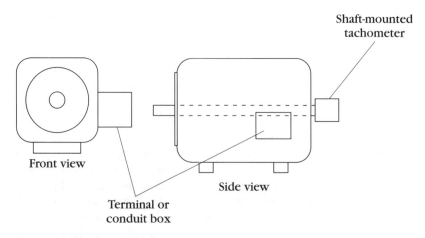

6-3 *Simplified front and side views of a typical square frame dc motor.*

from cast iron or even from aluminum. Other motor stators can be several steel laminations welded together. When selecting a motor stator, construction material should be carefully considered because of heat. After all, an electric motor is nothing more than a heatsink, with current running through it. As the motor heats and cools, the life of the motor is lessened. When comparing two motors of identical rated horsepower, the actual weights of each can help determine which motor can withstand heavier loads over longer periods of time. This weight comparison can also aid in determining which motor provides the best amount of metal for the price.

Motor cooling

Motor cooling is an important topic and worth discussing. As electrical current is passed through the electric motor, a buildup of heat occurs. The amount of heat produced is a function of the work, or loading, done by the motor, the type of waveform of the actual electrical signal to the motor, and the eventual changes from bearing wear and friction. Premium, or high-efficiency, motors provide better use of the electrical energy to get more work than losses out of the circuit. Rotor bar designs and cross sections, better conducting materials, and attention to air gaps are some of the areas where better efficiency is gained out of the electric motor.

A motor that runs fully loaded or sometimes overloaded naturally requires a greater amount of current. Extra current could affect its heat content, especially with respect to the duty cycle. Another factor that affects motor heating is that of the incoming power signal itself. A pure sine wave provides a known value when calculating motor losses. When that signal is subject to spikes, noise, or other line disturbances, however, the motor suffers as losses go up and less of that incoming power is used for torque production. In addition, bearings wear over time, as do other components in the drivetrain. This wear can cause the motor to perform extra work, and thus generate extra heat. Getting the heat away from the motor is the next challenge.

Many motors are sized so that the heat produced can be accepted by the metal content of the motor with normal convection and radiation to dissipate the heat. These motors could be classified as open drip proof (ODP) or totally enclosed nonventilated (TENV). Other electric motors incorporate a fan blade that rotates at the same revolutions per minute as the motor shaft. These motors are called totally enclosed fan-cooled (TEFC). Several common enclosure types for electric motors are shown in Fig. 6-4. If the motor heating is too great

TEFC—Totally enclosed fan cooled
Integral but external fan is coupled to the motor shaft directing cooled air over the motor frame. Suitable for use in dirty, dusty indoor environments.

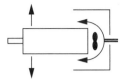

TENV—Totally enclosed nonventilated
Self-cooled by the internal circulation of air. Heat is dissipated from the frame by convection and radiation. Suitable for use in dirty, dusty areas.

ODP—Open dripproof, dripproof fully guarded
Self-ventilated with internal fan coupled to motor shaft. All openings are covered with screens or filters. Indoor use; not suitable for dirty, dusty environments.

TEUC—Totally enclosed unit cooled
Motor is cooled by closed loop cooling system by air-to-air heat exchanger. Heat exchanger is cooled by separate means. Suitable for both indoor and outdoor use.

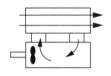

6-4 *Common motor enclosure types.*

compared to the rate of heat dissipation as with ODP or TENV designs, or if the motor rpm are not adequate to move the heat away in a TEFC design, auxiliary cooling measures must be taken. If not, the motor will either stop running if motor thermal protection is built in or damage itself. Either condition is undesirable, and up-front care should be taken to eliminate these circumstances.

One approach is to size the motor with a service factor. Another is to simply increase horsepower. More horsepower might put a motor into a larger frame designation, thus making it weigh more and allowing it to handle a greater amount of heat. A service factor of 1.15 means that the motor can deliver 15 percent more power when operating conditions are normal for voltage, frequency, and ambient temperature. This 15 percent extra power means that the motor is built and sized to handle that extra output continuously without damage to itself because of heat.

One phenomenon that occurs often is that a motor is selected for a particular load and duty cycle. Someone decides to play it safe and ask for a service factor of 15 percent. Another player bumps the horsepower rating up by a factor of two. Pretty soon, the application has a motor well oversized for the loading, energy is being wasted, and the initial cost of the motor was high. Granted a cool motor is a happy motor, but there are less-expensive, energy-efficient methods of achieving this state.

One method is the auxiliary blower. Basically, it is a separate fan and motor mounted on the main motor, as shown in Fig. 6-5. This auxiliary fan motor is much smaller and runs at full speed all the time, providing moving air across the larger motor to take the heat away. All that is required with the fan is a starter and motor. Another method of cooling is to duct cooler air to a heated motor. This method provides clean, cooled air when the environment around the motor to be cooled is harsh and a separate auxiliary fan motor is not attractive. Other means of motor cooling are possible but are usually more expensive. Water or liquid cooling is one method. Often, the environment where the motor resides dictates the type of cooling method chosen.

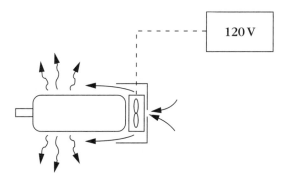

6-5 *Typical auxiliary blower scheme for a motor. A separately powered fan runs full speed no matter what speed the motor runs, thus cooling the motor at low-speed operation.*

Protecting the motor

In addition to cooling the motor, attention must be given to protection. The motor is an investment and should last several years if properly maintained. Motors should be equipped with thermistors that shut off power once a predetermined temperature has been reached. Often, a motor is running off an electronic drive, which can monitor the current sent to the motor. The drive can shut off power to the motor if it detects an overload condition exceeding a certain length of time. Thermal overload relays can also be installed in the circuit to add further protection for the motor.

Figure 6-6 shows three different circuits with the appropriate symbols for motor protection. In one, the electronic motor/speed controller (most often a drive) has an electronic overload circuit built in. It can monitor the current going to the motor and shut off power to the motor if excess current is sensed over a period of time. Another method of motor protection is by thermal overload relays. This adjustable overcurrent device is set at a predetermined current value to disconnect the load from the power supply when a specific condition exists. The device does not protect itself, but instead protects the load, which is the motor. The motor thermostat approach is also shown, which is the simplest form of motor protection. This contact shuts off power to the motor when opened. These contacts can be used in addition to other motor protection methods (i.e., using an electronic drive's fault circuitry).

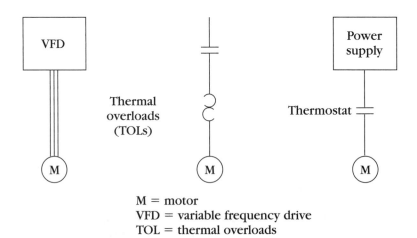

M = motor
VFD = variable frequency drive
TOL = thermal overloads

6-6 *Three ways to protect a motor: a motor controller (a VFD), thermal overloads, and a motor thermostat.*

The motor, as constructed, exemplifies a product whose design has evolved based on the need for extra protection and extended motor life. Care is taken to incorporate the windings, which are basically turns of copper or aluminum wire, into the motor package. Each turn of wire must be placed precisely or a premature failure in the motor can occur. Often, a motor is required to be *form wound*, which means extra special placement of the turns in the windings must occur. The term *random wound* describes the more common method of winding. As the windings are being wound, insulating tape is applied to further isolate the windings from each other. The entire network of coils is then coated with varnish or dipped to seal the insulation on the windings and add to the life of the motor. Any deviations or imperfect turns of wire can cause extra heating and degrade the motor's torque-producing capabilities, eventually causing overheating.

Another important piece of the motor package is the bearing arrangement. The front and back of the rotor must rest on sets of bearings. Premature wear at the bearings in a motor can cause a machine to shut down. Early failure of bearings can be mostly attributed to the actual mounting of the motor itself. If side loading occurs or the motor is coupled such that the shaft is somewhat off-center, unwanted forces begin to act on the bearings, thus wearing them out faster. Heavy-duty bearings should be requested when the motor mounting is overhung or side-loaded. In addition, good maintenance and lubrication can lengthen the life of the bearing. Some motors can be supplied with oil mist systems, which perform lubrication continuously. These systems are prevalent when motors must rotate at extremely high speeds.

A motor is often supplied based on its frame size. Motor manufacturers have their own designations when the horsepowers exceed a certain level. For three-phase ac motors, this level is usually beyond what is called the *NEMA sizes*. NEMA, or National Electrical Manufacturers Association, has provided standardization in motor sizes up to the 449 frame (some 500 frame designations and beyond do exist). Thus, for a given horsepower rating and a given base speed, most manufacturers adhere to standard frames. For instance, there is a certain foot-to-centerline of shaft dimension and a certain stack diameter (see Fig. 6-7). Dividing the first two digits by 4 indicates the foot-to-centerline of shaft dimension (447 frames have an 11-in. dimension). This dimension helps the machine designer locate the motor physically to the machine and compare a motor sized by one manufacturer

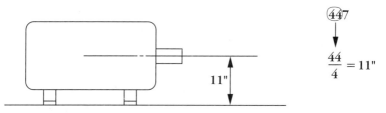

6-7 *The center line of shaft to foot dimension of standard ac motors.*

to another. Other issues also need to be considered, such as stack length, C-face mounting, duty cycle, service factor, ambient conditions, and torque requirements. Different frame sizes can be offered by different manufacturers for the same application and horsepower. Fully analyze what all the impacting factors mean and which are most important. This up-front analysis can save money, downtime, and headaches later.

Motors (ac)

All electric motors use some form of electromagnetism, induction, or repulsion. These principles of physics are the basis for motion in the factory today. Before these phenomena can occur inside an electric motor, electrical power must be supplied to make the necessary electrical changes and sustain motion. This power is either in the form of ac or dc. Once power is obtained, multiple ways of achieving the desired motion exist. This section discusses ac-supplied electric motors. First, however, we must go back to the 1830s to evaluate Michael Faraday's discovery of electromagnetic induction.

Faraday's law of induction has been a part of factory production for more than 160 years. Faraday discovered that by varying the magnetic field around an electromagnet, a current was produced and the magnet would actually move. By controlling the opening and closing of the circuit, he was able to gain control of the magnetic flux. Flux is the fundamental principle used in an induction motor today. Faraday's law also concluded that an induced electromotive force, or EMF, component and the magnetic flux's rate of change were proportional. That relationship makes induction motor discussion evolve around EMF, which is expressed as voltage.

Induction motors and repulsion motors are very similar in operation, and there are even overlaps in the function and naming of various types. For example, there are repulsion motors, induction motors, and repulsion-induction motors. Basically, the induction motor works on the principle of changing the state of electromagnetism around a magnetic field to achieve motion (induction). Likewise, the repulsion motor is close in operation. It contains two magnetic fields of like polarity that oppose each other and cause motion. A repulsion-induction motor is a hybrid and actually contains brushes, a commutator, and a wound rotor. It uses the brushes and commutator to get hard-to-start loads going. Once it reaches 60 to 70 percent speed, the brushes are lifted off the commutator by centrifugal force and the motor continues running as a squirrel-cage type.

Induction motors are the most common, and most use ac incoming power to operate. Most are three-phase supplied and sometimes called *polyphase* motors. The electrical action of this type of motor, particularly the squirrel-cage type, is likened to a transformer with a shorted secondary (see Fig. 6-8). The motor is much like a step-down transformer, in that the primary is equivalent to the windings of the motor, which is where the three-phase ac power is fed. The secondary is likened to the rotor, or armature. Fewer secondary turns, along with the gauge of the wire, equates to a higher current induced by the primary, with the maximum current created when the secondary is shorted.

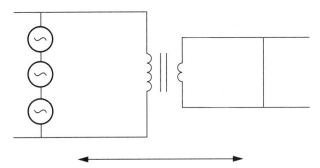

6-8 *The ac squirrel-cage motor acts like a transformer with three-phase input having a single-turn shorted secondary.*

The ac induction motors are also classified by their slip characteristics. Further evaluation of this characteristic can be found in the speed versus torque performance. When a load is applied, an ac motor has a tendency to slip. This phenomenon allows for the motor to request more current to maintain the given load at the desired speed. Some motors have more slip than others; typical values are 2 to 3 percent of synchronous speed common with NEMA A, B, and C design motors. Some are as high as 5 to 6 percent for special high starting torque applications. These higher-slip values are found in NEMA D and F motors.

Another issue with ac induction motors is insulation. Motor insulation is that nonconducting material that separates current-carrying components within the motor. This material is often insulating tape, which is applied prior to any dipping or coating of the windings. Motor insulation is rated based on temperature maximums versus a normal life expectancy of the motor. These values are shown in Table 6-1.

Insulation ratings are a good gauge when using motors with variable-frequency drives. The term *inverter duty motor* is thrown around occasionally. Variable-frequency drives do promote extra heating in the motor due to a nonsinusoidal waveform. This extra heating, depending on its severity, can lessen the life of the motor, although this phenomenon is exaggerated.

Many factors enter into the life expectancy of a motor, such as starting currents, quantity of starts, shock loading, and so on. Motor manufacturers deal with motors on variable-frequency drives by promoting class F or greater insulation systems for these applications. It is always better to get the better grade of insulation whenever possible and practical. Better insulation can help lengthen the motor's life, whether an electronic drive is controlling it or not.

Table 6-1. Temperatures for various classes of insulation

Insulation class	Temperature rating	
	Degrees F	Degrees C
Class A	221	105
Class B	266	130
Class C	>464	>240
Class E	248	120
Class F	311	155
Class H	356	180
Class N	392	200
Class R	428	220
Class S	464	240

The most common type of induction motor is the squirrel-cage motor. This motor is found in most ac applications in a wide range of horsepower ratings and is the workhorse of industry. Its construction resembles that of a squirrel cage, thus the name, and is shown in Fig. 6-9. The rotor is built up from steel laminations. Each lamination has a set of slots around its perimeter. Into these laminated slots are long rotor bars. These rotor bars are usually made from copper and have various cross sections to increase impedance by creating eddy currents. Traditionally, a squirrel-cage type motor has had a disadvantage in actual starting torque because of its fixed rotor design. Different rotor bar designs can provide extra impedance to better control the rotor flux at low speeds. At high speeds, this situation is not an issue.

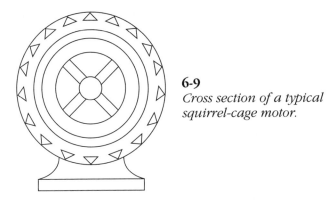

6-9 *Cross section of a typical squirrel-cage motor.*

When the stator windings have been supplied with ac power, a current is induced in the rotor. The difference in polarity between stator and rotor causes motion. This occurrence can happen in a three-phase motor because the phases are displaced by 120°. A single-phase motor could not start by itself. To get rotating, it would need a starting winding with a switch, which is sometimes called a *split-phase induction motor*.

The split-phase induction motor, shown in Fig. 6-10, requires a start winding and run winding. It is powered by single-phase, usually 120-V power. This type of motor typically is in the fractional horsepower range and has more commercial applications than industrial. Its starting and running operation is similar in principle to that of a repulsion-start-induction motor. It starts off the starting winding and, once it reaches 70 percent speed, changes to the run winding. The switch provides centrifugal force versus the centrifugal brush system in the repulsion motor.

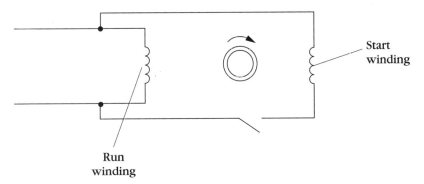

6-10 *Windings of a split-phase induction motor. The switch closes to provide current for start winding and then opens. The motor then runs off of run winding.*

Other types of induction motors include the capacitor motor, shaded-pole, wound rotor, synchronous motors, and the polyphase motors. The capacitor and shaded pole type are single-phase units limited in size. The capacitor type stores energy for starting and can be obtained in horsepower sizes up to 20 hp. The shaded-pole motor uses salient poles, which are copper loops that act as the starting winding. They are also called shaded poles, thus the name. The shaded-pole motors are very inexpensive and do not reach above 0.33 hp.

More industrial applications use wound rotor and synchronous motors. These types of induction motors can range in size from fractional to several thousand horsepower. The wound rotor motor contains groups of copper coils. The stator windings equal the number of poles to produce a rotating magnetic field. The wound rotor motor is the older brother of the squirrel-cage induction type machine. Wound rotor motors have been around longer than the squirrel cage and provide high starting torque.

Synchronous motors run at a constant speed regardless of the load changes. They run from a fixed frequency ac source at the same speed as the source frequency and as dictated by the number of poles in the synchronous machine. The synchronous motor's rotor has poles that synchronize with the rotating magnetic field. Many synchronous machines have low synchronous speeds and thus have 10, 12, and even higher counts of poles. These motors are used in compressor and pump applications quite often. They can be provided in very high horsepower configurations and are well-suited for steel and aluminum mill environments. Often, the synchronous motor must be started with a dc generator so the poles can lock in with the field.

Repulsion motors operate on the principle of using opposing magnetic fields to attain motion. They need an armature, a field, and a means of switching the polarity, which is done by the commutator. This motor also requires a set of brushes that ride on the commutator to transfer electricity to the armature. You can see the common dc motor is, in fact, a repulsion motor. Because dc motors have been a major force in the factory for so many years, these machines are covered in more depth later in this chapter.

When a larger motor is installed, the installation often needs to be evaluated relative to its mechanical drivetrain and electrical components. This evaluation is called a *torsional analysis* and is usually done before the actual installation. Design criteria help predict whether or not problems can occur with vibration, resonance, and shock conditions. All these conditions can be detrimental to the life of an installation. This analysis also helps to suggest what type of coupling should be used, if vibration isolation is required, or if the electrical pulses can be a problem.

The analysis requires data about the application: loading, speeds, motor mounting, type of electronic controls used, etc. Extensive calculations are then performed to determine as much about the operation of the system as possible. Engineering firms routinely perform these studies and can often predict where harmful operating points can be. Steps can then be taken to either eliminate or minimize the effects of these hazardous situations. For example, an electronic drive controller can vary the frequency to the motor to change speeds. At certain frequencies, an amplification of oscillations or resonance can occur, causing couplings or motor shafts to break or any other number of bad things to happen.

Motors (dc)

The dc motor has lost a lot of its prominence in the past few years. With ac induction motor technology getting stronger and manufacturing and maintenance costs a factor, the dc motor just is not used in the factory like it was years ago. It still has a place, however, because certain applications still require it. One important use is running from a portable power supply, such as a battery. Another common practical application of dc motor technology is in the winding and unwinding of material where holdback torque is needed. Still other applications we can call dc motor strongholds are cranes and hoists, traction motors requiring high starting torque, elevators, and large mill systems.

A major threat to the dc motor is ac vector technology, especially as these systems find cost-effective methods to handle regeneration. The dc motors have traditionally given high starting and ride-through torque capabilities along with excellent speed regulation.

The dc motor, as previously noted, is a repulsion motor. Figure 6-11 shows a dc motor in simplified block form. The main components are the armature, field, commutator, and brushes. These components provide an electrical-to-mechanical scheme, whereby dc power is supplied to the brushes, which transfer it to the commutator, which reverses the polarity in the armature, which opposes the magnetic property of the field, thus causing motion. As you can see, the dc motor's operation is straightforward. The equivalent circuit for the dc motor is shown in Fig. 6-12.

158 Electric motors

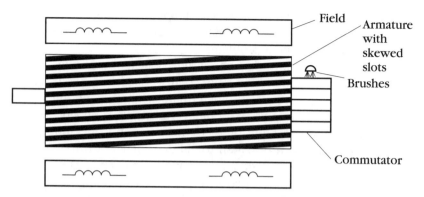

6-11 *Simplified dc motor with the four major components: armature, field, commutator, and brushes.*

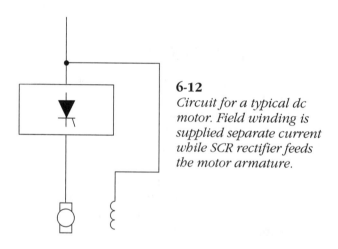

6-12
Circuit for a typical dc motor. Field winding is supplied separate current while SCR rectifier feeds the motor armature.

The current to the field windings maintains the field strength, which is half the equation to the dc motor's commutation. The other half is the armature winding's current and its smooth transitions from coil to coil. This current must reverse at each coil as a new pole is encountered. This reversal is actually the commutation, and the timing of this reversal must be synchronized by the brush position at the commutator itself. If the timing is off, flashing and arcing can occur. It is always imperative to have clean and properly aligned brushes with respect to the commutator gaps to provide smooth, effective torque output from any dc motor.

The typical frame construction of a dc shunt wound motor is more square than round (permanent magnet dc motors are usually round), making for a shorter distance from the feet to the centerline

Motors (dc)

of the motor. This design minimizes magnetic losses, thus getting greater use out of available current. The frame is typically made of iron laminations epoxied together with an oxide. The shunt field coils are mounted on the inside of the laminated frame and thus become the field.

The armature is mounted to the shaft within the housing of the frame and is supported at both ends by bearings. The armature is made of iron laminations in a round configuration so as to fit compatibly within the rounded field windings. The armature coils are fitted inside slots, sometimes skewed to reduce cogging, and insulated. The ends of these coils are at the commutator, which is connected to the armature. A skewed armature, although better for handling cogging conditions, is harder to manufacture. The amount of skew, or the amount of slot pitch offset, makes winding the armature more difficult and thus more costly. The commutator is a round, copper-based device with many individual wedgelike pieces. The surface is machined smooth to allow for a good, clean contact between the commutator and brushes. The brushes, which are made from carbon, ride on the commutator to pass current to the armature windings. The brush contacts are spring-loaded, and the number of brushes required in a given motor is determined by the current rating of the particular motor.

There are different types of dc motors: series-wound and parallel, or shunt-wound, along with permanent-magnet dc motors. The permanent-magnet dc motor is typically found in the fractional horsepower range. It usually has a lower armature voltage rating (90 V) than shunt-wound motors (180 V). The magnets are actually part of the stator and equate to the field. Thus, no power is required at the field, which is the permanent magnets. The magnets are usually ferrite type. Ferrite magnets are fragile and susceptible to breakage if the motor is mishandled. Ferrite magnets are attached to the motor frame by an adhesive bonding method, usually with epoxy, which means there is the possibility of dislodging them. Another common disadvantage with permanent magnets is that they can become demagnetized. An overcurrent condition in conjunction with prolonged overtemperature conditions at the motor can tend to demagnetize the permanent magnets.

Another, more expensive type of magnet used is the rare-earth version. This magnet is manufactured by a process known as sintering. *Sintering* is the fusion, or welding, of small particles of metal by applying enough heat to reach just below the melting point of the metal. This process allows the rare-earth magnet to be formed into an appropriate shape for later use in mounting to the rotor and is part of the reason for the added expense. In addition, these magnets are not magnetized until installed in the motor frame. They are attached with an adhesive that is somewhat resistant to high-temperature conditions.

Rare-earth materials are somewhat expensive, and the search is on for a good performing, inexpensive material. One acceptable synthetic magnetic composite is neodymium iron boron. Neodymium is one of the rare-earth elements (element 60 on the Periodic Chart with symbol Nd). Coupled with iron and boron, it has good magnetic characteristics.

Because the flux is basically constant in the stator of a permanent magnet motor, the speed versus torque is a linear relationship, even into extended speed range. Figure 6-13 illustrates this relationship. With a wound field motor, however, the torque diminishes as it goes into field weakening. Field weakening is illustrated in Fig. 6-14. To increase the speed of the dc machine, the stator's magnetic field intensity is reduced by reducing current. Less current means less torque is available to produce the current.

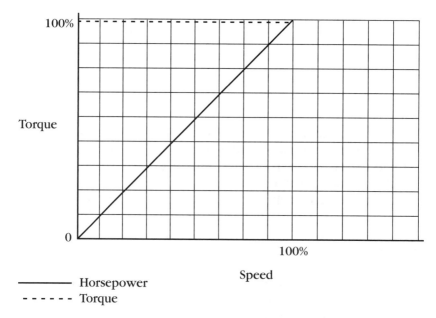

6-13 *Permanent-magnet dc motor speed versus torque curve.*

The field-wound dc motors compose the bulk of the dc motor installations. There are series and parallel wound versions. Figure 6-15 shows a simplified circuit of a series-wound dc motor. A series-

Motors (dc)

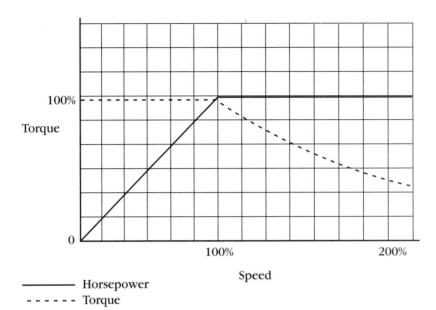

6-14 *Typical shunt-wound dc motor curve with the field weakening area beyond the 100 percent speed point. When the field weakens, torque diminishes rapidly.*

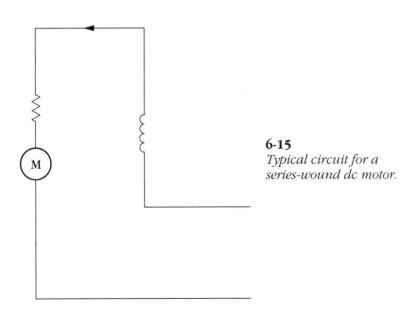

6-15
Typical circuit for a series-wound dc motor.

wound motor can deliver a substantial amount of starting and running torque at low speeds by developing magnetic flux in direct proportion to current but diminishing as the flux saturates. This motor is basically load-dependent and keeps generating torque as the load increases. If the load were to suddenly vanish, the series-wound motor would keep producing torque current, which could be hazardous. The field-wound dc motor has been used extensively in mill applications, especially on cranes and hoists.

The parallel, or shunt-wound, dc motor is a more common and less expensive dc motor on the market today. It can be controlled more easily by thyristor-based controls. Typical circuit configurations and curves are shown in Fig. 6-16.

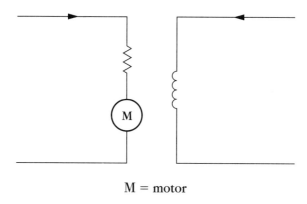

M = motor

6-16 *Typical circuit for a shunt-wound dc motor.*

A shunt is actually a short circuit and allows current to take a more direct path in the windings. The field windings are supplied a separate voltage, while the armature is provided its own. The armature rotates, creating an induced back EMF, which opposes the supplied armature voltage. This back EMF is a self-regulating, predictable quantity and allows for control by thyristor-based controllers.

Typical field voltage ratings for dc motors are 300 and 500 V although numerous 240- and 180-V shunt-wound field-rated dc motors are out there. These are harder to control. Typical armature voltage ratings are 120 V and 240 V, although plenty of nonstandard armature ratings exist.

A dc motor when operated from a silicon-controlled rectifier (SCR) power supply exhibits different characteristics. This fact is especially true when placing an older dc motor on a newer, SCR-based controller. This issue is called the *form factor* and is the ratio of RMS current to average current. It is an issue because the average current determines the amount of torque available. There can be effects from excess heating, as well as unwanted sparking and hard-to-commutate conditions.

Values have been placed on certain types of rectifiers. These values are selected based on the wave type and amount of SCRs in a given rectifier. For instance, at base speed and full load, a three-phase, full-wave, 6-SCR rectifier exhibits a ripple frequency of 360 Hz and a form factor of 1.01. A semiconverter with two SCRs and three diodes, single-phase, carries a form factor of 1.35 with a ripple frequency of 120 Hz. This current ripple is affected by line voltage, motor speed and load, the inductance in the circuit, and, most important, by the type of rectifier.

The form factor is used to base the realized current value, which determines heating. The current we are concerned with for heating is the RMS current, which multiplies the form factor by the motor-rated current. In this way, heating predictions are possible.

Commutating the motor can be difficult with certain rectifiers. Sometimes no commutation at all occurs, and the motor simply arcs or sparks. The motor can run adequately with sparking, but sparking should be avoided. Single-phase rectifiers can promote unwanted commutation problems. The lower the form factor, the better for heating, but this rule is not necessarily true with commutation. Smaller horsepower motors can commutate with a form factor of 1.35.

Servomotors

Another group of permanent-magnet motors are the *servos*. This motor uses a scheme of electronic control whereby electric current is changed to the windings, which are embedded in the stator. The name *servo* is often used to describe the motor, but the correct term should be *servomotor*. The servomotor is sometimes called an ac brushless motor or even a dc brushless motor. The bottom line is that the terms are basically interchangeable; the motor is brushless.

The motor is supplied three-phase rectified power and then furnished as an amplified power signal to the windings. Earlier servomotors used brushes and a commutator to actuate the rotor, but servomotors today are electronically commutated and thus brushless. Brushless means less motor maintenance and enhanced motor performance.

Figure 6-17 shows a brushless motor family tree. The family tree divides into two major branches of brushless motors: synchronous

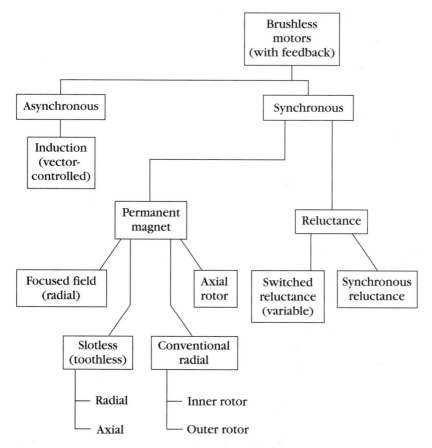

6-17 *Brushless motors family tree.* The American Institute of Motion Engineers

and asynchronous. This family provides a good look at the brushless motors' hierarchy and their relationship to each other.

The construction of a servomotor has made it relatively easy to implement from a physical standpoint. Servomotors have certain physical features. They need to be compact to fit in tight places on a machine, but this compact design should not sacrifice torque output. They also need to get rid of heat, just as all motors must, but an auxiliary blower is usually not practical. They must be easily retrofitted with a feedback device. Stack length and the mounting face of the servomotor also become issues. Mountings such as C-face and D-flange have become industry standards for mating to machines, gearboxes, and so on. A typical servomotor is shown in Fig. 6-18.

The external, or outside considerations of a motor are important, but it's what goes on inside that makes or breaks the application. A

Servomotors

6-18
Brushless servomotor.

view of the inside of a typical servomotor is shown in simplified form in Fig. 6-19. The windings to which current is supplied are located in the stator. Permanent magnets are located on the rotor.

A typical servomotor capable of rotating thousands of revolutions per minute might have six poles. This configuration is three N poles and three S poles. Correspondingly, six sets of windings exist for each pole. Three positive windings for phases A, B, and C, and three for negative, or reverse motion windings, phases A, B, and C. This motor design is sometimes called *inside-out* because of the permanent mag-

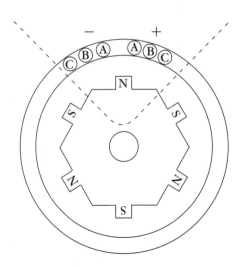

6-19
A simplified drawing of the inside of a six-pole servomotor. There are three positive windings (forward direction) for each pole and three negative windings (reverse direction) for each pole; this is typical for each pole, N or S.

net rotor and the wound stator (compared to the conventional brush-type motor).

The rotor of the servomotor can be many laminated iron pieces or one solid cylinder. The magnets are placed onto this cylinder. The magnets are permanent, rare-earth variety. They can be epoxied or press fit into place. This composite rotor is then carefully mounted into bearings that have been permanently lubricated. These bearings can be heavy duty if side or radial loading is expected when the servomotor is mounted. Attention must be given to axial loading, which can shorten the life of normal bearings from the standard 20,000 hours.

The next important component is the feedback device. How it is attached is paramount. This attachment can shut down a factory, so it is crucial that the device be straight, centered, and free of interfering wiring. The choice of couplings also matters. Will there be shock loading to the motor? Will there be a steady flow of back and forth, forward and reverse movement at high acceleration rates? A coupling must be tolerant of some (not much) misalignment, which actually means that the feedback device will run coupled to the motor shaft for a long period of time. The more the misalignment, the shorter the life expectancy of the coupling/shaft attachment. According to the Laws of Murphy, the coupling will fail at a period of high and crucial production.

Once a suitable coupling has been selected, it is important to ensure the integrity of the feedback device signal. Routing higher-voltage wire away from the feedback wiring and twisting and shielding the feedback wiring is good practice. Many servomotor manufacturers have developed standard housings to attach the necessary feedback devices.

The connection points for the power feeds to the motor and the feedback wiring are often in the form of what is called MS, or military style, connectors. A closeup of this style connector is shown in Fig. 6-20. The alignment of the pins and the screw-down aspect of the connection provides a true and sealed connection. This connector also eliminates time-consuming, individual wire connections when running the wire (assuming cables are made up ahead of time). No soldering or screw-driven connections are required with this type of quick disconnect.

Other facets of the construction of the servomotor have evolved over time and use. Motor environment has dictated how the motor is to be constructed. For instance, windings are insulated with a high grade of insulating tape, usually Class F or H. O-rings and shaft seals

6-20 *Military style (MS) connectors on a servomotor.*

are also provided to keep the environmental "nasties" out of the servomotor. Purging the servomotor and feedback device with air is a common practice for environments with hazardous gases. Consult local codes, the motor manufacturer, and the electrical code before purging.

Stepper motors

Another motor commonly used to position is the stepper motor. Like its servomotor brother, the stepper needs a drive package in which the electronic control, power supply, and feedback interface can be housed. This controller is looked at closer in Chapter 7. The stepper motor operation is simplified in the block diagram of Fig. 6-21. Basically, an indexer provides step and direction pulses to the electronic drive amplifier/controller. The current level for each phase to the motor is determined and subsequently output to the motor. An encoder is not necessarily required for step-motor operation, although it is needed in high-accuracy positioning applications.

A stepper motor works on the premise that pulses determine the distance for positioning systems, and the frequency of these pulses determines the speed. The stepper, or step motor, as it is often referred to, is inherently a digital device. It is a brushless motor that can deliver high torque at 0 speed and maintain stability. There is no drift and no errors due to the cumulation. These motors are fully reversible and can withstand numerous shock loads throughout their

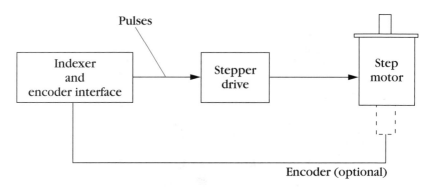

6-21 *Stepper motor for operation.*

life. Similar to the servomotor construction, the windings are in the stator for quicker heat dissipation, allowing for low inertia in the rotor. There are two windings in a step motor, and thus two phases.

Step motors sometimes exhibit oscillations with every step. Oscillation is common with this type of system and is most visible at low speeds. The goal is to match motor, amplifier, load, and mechanics to minimize the oscillations. Stepper motors tend to run hot because of the pulse-width modulation (PWM) waveforms from the controller, the high currents at 0 or low speed, and the duty cycle. A stepper motor also tends to jump when enabled, which is merely the motor seeking a tooth position. An air gap exists between the teeth on the rotor and the stator, thus they do not come into contact with each other. When flux is present, such as in a power-on situation, there can be an initial, internal movement. The stack length, or depth, provides the usable torque in a stepper or servomotor.

Conclusion

Electric motors are the workhorses of industry. They are today and will continue to be so for quite some time. Their construction allows for extended use and even abuse. They have minimal connections and are not necessarily hard to troubleshoot. The electrician's best friend when it comes to the motor is the motor nameplate, a sample of which is shown in Fig. 6-22.

Nameplates usually contain all the data required to troubleshoot, set up a drive controller, and even wire a motor. The frame, horsepower, full-load speed and current, and even wiring diagrams are sometimes provided right on the nameplate. If the motor can be wired for either 230 or 460 V, the connection information is usually

Conclusion

```
Frame   365TS        Design B      Serial number    34X909CK-01
HP      75           Volts  230/460    Phase    3     Hz   60
RPM     1768         Amps   192/96     S.F.     1.15
Insulation Class F   Duty   Cont               Encl   TEFC
Efficiency Premium   93.4   Ambient            40 degree C
```

6-22 *The ac motor nameplate.*

imprinted on the nameplate. Thus, the data follows the motor as the nameplate is screwed right onto the motor frame. The only way the motor data is lost is when the plant maintenance people paint over the nameplate. But paint can be scratched off, and the data is usually still there. Motor data is extremely important to the electronic drive application engineers if they are to implement the drive on that particular motor.

The controllers for the motors tend to be the challenge for the plant electrician. More wiring, some special, is required, along with programming a digital electronic drive. Attaching feedback devices and additional motor protection represent the remaining extent of extra wiring that has to be run or traced. Thus, in the basic sense, electric motors are relatively maintenance free. *Relative* because it depends on the type of motor and its use. Sometimes, motors must be rewound because of excessive high current conditions over a long period of time. Bearings also wear, and there are always those unwanted motor short circuits. Again, relatively speaking, the motor runs with little maintenance for many years.

That last statement is mainly for ac and brushless motors. Industry experts agree that dc motors are more trouble than ac brushless motors—hands down. The dc motor commutators must be cleaned, and the brushes often get dirty and need to be replaced. Nasty environments accelerate this maintenance, and eventually the dc motor is ousted in favor of the ac machine.

Replacing dc motors with ac machines is happening throughout industry and will continue to happen. There might come a time when the dc motor is no longer produced, but that is definitely many years away. The ac motor technology must continue to provide all the solutions that dc has for many years. Meanwhile, specialty and other brushless motors continue to provide solutions for position and other types of control.

Many controller manufacturers believe that the motor is the limiting factor in the technology. Motor manufacturers argue that the controller is the limiting factor. Both sides should get together and develop better standards. Some standards exist, and there are countless rules of thumb to follow when mating motor controller with motor. The controllers for motors will continue to change and so, too, will the motors. Some users would prefer the motor controller to be mounted integral to the motor housing.

The Department of Energy has set several standards to which electric motors must adhere to by 1997. The standard is lengthy, and many issues are covered. Motor efficiencies and power factor ratings are among the more important issues. These standards will affect not only industrial users but commercial and even residential markets. The days of wasting electrical energy are gone. It is far easier to make what we have installed more energy efficient and install new energy-efficient products than to build new power plants and diminish available resources. This issue alone will drive much of the future electric motor development.

Bibliography and suggested further reading

Electrical Apparatus Service Association. 1988. *Electrical Engineering Pocket Handbook*. St. Louis, MO.

Electrocraft Corporation, 1980. *DC Motors, Speed Controls, Servo Systems*. Fifth Edition. Hopkins, MN.

Early, Murray, and Caloggero. 1990. *National Electrical Code Handbook*. Fifth Edition. Quincy, MA: NFPA.

Tomal, Daniel & Widmer, Neal. 1993. *Electronic Troubleshooting*. Blue Ridge Summit, PA: TAB/McGraw-Hill.

7
Motion control

In the factory, if something moves, it is in motion, and if that motion is planned or predicted, it should be controlled. If the controlled motion happens automatically, without human intervention, we have industrial automation. This controlled motion is programmable, made possible by digital microprocessor-based devices. The tool and automotive industries primarily have driven this technology to the leading-edge environment it exudes today. Other industries have followed. We are in charge when motors and other types of prime movers respond to our stimuli. Thus, motion control is the essence of the production line.

Without motion there is no industrial automation! Sure, the programmable logic controllers (PLCs) can control some I/O in a process, or machine vision systems process graphical images, but unless a product is physically moved "out-the-door" in the factory, no true production exists. Controlling these motions is what this chapter is all about.

Many facets of automation are tied directly to motion control: electric motors, electronic drive controllers, feedback devices, computers, and programmable controllers. Vision systems also routinely interact with motion controllers. As communication capabilities grow, interaction between dissimilar pieces of equipment will become simpler and more commonplace. Even the basics of electricity, power transmission, and quality control relate to motion control.

The technology of motion control is constantly changing, which demands constant re-education for project engineers or designers. If these professionals are not continually updating themselves, they will learn too much via the school of hard knocks, because as much as we strive to engineer a project correctly, there are going to be mistakes. So many variables are involved; errors occur in data that is overlooked, new high-tech equipment, technical surprises, and unmet performance expectations can mean the project must be reengineered.

Reengineering resounds that old adage, "there was not enough time to do the job right the first time, but time was made to do it over!" Reengineering might be someone's thought at the time, but a motion control engineer, or the resident expert on these matters, must put advanced electronic controls on an old machine, perhaps replacing a few older mechanical parts, as well.

Mixing electronics with mechanics is always interesting. Simply electronically commanding a 2000-pound hunk of metal to move from point A to point B in 1.5 seconds does not mean it will! Generally speaking, we often find out why something did not work out as planned *after* implementing the original plan. We are constantly "chalking it up to experience!" Getting a good understanding of all aspects of automation, both educationally and through valuable experience, however, makes for a complete automation specialist.

Another emerging tool is that of data, process, and motion simulation. Although this technology has not yet become standard (actually a great need still exists for its penetration into many sectors of automation), software packages help predict performance under different conditions. These tools, plus the mechanical professionals who understand electrical concerns and the electrical people who understand mechanical concerns, might quite possibly help us win the battle against machines. The challenge continues as the technology of motion control keeps changing.

Motion control can be broken down into individual categories, such as speed or velocity control, torque or current control, and position control. High-speed microprocessors have allowed the industry to achieve major steps in controls. Coupled with the advancements in power semiconductor technology, designs have become dramatically smaller, faster, and more intelligent. Compared to the complementary metal-oxide semiconductor (CMOS), relay logic controls, and op-amps of yesterday, the technology of today is changing very quickly, maybe even too fast. For example, any manufacturer of a control product that introduces a new design or concept has some success with it before the competition improves on the original design. The first manufacturer then makes the next improvement, and the cycle continues. What happens is that the end user of the product attempts to standardize on a certain product (by the same manufacturer) but cannot. Everybody is changing so quickly that even spare parts inventories become obsolete in no time at all. What should be stocked?

Electricians must be constantly trained as products continue to become more complex. The free enterprise system is at work, and while traditional users continue to implement tried-and-true analog systems, one thing remains: motion control in some form or another must be achieved in the factory to compete nationally and globally. This topic is explored further in later discussions of quality control and ISO-9000 certification. For now, I discuss how we got here and what is happening now. Be prepared to keep abreast; the technology of motion control is constantly changing!

Motion control standards

As discussed earlier, motion in the factory is mainly actuated by either pneumatics (air), hydraulics (liquid), steam, or electrical power. The least common, steam, is sometimes used in larger plants to drive turbines to generate electricity for internal use. Pneumatic-, hydraulic-, and steam-actuated motion is controllable. Disadvantages are associated with each, however, which is why electronic motion control of electrically powered devices is most common.

Pneumatic and hydraulic systems tend to be maintenance-intensive. Hydraulic systems leak and are usually dirty, making them unsuitable for industries that need clean plant environments, such as the food industry. Pneumatic systems often get water in the air line and are noisy. Pressure losses equate to poor performance. Current pneumatic and hydraulic uses in motion control are either specialized or supplemented to machines controlled electrically. Using a pneumatic or hydraulic solution can still be appropriate, however, because of cost.

Older plants have air lines readily available and a compressor in place. An air-driven solution is more cost-effective than a new electrical scheme, but not in the long run. Costs for maintenance and loss of production must be considered. Interestingly, the compressor motor is electrically powered.

Electrically driven machines incorporate electric motors. These motors need a power source. This power source is either three-phase or single-phase ac electricity. Incoming power to factories start in this form. Before this power is fed to any electric motors, it is transformed, filtered, or converted to dc electricity. Some dc power is inverted back to ac power. In the production world, if the people who produce electricity all took vacations at the same time, our factories would virtually shut down.

Old technology

Early electrically powered machines mainly used constant-speed ac motors as the prime mover. Many machines, driven by ac motors that ran at their full speed, were energy wasters. This technology also did not allow for velocity or position control. Because the motor ran at full speed, gear reduction had to be incorporated to run the machine at an appropriate speed. It was mechanical solutions, such as brakes and clutches, that initially gave the plant personnel some control. The ac motors were a necessity in the factory, even though controlling them was another issue. Even motor generator (MG) sets used to control dc motors incorporated a constant-speed ac motor as the prime mover.

Motor generator sets

The MG set consisted of a motor and a generator. The motor, typically an ac-powered unit, was coupled to a generator. The net result was dc electricity that could be controlled to change the speed of a dc motor. The MG set's motor that ran the generator could be gasoline or diesel-powered, still powering the dc motor. This setup was ideal for remote and isolated locations. The MG set was the mainstay in industry for years, providing direct current power to dc motors. Figure 7-1 depicts an MG set configuration. With new ac and dc control technology, however, MG sets are no longer very popular. They are inefficient and use more energy than needed to perform the job. It is also hard to find replacement parts.

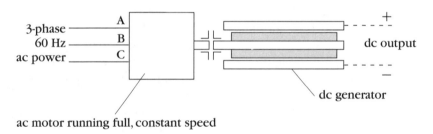

7-1 *Motor-generator (MG) set configuration.*

Eddy-current clutches

Figure 7-2 shows an eddy-current clutch. As the name implies, the eddy-current principle is used. Eddy currents, induced within the conducting material by the varying electrical field, cause changes in output speed. Soft starting and high-torque output, especially at low speeds, made this device a workhorse. The main disadvantage is that the clutch system is very inefficient and must be cooled by either water or air and there is still a lot of energy lost as heat. The eddy-current clutch was, in its heyday, probably the most widely used vari-

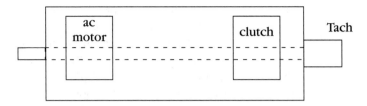

7-2 *Main components of an eddy-current clutch.*

able speed solution, but it still needed a constant-speed ac motor as the prime mover.

Static-speed control devices

Static devices were used mainly with dc motors to achieve variable speed control. Static devices could be equated to any power conversion system that did not require a motor driving a generator to achieve the desired electrical output. In other words, nothing dynamic was used to convert ac to dc or dc to ac. The power conversion was handled through devices that took electrical energy in one form, and the energy emerged out in another form. Vacuum tubes (or electron or thyratron tubes) were used to change the dc voltage to the dc motor's field or armature. These systems used a lot of space on the factory floor and gave off a lot of heat. It is difficult today to even purchase replacement parts for these systems. Many are being replaced with digital drives.

New technology

With advancements in power semiconductor technology, newer feedback technology, and the more frequent use of microprocessors, major strides have been made in motion control. All we are trying to do is control the speed and torque of an electric motor. How to achieve it and how to keep costs minimal drive the technological advances.

Because motors come in two basic packages, alternating or direct current, the controls for those motors can be simplified, right? Not so, because there are different ways to get the desired ac and dc output to the motors. Motors also come in many shapes, sizes, and configurations. Common input, or operating voltages, for most three-phase motors in the United States are 208/230 V, 460 V, 575 V, 2300 V, and 4160 V. Some smaller motors are single phase and run off voltages of 115 V and 24 V. If the motor was supplied with these voltages directly, it would run at its maximum rated speed, or synchronous speed (synchronous with a 60-Hz supply voltage).

Controlling the speed of a motor is the beginning of motion control. Torque, motor shaft position, and fast reversing become ancillary concerns. Let's first look at speed control. Industry often needs to reduce the speed of a motor that typically runs full speed, such as for pump and fan applications where demand is not always a criterion. In addition, a process might require that different speeds be available for process control. This control must be automatic and immediate.

One means of speed control is using electronic drives. These motion-control devices allow for the control of power to a motor. This control can be elaborate, involving positioning and precise torque control. The ac or dc motors can be controlled, and many versions of these motors are available. Presently, there is a gradual movement toward ac systems. The ac drives and ac motors appear to be more cost-effective, better performing, and more standard.

Drives (ac)

The term *drive* means many things to many people. Some might interpret it as the device on a computer for accepting floppy disks. Others might consider a drive to be all the mechanisms required to move part of a machine, while some might call this the drivetrain. When referring to a drive that electrically changes the electrical input to a motor, it should be called an *electronic drive*.

Prior to using electronic drives to control ac motors, methods for slowing the operation's output were limited. Many were mechanical. Various methods are described in Table 7-1 and illustrated in Fig. 7-3.

Table 7-1. Methods for slowing operation

Variable speed pulley (Fig 7-3A)	Needs a constant-speed motor, usually ac. The pulley is adjustable in the x direction to allow the V-belt to move up and down in the groove y, thus varying output speed. Disadvantages include belt wear, limited turn down, and belt slippages from fast accelerations and decelerations. No real soft start. Low horsepower. Advantages are simple design and less expensive.
Eddy-current clutch (Fig 7-3B)	Needs a constant-speed motor, usually ac. Soft start and variable speeds via controller. Reversible. Disadvantages

Drives (ac)

	include inefficiency and need for water for cooling. Advantages include high torque at low speeds and a good speed range.
PIV, proportionately infinitely variable, transmission (Fig 7-3C)	Needs prime mover. A complex, multiple-gear transmission box. No real soft start. Reversible after stop. Expensive to replace. Lower horsepower because of manufacturing costs.
Fluid-based speed variator	Needs constant speed input device, usually an ac motor. Soft starting and reversible. Disadvantages include high maintenance, leaks, and limited horsepower. Advantages are high torque at low speeds and wide speed range. Basically a fluid coupling in a box.
dc drive (Fig 7-3D)	Requires a dc motor and electronic converter. Advantages include high-torque speeds, wide speed range. Disadvantages are that the dc motor is not very efficient and that the dc motor is a maintenance item and not suitable for harsh environments.
MG sets (Fig 7-3E)	Requires ac motor, dc generator, and dc motor. Old technology. Disadvantages are that parts are hard to get and it is not efficient. For its time, it was a good variable-speed controller.
Fluid coupling (Fig 7-3F)	Needs a constant-speed prime mover. Variable-speed control by controlling the flow of liquids not the impellers. Soft start device. Disadvantages are leaks, efficiencies, and maintenance. Advantages are high horsepower and low initial cost.
Variable-pitch fan (Fig 7-3G)	Requires a constant speed motor, usually ac. By mechanically changing the pitch of fan blades, variable speed is achieved. Disadvantages are its limited speed range, expense and lack of energy savings.
Inlet vanes, dampers, valves on constant speed systems	Much of industry is still trying to overcome these conditions. They offer no energy savings!
ac drive	The electronic, cost-effective, efficient, soft starting, very wide speed range (extendible beyond base speed), low-maintenance solution. Low and high horsepowers.

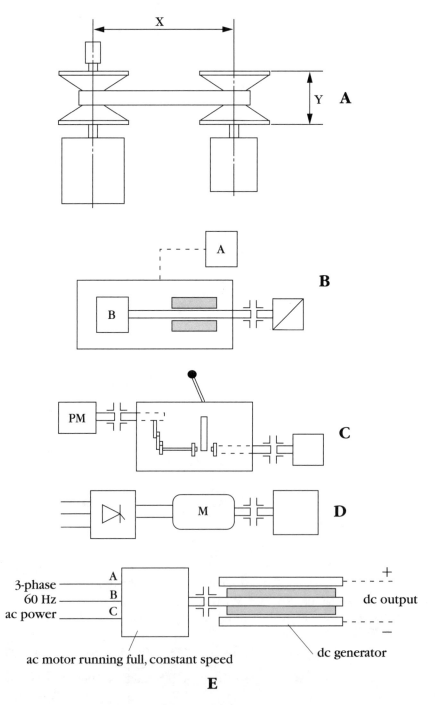

7-3 *Common methods for varying the speed of a given load.*

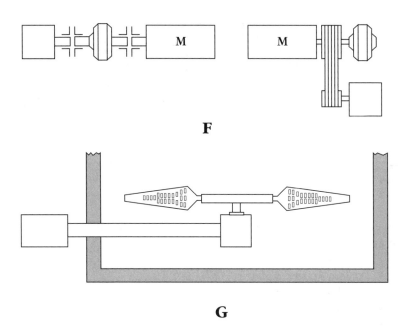

7-3 *Continued.*

One way to slow airflow on a fan was to install dampers downstream from the fan motor. Energy was wasted because the motor ran at full speed. In pumping applications, valves were installed to reduce liquid flow, while the pump motor ran at full speed. Other methods of speed control incorporated other mechanical solutions, such as the variable-pitch pulleys, fluid couplings, eddy-current clutches, and so on. Variable-pitch pulleys used a belt, as shown in Fig. 7-3(a). This belt could break, slip, or simply wear out. Fluid couplings could develop leaks. Eddy-current clutches tend to be inefficient. Most of these methods of speed control are high in maintenance, while others are energy wasters. The ac drive was a very welcome solution.

The ac drive has many names in industry:
- variable frequency drive (VFD)
- variable speed drive (VSD)
- adjustable speed drive (ASD)
- volts-per-hertz drive
- frequency drive
- inverter

Figure 7-4 shows the main components of an ac drive. As can be seen, the drive is actually a converter and an inverter of three-phase power. Many types of devices are used in the converter and inverter

Motion control

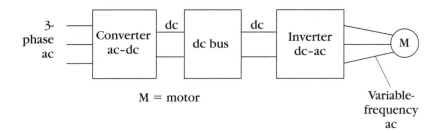

7-4 *The three main parts of an ac drive: the converter, dc bus, and the inverter.*

power bridges. These different devices contribute to further classifying ac drives, for example
- pulse-width modulating (PWM)
- current source inverter (CSI)
- variable voltage inverter (VVI)
- six-step inverter (SSI)
- insulated gate bipolar transistor (IGBT)
- pulse amplitude modulating (PAM)

One reason ac drives are so popular is because of their ability to provide speed control to a standard NEMA design-B squirrel-cage motor. This ability is attractive not only for new installations, but also because of the huge, already-installed base of existing ac motors that can be retrofit. The squirrel-cage motor is attractive for many reasons.

First, it is usually less expensive to purchase an ac squirrel-cage motor. It costs less than a similar horsepower dc motor because it does not have brushes or commutator. It is also smaller for a given horsepower, thus making its metal content lower and less expensive. A motor's enemy is heat buildup and an adequate means of dissipating that heat must exist. Auxiliary fans must be used occasionally to provide cooling. These fans are usually cycled on and off by the ac drive. Motor heating as it affects motor construction must always be considered, which affects cost.

Another benefit of the ac motor over the dc is lower maintenance. Because there are no brushes or commutator, there is no need to clean or replace them. The motor can go into dirty, hostile environments and run, whereas the dc motor and its brush design would falter. These maintenance and downtime costs attributed to the dc motor make it easy to consider the ac motor.

Other advantages to going ac include size and environment. The ac motor, being smaller, can fit in confined spaces on a machine, and that same ac motor can be used in environments where nonsparking

Drives (ac)

motors is a requirement. The ac motor can be overloaded within reason. Multiple ac motors can be run simultaneously, and the ac motor can be run at full speed in an emergency, as shown in a bypass scheme in Fig. 7-5. The ac advantages are so many that most applications consider ac motors first. All that is needed is an electronic means of controlling it.

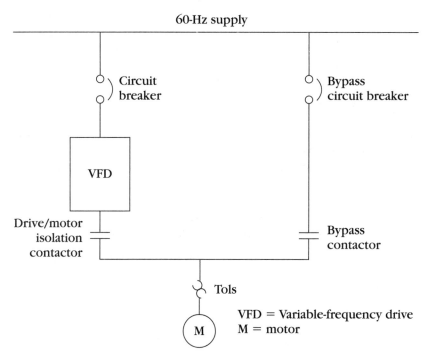

7-5 *Across-the-line bypass scheme.*

Technology for ac drives has existed since the early 1960s, but ac motors did not become popular until the 1970s due in part to the energy crisis experienced during that time period. It is amazing what we humans can accomplish when we have to!

Other factors that drove the drive development were power semiconductor advancements, the need to increase production and control maintenance costs, and the ability to interface with process control equipment in the plant. Early ac drive designs were awkward, complex, and costly. Physical units were large and took up valuable floor space in plants. As microprocessors, logic controls, and advanced power semiconductors became available, so too did a simplified, smaller ac drive.

To understand the application and operation of an ac electronic drive, one must first understand motor speed concepts. Speed is equivalent to the following:

$$\text{Speed} = \frac{120 \times \text{frequency}}{\text{number of poles}}$$

The value 120 is a constant. For a given motor, the number of stator poles is constant; thus, to change the speed, the frequency must be changed, which is exactly what a variable-frequency ac drive does. In the equation, if the frequency is 60 for normal 60-Hz supplied power, the motor speed is the maximum that can be achieved for a specific number of poles. This speed is referred to as the *synchronous speed*. If the frequency is 30 Hz, the resulting speed is half.

The number of stator poles refers to how the coils are wound in the stator, which becomes a fixed number and thus determines the motor's revolutions per minute. Table 7-2 shows the corresponding speeds for a given number of poles in a motor. Those examples shown are for the most common pole configurations in ac motors.

Table 7-2. An ac induction motor speeds/poles chart

Number of poles	Synchronous speed (rpm)
2	3600
4	1800
6	1200
8	900
10	720
12	600

When driving an actual load, the shaft speed of an ac squirrel-cage motor is slower than the synchronous speed. This difference is called *motor slip* and is typical of all induction motors. Because the motor dynamically corrects itself to maintain speed when loaded, it lags behind in actual rpm. The amount is usually 2 to 3 percent of the synchronous speed. For example, an 1800-rpm synchronous speed motor, when supplied three-phase power at 60 Hz and no loading, can run at 1800 actual revolutions per minute. When the motor shaft is applied to a load, however, the actual speed is 1750 revolutions per minute. The speed at which a motor can run fully loaded is called *base speed*. A motor can run above or below its base speed by increasing or decreasing the frequency.

Drives (ac)

Torque can be said to equate to load, which is equivalent to current. Given relationships exist for speed and torque. Typical speed/torque curves are shown for NEMA design motors A (Fig. 7-6), B (Fig. 7-7), C (Fig. 7-8), and D (Fig. 7-9). Slip always remains constant anywhere on the curve while frequency is reduced. The different motor designs have different values, or percentages of slip, as noted.

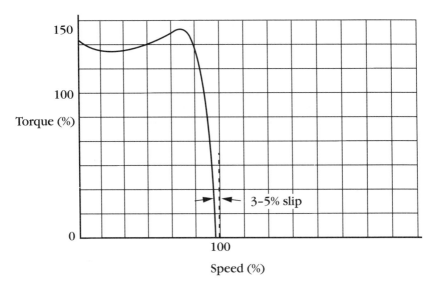

7-6 *Speed vs. torque curve for a Design A motor.*

Slip is an important element in controlling an ac motor, especially at low speeds. In essence, controlling slip means the motor is under control. Some motors reach up to 8 percent slip. In addition, each design has different starting and maximum torque characteristics. The most common design is the NEMA design B, but a NEMA A, C, or D design can be selected for its particular torque output.

When applying a variable frequency drive (VFD) to a motor other than NEMA design B, some difficulties can be encountered in matching the two. Depending on the drive type and motor parameters, special attention might need to be given to these applications.

As the induction motor generates flux in its rotating field, torque is produced. Flux must remain constant to produce full-load torque, which is extremely important when running the motor at less than full speed. Because ac drives are used to provide slower running speeds, there must be a means of maintaining constant flux. This method of flux control is called the *volts-per-hertz ratio*.

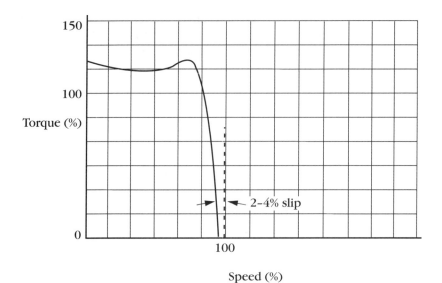

7-7 *Speed vs. torque curve for a Design B motor.*

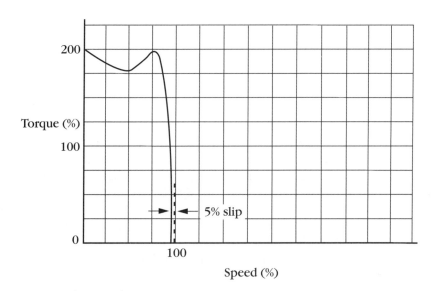

7-8 *Speed vs. torque curve for a Design C motor.*

When changing the frequency for speed control, the voltage must also change proportionally to maintain good torque production at the motor. The nominal volts-per-hertz ratio is 7.6:1 for a 460-V, 60-Hz system (460/60 = 7.6). The variable-frequency drive tries to maintain this

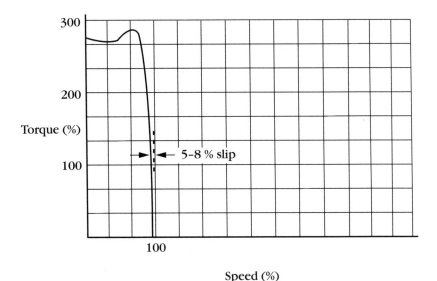

7-9 *Speed vs. torque curve for a Design D motor.*

ratio because if the ratio increases or decreases as motor speed changes, motor current can become unstable and torque can diminish. Thus, variable-frequency drives have control troubles below 20 Hz.

The flux-vector design of a variable-frequency drive can maintain better control of the volts-per-hertz pattern at very low speeds. This drive is discussed later. Another method of increasing the voltage at low speeds to produce adequate torque is by incorporating a voltage boost function available on most drives. If the motor is lightly loaded and voltage boost is enabled at low speeds, however, an unstable, growling motor might result. Voltage boost should be used when loads are high at low speeds.

Variable-frequency drives also allow for motor operation in extended speeds, or overspeeding. The application might require that the motor be run beyond 60 Hz. Frequencies of 100, 200, even 400 Hz are possible with higher switching inverters. Even higher speeds can be achieved, but torque diminishes rapidly as the speed goes higher. High-speed applications include test stands and dynamometers. While high-speed capabilities are an attractive feature of the variable-frequency drive, care must be applied when using this function. Many applications cannot tolerate extremely fast speeds, which can be dangerous.

Alternating current drives are usually classified by their output. The object of the ac drive is to vary the speed of the motor while providing the closest approximation to a sine wave. When an ac motor runs directly off 60-Hz power, the signal to the motor is a sine wave (as clean as the local utility can provide). To get the desired speed, you need to put a variable-speed drive in the circuit and vary the frequency. This procedure sounds simple enough, but the industry continually strives to address the side effects and provide a pure system.

As discussed earlier, the variable-frequency ac drive consists of a converter section and an inverter section (the name inverter has stuck with the product). Many converter designs exist, as do many inverters. Generally, the converter section, or front-end, of the variable-frequency ac drive is the dc drive for a dc motor with some modifications. The dc drive is discussed later in this chapter.

The more common designs for the converter use diodes or thyristors to rectify the ac incoming voltage into dc voltage. One design is called the *diode rectifier*. In Fig. 7-10, the incoming, three-phase 60-Hz power is channeled into three legs of the converter circuit, with each leg having two diodes to create a constant dc voltage. From here, the constant dc voltage goes through the dc link to the inverter circuit where it is changed into variable-frequency ac supplied to the motor.

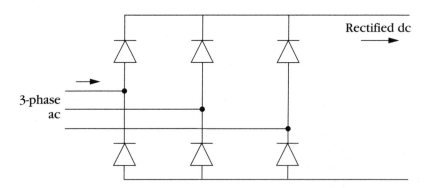

7-10 *A diode rectifier circuit.*

The diode rectifier is the most popular design because it is simple and the least expensive. Advantages of this rectifier include a unity power factor, less distortion backfed to the supply, and resilience to noise in the converter itself. The biggest drawback to using diodes is that no voltage can return to the source; therefore, no regeneration is allowed. Another separate power bridge must be added to the converter section if regeneration is required.

A diode rectifier with a chopper is similar to the aforementioned simple diode rectifier. The difference is shown in Fig. 7-11. A thyristor is incorporated, usually a silicon-controlled rectifier (SCR), to act as a valve to allow the dc voltage to rise. Once the voltage reaches a predefined level, the valve, or chopper switch, opens to de-energize, and the voltage decreases. This process controls the high and low levels of dc output voltage to the inverter and allows for better control of a constant volts-per-hertz ratio. The cost to produce this design is slightly higher than a straight diode rectifier.

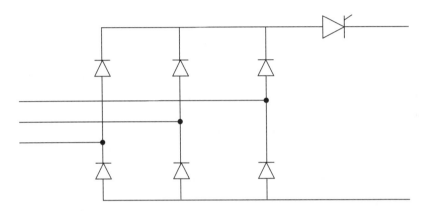

7-11 *A diode rectifier with a chopper circuit.*

Figure 7-12 illustrates the robust SCR converter. This design is also a full-wave type rectifier. Six SCRs are used to control the gating of the device. *Gating* is the term given to controlling the SCR's time of conduction by turning the SCR on and off. The SCR cannot be turned on until it has de-energized after being turned off, which is sometimes referred to as the *zero-crossing* of the current. Many SCRs have different turn-off times, and it is sometimes necessary to get all six SCRs to match in one drive's circuit to assure proper, smooth gating. The drive's logic circuitry provides the control for this gating sequence and thus controls the output voltage to the inverter.

The SCR, or thyristor-type converter, unlike the diode rectifiers, has distinct disadvantages and advantages. As speed is decreased, so too is the power factor of the system. Distortion fed back to the supply is a major concern, and a choke or other special circuit must often be added to minimize disturbances. In addition, these SCRs are more susceptible to line disturbances, which can result in nuisance drive trip-

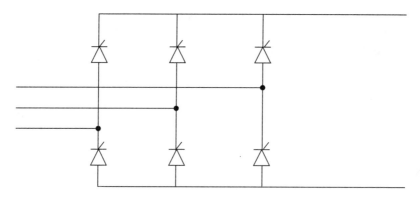

7-12 *Silicon-controlled rectifier (SCR) converter.*

ping. If manufacturers take measures to protect against these problems, this type of rectifier is attractive because it can regenerate power back to the ac supply simply by gating the SCRs in reverse order.

The heart of the variable-frequency drive is its inverter section. It is the most complex portion of the ac drive and is probably the reason that ac variable-frequency drives are sometimes just called inverters. Many designs exist and discussion has ensued on which drive is best. The answer is they all are, provided they are selected for the right application. Some designs inherently cost more than others, while other designs have horsepower limitations or address harmonic distortion. Basically, most lower-horsepower inverters incorporate high-frequency transistors, while higher-horsepower inverters do not. Individual power devices and the paralleling of devices to get higher currents can be expensive.

One of the latest inverter designs is the insulated gate bipolar transistor (IGBT) type. This transistor is a combination of features provided by the MOSFET transistor and the bipolar transistor. The IGBT has good current conductance with lower losses. It possesses very high switching frequency and is easy to control. Figure 7-13 shows its simplified circuit. This technology has gained much momentum as the IGBT can be used in powers of up to several hundred horsepower.

Other inverters using high switching frequency transistors incorporate MOSFET, bipolar transistors, or Darlington transistors. The transistor's basic advantage is that it can be switched at will from conducting to nonconducting. It does not need to wait for a zero-crossing condition as do its diode and SCR counterparts. Because higher current ratings of transistors are now available, higher resulting horsepower drives are now being built. Transistors are even being put in parallel to get the desired output to the motor. These transistors can

Drives (ac)

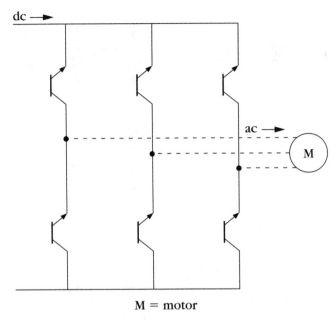

M = motor

7-13 *Insulated gate bipolar transistor (IGBT) inverter.*

switch at several kilohertz, virtually eliminating audible noise at the motor, which was an earlier objection to using high switching frequency transistors. The current waveform to the motor is close to sinusoidal, while the voltage can be modulated much easier.

Other types of inverters use diodes and SCRs. The six-step or variable-voltage inverter needs a chopper in front of the diode bridge or individual SCRs to switch on and off to produce the desired six-step voltage waveform. Each power-switching SCR's conduction time is controlled to get the desired change of each individual step, thus changing the frequency output to the motor. Commutation, or turning semiconductors on and off, is accomplished using an extra circuit, usually consisting of capacitors, to provide power to switch off devices. Transistors and gate turn off (GTO) devices do not need commutation circuitry.

Another type of inverter is the current source type. This drive can be called a CSI, or current source inverter. The equivalent circuit for this inverter is shown in Fig. 7-14. This inverter typically uses SCRs as switches to gain a six-step current waveform output. Here, the conducting time is changed up or down for each individual step, resulting in a longer or shorter cycle time.

Another type of ac drive inverter is the pulse width modulated (PWM) output version. This type of inverter is a constant-voltage

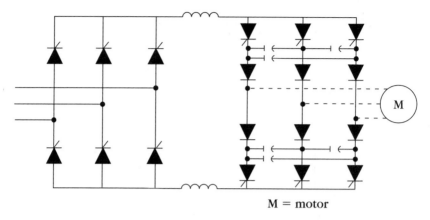

7-14 *Converter and inverter circuitry of a current source drive.*

source drive and is very common. Its equivalent circuit is shown in Fig. 7-15. This inverter is typically combined with a diode rectified converter. It can use GTO thyristors or transistors because a PWM inverter requires fast switching devices. IGBTs are widely used, as they incorporate very high frequency switching. This device has also practically eliminated noisy, whining motors because the frequency is out of the human audible range. The pulse amplitude modulated version varies its output voltage to create either 6 or 18 pulse signals to the motor. This 18-pulse design is good for shaping the motor current more closely to a sine wave and for less harmonic distortion at the motor. Six-pulse designs promote cogging at low speeds and increased motor heating.

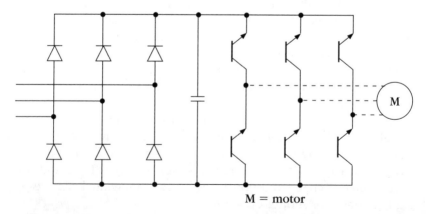

7-15 *Converter and inverter circuitry of a pulse width modulated (PWM) drive.*

Drive selection (ac)

So why use a particular inverter over another? This is a hard question to answer. Each application dictates what is required. The first issue is whether a dc motor or an ac motor will be used. Will there need to be regeneration? What is the horsepower? Voltage? What is the speed regulation? Torque regulation? Is cost a factor? Is plant floor or wall space at a premium? Is a digital or analog drive preferred? What will the maintenance person find when a problem occurs? What type of duty cycle, or loading, is predicted? Does the plant have a "clean" voltage supply? Do we need an efficient drive? The questions seem to be endless. Let's look at the issues.

Once it has been decided that an ac motor is to be used for the application, it must be decided what horsepower, voltage, and enclosure will be used, along with whether or not the motor selected can run properly with any inverter's output. In other words, is there a good match between motor and drive?

The Do's and Don'ts of ac drives have created many arguments among professionals over the years. Misapplying drives is always a potential problem. Manufacturers of drives might have biased reasons to support or degrade a particular design. The rules stretch into the dc sector, along with the servo- and stepper-drive worlds. The best suggestion to be given is this: take the time and perform an in-depth analysis of the current equipment available, and completely understand the application from both a mechanical and electrical vantage point.

One manufacturer's variable-frequency drive might be well suited for one application but not for another. Hundreds of considerations need to be pondered when selecting a variable-frequency drive. Section 15 of *Architectural Specifications*, Electrical, usually contains written descriptions of drives to be used for building and construction projects. Try to secure a copy, and talk with as many technical vendors as possible because the industry is constantly changing. These professionals can advise you of whether your application is right for their drive and vice versa. Also consult the bibliography at the end of this chapter for a more in-depth review of certain drive subjects. The following is a list of issues and concerns for ac drives.

Circuitry Determine the complexity of the ac drive's circuitry. The more components, the greater the risk of component failure. Also consider how long a particular complex drive type has been in production. Reliability is the key in an ever-changing industry.

Digital or analog Determine if a digital or analog drive is necessary. While most manufacturers promote digital designs, analog types are simpler. Digital drives offer plenty in diagnostics, protection, and communications.

Speed What is the application? What is the speed range and loading at various speeds? Fully loaded low speeds are the most difficult for ac drives to control. Similarly, hard-to-start loads require high breakaway torque. Is speed regulation tight? Typically, a drive's slip compensation function can provide adequate adjustment. Check if a drive design is retrofittable with a feedback device for better speed regulation. Determine whether or not the load is variable or constant torque. Figure 7-16 illustrates the different torque at various speeds. Attention might need to be given to the motor if loads, or motor currents, are high at low motor speeds. A separate blower for auxiliary cooling might be required.

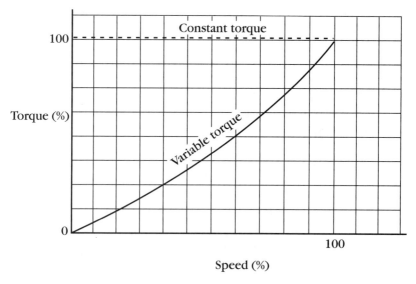

7-16 *A variable torque and constant torque comparison.*

Braking Can the load being driven regenerate? Is braking required or can the motor coast to a rest via the friction in the drivetrain? What will happen in an emergency stop (E-stop) situation? The drive has no control when it is not being supplied power.

Power supply How good is the power supply furnished to the drive? Some drives are more sensitive than others. Some are phase-sequence sensitive.

In-rush currents The ac drives generally limit the amount of in-rush current to the motor thus lengthening its life while providing softer, smoother starting.

Drive selection (ac)   193

Motor quantity Does the application require several motors to be run from one inverter? If so, a current source drive will have limitations because it is load dependent. Only a few, if any, motors can be taken offline when running the entire set. A voltage source drive is more suitable in this situation.

Ground fault and short-circuit protection Does the drive manufacturer require an isolation transformer? How will the drive handle a ground fault or a motor short circuit? What happens to the inverter in an open-circuit situation?

Power factor Discussed at more length later in this chapter, must be considered because of utility penalties. A drive with a constant power factor throughout its speed range is attractive.

Harmonic distortion, or content Variable-frequency drives create disturbances back to their supply and out to the motor. Determine what your particular system can tolerate. Harmonics is also discussed later in this chapter.

High horsepower Motor horsepowers can sometimes dictate the type of inverter required. Some transistors are not available in higher current ratings. Explore the variable-frequency drive manufacturer's experience in high horsepowers.

Efficiency Variable-frequency drive efficiency is just as important as cost. Analyze each using the motor and actual drive. Consider efficiencies at all speeds but mainly for the predicted operating speeds to be operated. Remember, the variable-frequency drive is supposed to be an energy-saving device.

Coolants VFDs are usually air-cooled; some are water-cooled, and some are air-conditioned. Air-cooled drives tend to be noisier than others mainly because of the cooling fans inherent to the package. Ensure the plant can tolerate higher decibel levels.

Ventilation A cool drive is a happy drive. When physically installing any drive, whether ac or dc, special attention should be given to the heat generated by the drive. The drive has current running through it. This heat must go somewhere. It can naturally dissipate if there is a light duty cycle. If loading is heavy, provisions for cooling or ventilating are necessary.

First, look at the ambient environment around the drive. Is the room in which the drive is located naturally warm or hot? What is the temperature on the hottest summer day? Is the drive going into an enclosure? Is this enclosure completely sealed? Determine if the drive will be heavily loaded. Will it run 24 hours per day or intermittently? Ventilation fans that pull ambient air into the enclosure and up across the drive, can suffice. Figure 7-17 shows a typical drive installation with ventilation.

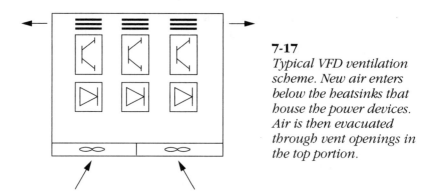

7-17
Typical VFD ventilation scheme. New air enters below the heatsinks that house the power devices. Air is then evacuated through vent openings in the top portion.

How clean is the ambient air? If dirty air is brought into the enclosure, a new problem can emerge. Dust and dirt collect on the drive and can virtually suffocate it. Eventually, no heat will be able to escape from the drive, and it will overheat. Most newer drives will trip, or fault, on an overtemperature fault, protecting the drive but being a nuisance because the drive must cool off before starting again.

Another way to handle drive-enclosure heat is using air conditioning in the cabinet. Air conditioning is more expensive but sometimes the only answer. When the ambient air is too warm or too dirty, air conditioning makes sense. Many manufacturers specialize in small, compact air conditioners that attach directly to the wall of a drive enclosure. These units are self-contained, closed-loop units that keep the inside of the drive enclosure completely cooled.

If the ambient area is dirty, the air conditioner must be kept clean to operate efficiently. If the air conditioner stops running for whatever reason, the drives will trip quickly because they are now in a completely sealed enclosure with no way for heat to escape.

Many air conditioners today work without any chlorofluorocarbons, or CFCs. They are green products, environmentally safe, although they probably need cooling water supplied to them and require a drain. They are worth looking into.

High altitude High-altitude locations can be a problem for a drive installation. Knowing the elevation conditions ahead of time can prevent drive-related problems. As the altitude increases, so too does that air's inability to dissipate heat. Thin air, which you might get at 3300 feet above sea level (or approximately 1000 meters), cannot hold as much heat as an equivalent amount of air at sea level. Therefore, transferring the heat from our heat source, the drive, is more difficult at higher altitudes. For a given elevation, drives often have derating values for equivalent horsepower, or continuous current output.

Drive selection (ac)

High humidity Typically, high humidity is a problem for a drive if the excessive moisture in the air condenses on the components. Water will conduct electricity and, if enough water forms on the drive, short circuits can occur. The maximum level of moisture in the air is a relative humidity of 95 percent. High humidity causes nuisance tripping, and downtime will occur until the drive is dried.

Another moisture-related problem is that of corrosive gases in the air, such as acid mist, chlorine, salt water mist, hydrogen sulfide, and others. Problems can range from slow deterioration of the printed circuit boards to actual corrosion of the bolts that hold the drive together. Protective coatings on the boards can help, but the better solution is to keep the contaminated air from getting to the drive at all.

Distances between motor and drive Distance can be a factor depending on the type of drive used. A voltage-source drive tries to maintain a constant voltage in the motor/drive circuit. Long runs of cable can create voltage drops, which can affect the motor's ability to maintain speed under heavier loading. Cable length and thickness determine the maximum output current. The longer the cable is, the more heat generated in the VFD because of lower capacitive reactance. The Dv/Dt phenomenon, or the derivative of voltage and time, and the capacity of cable can be tolerated with a filter on the output of the VFD.

One solution to minimize drops in voltage is to increase the gauge, or diameter, of the cable. Increasing the gauge might need to be done if the actual supply voltage is lower than nominal. For instance, if the supply is supposed to be 460 V, most drives can handle a 10 percent range above and below that. If the supply is lower than the projected 460 V, a long run of cable can further reduce that value to the motor in less available volts per hertz, limiting speed and torque capabilities. Typical distances, though, are less than 300 feet and usually present no problems.

On occasion, it is necessary to locate the actual electronic drive farther than 300 feet away from the motor. If the drive is a voltage-source drive with very high switching transistorized output, a different phenomenon can exist called the standing wave condition. This condition can make for high peak voltages and possibly damage motor windings. This condition is more prevalent with these types of drives and runs of cable greater than 300 feet.

One solution is to install output reactors between the motor and drive. The reactor smoothes the voltage but adds impedance to the system, adding to the overall voltage drop. This disadvantage must be considered when evaluating this option as a solution. If the voltage supply is steady and higher than nominal, all should be fine at most speed and load conditions. Two other considerations relative to this standing wave condition are: 1) the longer the distance between motor and drive, the higher the impedance output reactor be used, and 2) a higher class of insulation in the motor might be better able to withstand voltage peaks without substantial degradation.

Output contactors A motor can be located very far from the variable-speed drive, even completely out of sight. If a maintenance person wants to work on the motor, he or she must ensure that no electricity is going out to the motor. One common practice is to install a contactor or disconnect near the motor in the circuit between the output of the drive and motor. This contactor is fine from a safety point of view but can be harmful to the drive.

For example, the maintenance person goes onto the roof of a building where a motor is located that drives a fan. The variable-speed drive is located two floors below in a mechanical room. The fan is running at full speed and full load when the maintenance person decides to kill power to the motor by opening the contactor. This action could cause a high-energy spike back to the drive, blowing an output device. Rather than assume all maintenance personnel are trained to never open an output contactor under load, it is more practical to interlock a contact that faults the drive first and then opens the output contactor, ensuring no power is flowing through the drive and thus eliminating the possibility of a spike. This contact must be an early auxiliary contact.

Input reactors Do variable-speed drives require isolation transformers? What needs to be accomplished? Are we trying to minimize noise in and out of the drive at the supply point? Do we need ground-fault protection? The isolation transformer can provide all of this, but the user should understand the type of drive being supplied before requesting input line reactors.

Output reactors Certain types of drives sometimes need output reactors to reduce ringing at the output to the motor. The major reason, however, to use output reactors was discussed in the subsection on distances between drive and motor.

Flux vector drives (ac)

The ac drive technology has progressed to the point that more performance is demanded from the ac drive and motor. Because the standard open-loop ac drive cannot hold speeds as the speed decreases, a better design has appeared. This drive is called the *flux*

Flux vector drives (ac)

vector drive. It takes complete control of the motor slip and air gap flux within the ac induction motor by using special algorithms, high-speed microprocessing, and digital feedback from the motor itself. An ac flux vector system (motor and drive controller) can perform similar to an equivalent horsepower dc system.

Why is this drive called a flux vector? Why not a magnetic field vector, or a flux corrector? The name *flux vector* was coined, and it stuck. Other names for similar drives are space vector, voltage vector, and simply a high-performance drive. Vector is the key, and it is defined as a quantity of magnitude in a specific direction. What the drive controller tries to do is constantly control the flux angle and magnitude of flux within the motor's air gap. The flux vector is thus created by the supplied currents. The flux vector drive has incorporated the basic ac power bridge, as well as some extra control.

By attaching a pulse generator to the motor's rotor and feeding the signal back to the drive, slip frequency can be determined. The signal is compared to the required speed. By knowing what the motor slip is, a precise speed correction can be made using new voltage output. The precise speed connection is made possible by using high-throughput microprocessors that crunch the data and continuously create a newly corrected output. Thus, the flux vector drive is constantly correcting as the motor runs. This feedback makes the flux vector drive a closed-loop drive.

The driven load and the control of the torque to drive the load also enters into the equation. Torque control is accomplished by maintaining a constant value of air gap flux, which means that the volts-per-hertz ratio must be held constant. With rotor speed completely controlled along with the magnetizing- and torque-producing currents, we have flux vector control.

As previously mentioned, the microprocessor-based drive must be programmed with values of motor slip (the difference between base speed and synchronous speed), full load and no load (or magnetizing current), and other parameters. Once the drive knows what to expect, it can correct for deviations while running. Speed regulation to 0.01 percent of base speed is now possible. Speed regulation can be held even at very low speeds while loaded, which has been the dc drive's sole domain.

Attention must be given to the ac motor being controlled by the flux vector drive. Its nameplate and test data should be available so the correct parameters can be entered into the drive. The motor needs to be equipped with a pulse generator as well as an auxiliary blower for cooling at low motor speeds. It is often better to procure the motor from the same supplier as the ac flux vector drive to assure compatibility. In some instances, a standard ac drive can be physically modified in the field to become a flux vector drive, but this situation can be avoided if the up-front application evaluation is done.

Do's and don'ts of flux vector drives

- A flux vector drive provides much better torque control but cannot create more current for a given horsepower-sized drive. A 20-A rated ac drive produces 20 A continuously to a motor (and sometimes more for short periods), whether a flux vector drive or not.
- Flux vector drive systems cost more, so be sure the application justifies the extra cost. The major contributor to the cost increase is the motor and its need for a pulse generator and auxiliary blower.
- Affixing the feedback device, which is the vital component in the flux vector drive's feedback loop, is most important. Take special care when mounting these devices.
- Determine beforehand the speed range and actual torque requirements, especially at the low speeds. Does the load change frequently? The answers to these issues can help select the proper size and type controller to use.

Many other issues and concerns are similar to that of the standard ac drive. Refer to those listed previously.

Drives (dc)

The speed of a dc motor must be controlled under different load conditions. Some loads are hard to start, while others must have regeneration or reversing capability. A dc electronic drive must be able to dynamically change output levels of both voltage and current to a dc motor to control speed and torque. Today's technology uses solid-state electronics to accomplish that.

The dc drives can be classified into two groups, transistorized and thyristor-, or SCR-based. The workhorse of the dc drive world is the SCR-based unit, as it is available in a much larger power range. The transistorized dc drive usually is used in specialized dc motor applications, such as running a permanent magnet motor. Regardless, the dc drive controls the speed of a dc motor. The dc drive is less complex than the ac drive; it is actually the ac drive without the link and inverter circuits. Therefore, the dc drive is usually less expensive than its ac counterpart. Both the motor and the drive packages must be compared, however, not just the drive.

Figure 7-18 shows an equivalent circuit and the associated waveform for a three-phase, full-wave power bridge using SCRs. The dc drive mostly varies the voltage of the motor armature while the motor field has a constant voltage. The SCRs are configured in a power bridge. The dc power bridges can accept either single- or three-phase ac or dc (as from a battery source) input supply. The smoothest output from a dc drive would be that of a full-wave, three-phase supplied input to the bridge. The gating of the SCRs creates pulses of current. The more pulses per second, the smoother the torque output of the motor. Today's dc drive technology allows up to 12 pulses of dc current for every ac cycle (there are 60 cycles per second). The dc drives can reach 720 pulses per second, which is not bad for a dc system.

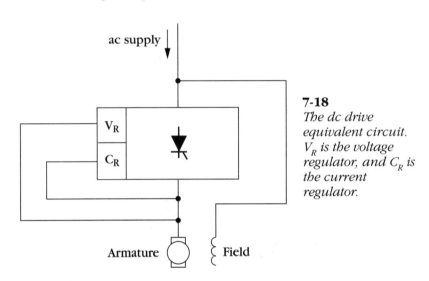

7-18
The dc drive equivalent circuit. V_R is the voltage regulator, and C_R is the current regulator.

The speed of the dc motor is typically in direct relationship to the voltage supplied. Today's higher voltage rated shunt-wound dc motors have 500-V fields and 300-V armatures. Lower ratings are also available at 300-V fields and 150-V armatures. If 100 V is supplied (20 percent), the motor can run at 20 percent speed, or 350 rpm (with a base speed of 1750 rpm). At full voltage, the motor runs at full speed. The real feature of the dc motor and drive system, however, is that it can deliver 100 percent torque virtually anywhere in its speed range. It produces as much torque as the load requires, even to where the drive is overloaded with current and can damage the motor. Thus, motor overload protection must be built into the drive and motor.

The type of feedback circuit used is armature. In it, the dc drive compares the commanded speed reference to the armature voltage being produced to run the motor at that given speed. Constant correction of this loop is the dc drive's chore. Speed regulations of 3 to 5 percent can be achieved while running in this manner. For better speed regulation, an analog tachometer of 50 or 100 V per thousand motor revolutions per minute can be attached to the motor's rotating portion. Regulation below one percent is common with an analog tachometer, although impact loading and device age can deteriorate this accuracy. More common today is a digital tachometer, or encoder, which can provide motor speed regulation of less than 0.01 percent.

A dc drive with two SCR bridges can regenerate power back to the main input supply. It can also reverse directions. The dc drive can provide 200 percent of full load current for short periods and 150 percent for longer periods. These features make it attractive for many applications because of its simple design. Its ruggedness has kept it around for many years and will keep it around for many more. Contrary to the prediction that all dc applications will eventually be replaced with ac drives and motors, this transition is happening much slower than anticipated. There are several reasons to switch from dc to ac, but there are still those applications that require the performance of dc. Because dc drives have kept pace with digital and solid-state electronics, many extra functions are now possible.

Digital dc drives can accept various inputs and can send certain outputs. Start and stop logic using low-voltage discrete inputs, as well as the ability to accept 0 to 10 V analog signals, have become standard in dc drives. Some dc drives can carry a software program that can define application functions and send both discrete and analog outputs when required to another computerized device. With microprocessor technology, even high-speed data transmission to other smart drives and host controllers is possible. We are now on the cutting edge of automation with dc drives.

Digital dc drives can also provide diagnosis for faults when they occur. The drive acts as the protective portion of the drive and motor circuit. The dc drive will trip if output current remains at a harmful level for too long, preventing the motor from overheating and breaking down. Thermostats can be included at both the motor and drive heatsinks to monitor excessive levels of temperature. Thermostats are another protective measure that the drive can monitor and alarm accordingly. As most drives read a tachometer feedback signal, they should shut down if that signal disappears for any reason. Better for the motor to shut down than run away. The electronic drive can even detect a blown fuse, shorted SCR, or loss of motor field.

Like ac or servo digital drives, digital dc drives afford the user the luxury of programming the drive for the application much faster. By entering numeric parameters into the drive's program, hundreds of drive characteristics can be set. Acceleration and deceleration rates, minimum and maximum speeds, plus several other application-dependent requirements can be set simply by pushing a few buttons. The gains, or stability and response, of the motor/drive system can be tweaked digitally by entering new values for proportional and integral drive math functions. These parameters are those special ones that need adjusted, usually in the field, for hard-to-start loads as well as those surprises not covered at the engineering review.

The dc drive and motor system needs to isolate the motor mains from the drive. Isolation is accomplished by using an M or loop contactor. This contactor must be sized for the current-carrying capacity of the motor and drive.

The dc drive is typically more efficient than its ac drive counterpart because the equivalent nonregenerative dc drive has only one power conversion section, whereas the ac drive has two, the converter and inverter. This fact is also true when comparing the ac and dc motors.

Figure 7-19 shows four-quadrant operation, which is common in dc drives with two bridges of six thyristors each. Four-quadrant operation means that the motor can be controlled in the forward direction with positive torque to drive the load or negative torque to regenerate the load. The other two quadrants perform similar functions in the reverse direction. Similarly, an ac drive can become a four-quadrant drive with a matching bridge design on its front end. Four-quadrant ac means that regeneration and reversing are possible with the motor.

Field weakening and constant horsepower are two issues often attributed to dc drive and motor operation. These conditions were illustrated in Chapter 6. Field weakening, the running of a dc motor beyond base speed reducing the current to reduce the stator magnetic field, does not allow for as much usable torque. Trade-offs exist when running a given motor faster than its base speed. All motors have maximum torque limitations. A winding application is a constant horsepower application.

Specialty electronic drives and soft starts

Other application-dependent electronic power-converting devices exist that control specific motors. These drives are called amplifiers

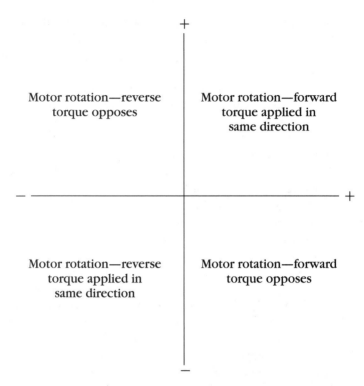

7-19 *Four-quadrant motor operation.*

and power supplies, but they control the motion of the motor. Some control high speeds, while others control shaft position. They usually start with three-phase, 60-Hz supply power. What they do after that is the interesting part. Feedback systems play a major role in the success of a given application. Several types of specialty drive systems exist, such as servos, steppers, and spindles, which are the more prevalent, and soft start devices, which are used for controlling the voltage to a motor at start-up.

Servo drives

Perhaps the most versatile drive system that can be used on the widest variety of applications is the servo. It is also the most expensive and most complicated for comparable horsepowers. Current servo systems do not allow for horsepowers above 50 hp. A servo can be defined as an automatic closed-loop control system that uses feedback to electronically control a mechanical motion event.

One of the first rules in speaking servo is to discuss power requirements in terms of torque rather than horsepower. This rule should exist for all applications, but most often people equate power, size, and capability of a motor by its horsepower or lack thereof. Torque is the force that does the work as we rotate the motor shaft. Servo people talk in inch-pounds of torque. The analysis can also be made in foot-pounds of torque, but this definition is less common because servo systems are smaller, and foot-pounds of torque must be converted into inch-pounds later anyway. The upper limit in servo motor construction is approximately 800 in-lbs. This limitation is due in part to motor construction, the magnets, and the fact that it is less expensive to use a high-precision gearbox to get the output torque rather than a larger servomotor and amplifier.

Servo drive systems can be used to control speeds of motors down to a virtual crawl, with turndown ratios of 10,000:1 and greater. Speed accuracy, or regulation, can be held to 0.01 percent or better. The response of the overall system or bandwidth is extremely fast. The servo system can hold to very tight torque accuracy, meaning that as the load changes so too does the servo torque, or current regulator. It corrects quickly and can also provide more-than-adequate values of stall torque when its amplifier is sized accordingly.

The servo drive can be likened to the ac and the dc drives' smart little brother. It is limited in high horsepower capability but can be equipped to control the actual position of the load it is driving. It has full, fast torque control from full speed down to stall. It can control the acceleration and deceleration rates of its associated motor with ease. It is providing the same control aspects of an ac or dc controller plus more.

The servo drive can control a brush-type motor or a brushless motor, which results in confusion, especially in terminology. There are brushless ac and brushless dc systems. Both are basically the same: the motor is a permanent magnet type motor with no brushes. A better description might be that the motor is a brushless synchronous motor with permanent magnets. The brushless ac or brushless dc amplifiers take feedback from a motor, which uses no brushes for commutation, and linearly amplifies the waveform out to the motor. Servomotors were reviewed in Chapter 6 and a family tree of the brushless motors was shown in Fig. 6-17, for reference. Servomotor function and commutation was also shown in Fig. 6-19, which illustrated the motor construction and clarified servomotor commutation. Some servomotors use permanent magnets, while others use the switched reluctance type. Whatever the name, it is the function and interactivity with the servo controls and feedback devices that make the servomotor work well.

A servo system is not just the servomotor, amplifier, controller, and feedback device. It is these components working in complete unison with each other quickly. There are several other factors to consider when looking at the overall servo system.

Bandwidth, the response factor of any drive system, is most important in a servo drive system. Bandwidth is often described as the time a motor controller takes to correct from no load to a loaded condition. It is usually expressed in radians per second, and while ac and dc speed control drives must use bandwidth in their control algorithms, a servo system lives and dies by its bandwidth. It is a drive system installed by virtue of its performance. It usually costs more initially and is more complicated than a dc or ac drive system. Thus, a higher bandwidth system is more immune to disturbances and responds better to produce better. One important factor associated with bandwidth is that of inertia, particularly load inertia and motor, or rotor, inertia. Other factors that affect system performance and bandwidth are backlash, belting, gearing, acceleration rates, overshoot, and torque ripple.

Figure 7-20 shows a block diagram of a servo system with three control loops. The fastest loop is the current loop, which solves the new output, in amperage to the motor, based on actual motor current sensed in the servo amplifier by current transformer (CT). The next fastest loop is the speed, or velocity loop, which needs speed feedback from the motor rotating element, or rotor. Either a separately mounted tachometer or similar device that can provide the appropriate pulses to translate into speed is needed. The slowest loop is that of position. Good servo systems have position loop updates in the 5- to 10-millisecond (ms) range to provide very tight positioning accuracy. Of course, many variables can affect each loop's performance,

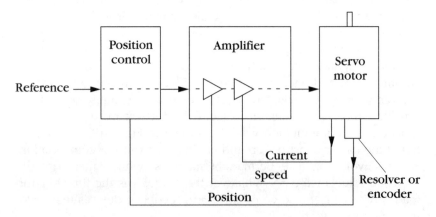

7-20 *Three servo loops: position (slowest), speed, and current (fastest).*

such as temperature or sloppiness of the system mechanics, but a high-response amplifier and controller is paramount.

The type of feedback devices used can vary. In a later chapter, feedback devices are explored in greater detail. As for servo systems, the feedback device is an integral part of the system. Some devices can provide feedback data that can be used to both calculate new speed and new position requirements. An encoder can provide output in the form of pulses, which can be used to tell rotor position and can count those pulses for a given period of time for speed correction. Some encoders are absolute, knowing where they are upon power-up, while others are incremental. Another device is the *resolver* and its feedback must go through a resolver-to-digital (R-D) converter to be useful in the controller or amplifier. R-D conversion might be a limiting area in the servo system's resolution. You need to check the R-D's resolution as to how many bits that conversion solves to. Encoder, also called tachometer, or resolver feedback can be accepted into a given amplifier or controller, but not all at once. Different manufacturers have different standards for their set algorithms, and expect certain types of feedback in their products.

The stability of a servo system is dependent on how well the crucial components are tuned and matched with each other. The system's crucial components are the servomotor, amplifier, power supply, and controller. These components receive an input signal and respond with an output signal. This relationship to signals can be referred to as the *gain* of the system. Gain is adjustable and makes the system's performance good or bad. Often called *tuning a system*, adjusting the proportional and integral gains determines performance issues such as smoothness, amount of overshoot, ringing, and oscillation. *Proportional gain* is that adjustment that acts directly and in proportion to its input. *Integral gain* takes adjustment further and uses an integrator algorithm to adjust its output based on a timed calculation. Adjusting the gains is one method of getting the most out of a servo system; however, application sizing and requirements must be attended to as well.

If a robot, whose individual axes are servo-driven, tries to move a delicate part from point A to point B, instability and unwanted mechanical vibrations are hazardous. The servomotors will overshoot, as shown in Fig. 7-21. This situation can lead to a condition called *ringing*, whereby the servo system tries to correct itself but never quite does. A well-tuned servo system means that the crucial components have been sized and matched to each other, proper gain values have been programmed, and the user runs the system within the specified operating guidelines.

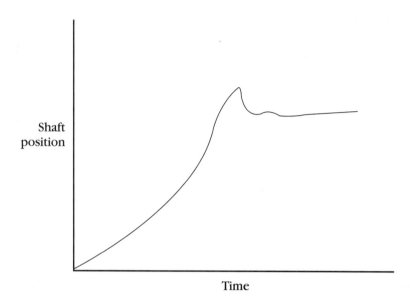

7-21 *Oscilloscope image of a servomotor move. This move over-shot after the acceleration ramp.*

Matching components in a servo system is always a challenge. There seem to be many variables for a given application. Careful attention should be given to the inertia of the load being driven and the motor selected. This variable is called *inertia matching* and involves numerous calculations. The actual inertia of each component in the mechanical drivetrain must be calculated to match the servomotor to provide good acceleration and deceleration performance. It might be necessary to incorporate a gearbox to greatly reduce the reflected inertia of the load back to the motor. As was discussed earlier, inertia through a gear reduction is a square function. Thus, whenever the inertia value is high, a gearbox should be considered.

Likewise, when sizing and matching system components, torque requirements must be considered. Peak torques and accelerating torques must be calculated along with the inertias. Once these values are known, a servo system can be pieced together. It might be better to leave component selection and the lengthy calculations to the manufacturers, as they routinely perform these tasks and can thus eliminate any glaring problems.

Applications in which servo systems are prominent are in the machine tool and robotics industries. Grinders, lathes, milling, and boring machines are good places for servo control to hold position for tight tolerances. The same can be said of robotic applications. A servo system can be used for simple speed regulation but much of its capability is then wasted. The cost of the complete system along with all its components dictates where it will be used.

Stepper drives

Some motor/drive applications require inexpensive positioning and lower precision. Stepper-motor systems allow just that. A stepper-drive system is basically similar to a servo system but without the feedback. A step-motor system, shown in Fig. 7-22, contains a step motor, an indexer, and the drive. The indexer can be substituted with a programmable motion controller, and the drive can be called an amplifier. For higher-accuracy requirements, a feedback device can be attached to the motor and better precision can be attained.

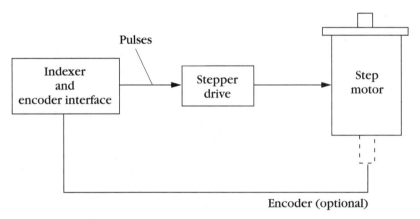

7-22 *Stepper-motor system block diagram.*

Functionally, the step-motor system works like this: the indexer provides pulse signals to the drive, or amplifier. The number of pulses represents the distance, and the frequency indicates speed, or velocity. The pulse signals are then amplified and sent directly to the step motor. The motor, constructed specifically to convert electrical pulses into discrete moves, turns the load at the desired speed to the predetermined location. The step motor's construction was described fully in an earlier chapter. The indexer, or motion controller, does not accumulate error or drift; thus, for every pulse output, the step motor should move the load to the desired location each time. Of course, closing the position loop with an encoder assures moves to the exact location every time.

Stepper motors have two windings, or phases, which are located in the stator for maximum heat dissipation. Current is switched on and off in these two phases, which creates an electromagnetic field for rotation. The position of the motor shaft is dictated by the number of current switches and which phase was turned off last. The frequency of the switching dictates the speed of the motor. Some oscillation occurs in a step motor with each step, which is more noticeable at low speeds. Microstepping is one method of minimizing resonance due to oscillation. In microstepping, the increments are much smaller, thus smoothing the operation. A standard of 25,000 pulses per revolution is very common.

Spindle drives

A spindle is often referred to as the shaft in a lathe or capstan. This spindle, if it is to turn, must be motor-driven. The motor must be controlled by a specialized device called a *spindle drive*. This drive has a niche in the machine tool industry. A spindle is typically required to turn a part at high speeds, which is called machining, or working, the part. When turning the part at speeds to 7000 rpm, power requirements are constant. The part spins at high enough rpm to remove metal from it. It is typically necessary to run at lower speeds to provide constant torque, resulting in the spindle system. The duty cycle of the spindle is contingent on the associated motors and drives coordinated with it. As the spindle spins at high speed, the other axes of motion are either moving in or out and traversing the part. Depending on the machining, the spindle can slow or remain at a constant speed when not doing any work.

In the machine tool environment, smaller, compact designs are needed because space is usually a premium on the plant floor and on the machine itself. Spindle motors are usually of a longer stack length, square, and completely enclosed with a separate fan for cooling throughout its speed range. The industry has, over the years, stressed low maintenance, and therefore no brushes or commutators are incorporated into the design. The most common sizes of typical spindle motors go to roughly 30 hp and can be handled by ac drive technology. Larger spindle drives have traditionally been left to the dc drive, but this scenario is rapidly changing as ac drive current and switching capabilities increase.

Drive technology has evolved in other ways, especially in the ac inverter realm. Inverters can provide outputs as high as 800 Hz, which is over 12 times the base speed of 60 Hz. The ability to provide constant torque to lower speeds also makes the ac drive attractive to the machine tool user. The emergence of flux vector ac drives has also impacted this industry. The enhanced speed and torque regulations of this controller have allowed the vector drive to become a standard.

Reduced voltage starters

A particular motor might need to be capable of starting with a lower starting current. It might be necessary to start the load softly so as not to disturb the process. Another reason to soft start would be to minimize the in-rush current to the motor, thus lengthening its winding life. Starting a large motor can also demand so much current that the lights in the plant might dim and voltage drops might occur, causing unwanted disturbances. When variable-speed control of the motor is not required but limiting in-rush current is, it makes sense to use a reduced-voltage soft starter. These devices are often simply called *starters*.

An ac induction motor, when started directly across the line, requires 5 to 6 times its current rating to start a load. This in-rush current, after many hundreds of starts and restarts, causes premature wear in the motor. In addition, the other components in the mechanical drivetrain, such as the couplings, belts, and gearboxes, take a beating with every hard start. This shocking of the entire system eventually takes its toll on both the electrical and mechanical parts of the machine. The shock effect is also undesirable to the driven load. A conveyor motor loaded with product, when shock started, can actually damage the product if it fell over or even off the conveyor belt.

The electrical supply system must be compatible with the direct across-the-mains, or line-starting, of the ac motor. Larger ac motors with high current ratings, being started at remote sites where generators are the power source, must be started when power is available from the generator, which requires coordination and procedure. Many smaller ac motors, running or starting at once, can cause voltage starving problems. A facility with adequate power can still have a voltage dip when a large motor is started across the line. A device is needed that can start the motor with a reduced voltage and then allow voltage to increase slowly as the starting procedure advances. This device is the *reduced-voltage soft start*.

The reduced voltage starter is an electronic power-converting device that contains a power bridge consisting of thyristors and a logic circuit. When three-phase power is applied to the bridge and a start contact closed to the control circuitry, the thyristors begin firing. The control circuit governs how long and at what angle the firing, or gating, of the thyristors occurs, which controls the amount of voltage seen at the output of the bridge. The controller is also programmed to know in advance the acceleration, or ramp time, to control the voltage as speed increases over time.

Other means of starting an ac motor and limiting the in-rush current somewhat are by using an auto transformer or configuring the motor windings into a star configuration. Either of these methods can greatly reduce the in-rush current compared to across-the-line starting, but they are still not as easy on the motor and drivetrain as the reduced voltage starter. Both the star-delta and autotransformer method incorporate repeated instances whereby the motor windings are reconnected to the supply but at an increased, stepped-up voltage. While this is better than shocking the motor windings with 600 percent current, it is still not a soft start.

The auto transformer, having close primary and secondary voltages, allows for the motor windings to connect to the lower voltage tap of the secondary. As the motor ramps up in speed, the windings are open-circuited and then reconnected at a greater voltage level than before open-circuiting. This procedure is repeated until full speed is attained and the motor windings are transferred to the three-phase, 60-Hz supply. This is also a step function procedure, as the star-delta configuration of the motor windings. The motor can see far greater values of in-rush current than even across the line-starting with this method, but for substantially shorter periods.

For those applications requiring a controlled, reduced-voltage soft start, the starter is the answer. The reduced-voltage starter has, however, been misapplied more than once. It can provide some speed control but only on very simple loads. Consult the starter manufacturer before applying it to an application other than soft starting. The variable frequency drive can and does provide soft-start capability while also controlling the speed, which is one of its extra benefits.

Braking and regeneration

One of the most misunderstood topics in motion control and electronic-drive applications is that of braking and stopping an electric motor. Which method is the fastest at stopping the motor? Which is the safest? Does the application require braking? Can the electronic drive provide braking? Should a mechanical brake be used? Many good questions with many good answers.

An electric motor, whether dc or ac powered, moves its desired load and demands that amount of power to do its job. When the load decides to drive the motor, the motor now is a generator, a generator of power. Where can this power go? This energy coming from the motor is called *counter EMF* (CEMF), also known as *back electromotive force*. If this CEMF has a channel to get fully back to the source of supply power, it is called *regenerative*. If this channel is used to slow or stop a motor and its associated load, it is called *regenerative braking*. A high-inertia, low-friction load, an overhauling load, or a crane or hoist all are applications in need of stopping. Choices are a controlled stop, dynamic braking, or regenerative braking of the motor.

Regenerative braking assumes that there is a means of getting the load-generated energy back to the supply mains. It also assumes that this means is operating. If a current source ac drive or a regenerative dc drive has faulted and is no longer in control of the motor because it has lost its control power, regeneration through its power bridges via the firing of SCRs cannot happen. Thus, it is customary to include dynamic braking as a safeguard in those applications needing guaranteed stops. A circuit (contactor-activated) must be added that will dissipate the motor-generated energy to a resistor bank upon power loss to stop the motor fast.

Dynamic braking takes mechanical energy that is being backfed through the system as electrical energy and dissipates it as heat at a resistor bank. Voltage source drives, diode rectified drives, and PWM inverters use this approach to braking an electric motor. This method of dissipating regenerative energy is also used to slow a motor to provide back tension, or holdback torque, as in winding and unwinding applications. This method is used when regeneration back to the mains is not possible. When a drive that is not fully regenerative tries to control a motor in a generating state, the dc bus voltage rises quickly, and an overvoltage fault occurs. If the regeneration is tremendous, devices will most likely "pop!"

Obviously, dynamic braking wastes energy as heat. If the application brakes often and quickly and the loading is heavy, alternate methods should be explored. The regenerative drive is one. It allows electrical energy to go back to the mains. This type of drive must have the appropriate number of power semiconductors and the right type. There is a premium for this extra hardware, and it must be decided if the energy losses in heat outweigh the up-front costs of the hardware in a regenerated system.

Another method used as an alternative to dynamic braking is that of common busing. In this scheme, regenerative power can be used to power another motor that is in a motoring state. This power is resident on the bus for the appropriate use. It can only be used in those instances where the machine has a motoring and generating component, which can typically be found on a line that has an unwind and rewind motor.

To the electric motor, it does not matter if braking is accomplished via resistors or through regeneration. The motor has become a generator and that energy must go somewhere. The motor controller must have the necessary logic and means to divert this energy. Today's dynamic braking uses solid-state componentry and drive-integrated logic to ensure proper sequencing. The conventional contactor to remove the armature power supply in a dc system can now be replaced with an SCR.

Remember that we are trying to take this motor-generated current and reuse it as braking torque. Deceleration is a form of regenerating motor energy used to achieve slowing and stopping. Deceleration can be described as controlled stopping. It is common to see an electrically actuated holding brake when no motion occurs at the motor. Once in normal running mode, these brakes should not be used to allow the CEMF dynamic braking through resistors or back to the mains to be optimized.

Sizing dynamic braking resistors usually is left to the drive manufacturer. It is easily calculated if you know the following information:
- duty cycle—braking times and how often
- maximum speeds
- power rating of motor and efficiency
- ac drive, dc bus voltage

Another form of braking is *dc injection braking*, sometimes simply called *dc braking*. A dc voltage is forced between two of the phases in an ac motor causing a magnetic braking effect in the stator. This type of braking is commonly found in systems that can tolerate braking at frequencies of 2 to 3 Hz or lower. The reason for not applying this type of braking at higher frequencies is because the brake energy remains in the motor and could quickly cause overheating.

Harmonics

One of the emerging mysteries of power conversion and high-frequency switching is that of harmonics and harmonic distortion. Talk to anyone in academia or in the plant and they will have a story on what harmonics is. First of all, it is another misnomer. *Harmonious sound waves* are generally those pleasing to the ear. *Harmony* is said to be in agreement with or to blend with. *Harmonic distortion* is those waves blending unwantingly into a system. Why do they occur and how to handle them seem to be the most popular questions. If we did not have nonsinusoidal circuits, we would not have harmonics.

Better defined by IEEE, the *harmonic* is a sine wave-based component of a greater periodic wave having a frequency that is an integral multiple of the fundamental frequency. We start with a sine wave and evolve into a nonsinusoidal condition. *Fourier analysis* is the term given to the study and evaluation of nonsinusoidal waveforms. In 1826, Baron Jean Fourier, a mathematician, developed a series of formulas and terms to work with these types of waveforms.

There must always be a fundamental component that is the first term in the sine and cosine series. This component is the minimum frequency required to represent a particular waveform. From there, extend integer multiples of the fundamental. The component called the *third harmonic* is three times the frequency of the fundamental frequency. Four times is the fourth, and so on. In a six-pulse system, typical of a drive's converter, the real harmonics of concern are the odd-numbered harmonics not divisible by 3. The most important harmonics to filter usually are the fifth and the seventh. In a 12-pulse system, the lowest producing harmonic is the eleventh. What is created from these unwanted waves is the actual issue in dispute. Maybe these waves could be renamed, but the name harmonics has been attached and will stay. The real issues are what causes harmonics and what are the side effects of those distorting waves.

Harmonic distortion can also be called electrical noise. It can be said to be garbage, hash, or trash on a given electrical line. It is unwanted and nondesirable because of the complications it creates with computers and other sensitive equipment now used in many process control environments. More is understood today about what causes the distortion and what can be done to correct the problems, yet manufacturers of power conversion equipment are always looking at new ways to solve the problem, all the while creating new problems. For now, industry is working together to solve this problem.

When static power converters convert ac power into dc power and vice versa, a disturbance is created both to the incoming and outgoing supply power because the waveform is being distorted from its original sinusoidal state. This distortion can have detrimental effects on other electronic devices, depending on the severity of the distortion. Studies have given industry standards by which we can minimize or predict the harmonic content of a given system and thus take corrective measures. One such standard is IEEE Standard 519. The 1981 version of the standard was referred to for many years and finally revised in 1992. In its new form, the standard has become more than a guideline, it now makes recommendations.

A user with a harmonic problem now has some options. Prior to these recommendations, a facility had to live with harmonic distortion problems, which is how the vicious rumors got circulated and amplified. No one single solution exists to a harmonic distortion problem, and it is not practical to think that by just meeting an IEEE standard for allowable distortion levels will keep your plant from having problems. This issue of harmonic distortion definitely takes study, cost, and expense for correction, and, let's face it, loads change in the plant, thus making it hard to predict future distortion levels.

Harmonic analysis should neither be oversimplified nor taken so seriously that no other productive work is done. Many entities play a role in the analysis. The electric utility has input. The factory or plant has a lot to say, especially concerning the other sensitive pieces of equipment. The supplier of the phase-controlled converters or rectifiers also has much to offer. It is important to include all parties along with someone who understands the issue of harmonics and can act as the consultant, or even the mediator, in solving problems and disputes.

One fact that is clear is that there are definite misconceptions of what harmonics are. For instance, when a large electric motor is line-started in a factory, the lights can dim or flicker. Many might imply that this is a harmonic distortion problem, when actually the electric motor's voltage and current demands simply lower the plant's available electricity for a brief moment before voltage levels recover. This situation does not preclude that the sine wave has been distorted in the area where the lights flickered.

In any electrically supplied installation, there are linear loads and nonlinear loads. A linear load can be defined as a predictable sinusoidal waveform generated by an electrical load. Examples of this type of load include a facility's lighting system and its other resistor and inductor loads. This load is termed "predictable" because a relationship exists between the voltage and current whereby the sine wave is "cleaner" and smoother. With the advent of rectifier circuits and power conversion devices came nonlinear loads. Other nonlinear loads have been introduced over time for one energy-saving reason or another.

Lighting ballasts, which help save energy in lighting systems, exhibit a certain hysterisis that contributes to magnetic saturation, thus causing the load to be nonlinear and the wave to be nonsinusoidal. Other nonlinear devices include metal-oxide varistors (MOVs) and electrical heating equipment. With these nonlinear loads came waveforms that were no longer nice, clean sine waves. The waves now contained many portions of other waves, thus making the resulting wave notchy and distorted. The challenge becomes discovering what problems this condition causes and what can be done to correct the situation. But first, more basic information.

Harmonic distortion affects many components of the electrical system. Voltage, current, and sometimes both types of waveforms can be simultaneously distorted. Voltage distortion can subsequently affect many other sensitive devices on the same electrical system, whereas current distortion tends to be more local to the distortion-causing load. Consequently, the attention and correction has been focused on voltage distortion. Current distortion should not be ignored, however, and newer standards even address it.

Because static power converters have become commonplace in industry, a significant rise in the awareness of harmonic distortion has occurred. Variable frequency drives, UPS systems, and electrical heating equipment all convert ac to dc or dc to ac and thus create changes to the sinusoidal supply and cause definite interference problems with the communication and computerized equipment in the industrial facility. A typical electrical circuit in a factory can have motors, drives, sensitive computers, lighting, and other peripheral equipment on it. Every factory is different, and every circuit within the factory is also different. A complete analysis must be made of the system in place to properly identify the magnitude of the harmonic distortion and the corrective action required.

A relationship exists between the amount of harmonic filtering and the apparent costs involved, which must be considered at some point, along with predicting load changes on the particular system in the future. It might be simpler and more cost-effective to place a filter ahead of the only computer in the circuit rather than installing a more elaborate filter upstream.

For any harmonic analysis, the point of common coupling (PCC) must be selected. PCC is an important location in the electrical system, as it is the point where the harmonic distortion measurement is made. This point is usually at the secondary of a transformer, where many parallel loads of the same electrical system come together to connect to the main power supply. Many people believe that isolation transformers eliminate harmonics. This is not so.

Harmonic currents pass through a transformer. Voltage distortion can be affected by the impedance of the transformer, but it can also be accomplished with a less expensive line reactor if the line reactor has an equal impedance value. Transformers are designed to operate at 60 Hz. Harmonics tend to be present at higher frequencies, thus creating losses in the transformers in the form of heat. A transformer can overheat if subjected to currents containing high levels of harmonics, which has given rise to what is known as the *K-factor element* in transformer sizing in converter/inverter applications.

Not only will power conversion affect the supply power system, but it will also affect the output waveform's shape, most often to a motor. The occurrence is the presence of higher frequency disturbances that now tag along with the useful voltage and current. Figure 7-23 shows voltage notches in a converted sine wave's current. These notches are called *commutation notches* and their depth is of great significance.

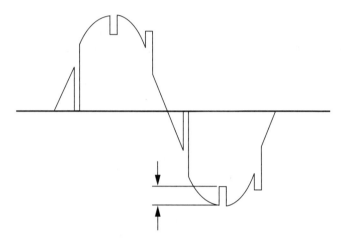

7-23 *A commutation notch. The depth of the notch is relative to the amount of distortion seen.*

The frequency with which commutation notches are seen is important. Basically, because controllers change sine wave ac into pulses by delayed conduction, harmonics are produced. Solid-state devices are the culprits in these rectifier systems, with SCRs and other thyristors being exceptionally notorious for high amounts of distortion because of their intermittent conduction. They appear as a short circuit to the system because they do not dissipate any energy when they conduct; they are good for efficiency, but bad for harmonics and the power factor.

Windings in an electric motor exhibit similar characteristics to the induction in a transformer. Should harmonics be present in the output to a motor, excessive heating can occur. This heating is lost power and more audible noise. Knowing ahead of time what levels of harmonics might be present can help select proper-sized equipment for the desired performance.

Distortion limits are classified by the depth of the commutation notch, the area of that notch, and the total harmonic distortion (THD) or in terms of voltage. Limits are also being set on current distortion. Thus, manufacturers of power-converting equipment must provide filtering where required, resulting in higher system costs. The harmonics themselves are not necessarily harmful, except to other sensitive electronic equipment on the same electrical system.

In summation, harmonics can and are blamed for a variety of problems in the plant. Blown fuses, burnt motors, and hot or burned connectors all point to a potential in-house harmonics problem. Consequently, it is worthwhile to make an analysis and take corrective action. It is also good to determine whether or not harmonics are present and to what extent, rather than assume them to be the problem and proceed down the wrong path.

Data required for harmonic analysis

When considering installing a device that can convert ac to dc and vice versa, it is recommended that the area of the plant where the device is to be located be investigated. Analyze all the equipment on the same circuit. Calculate the total linear and nonlinear loads. Look at the impedances of transformers and reactors in the circuit. Obtain the short-circuit current data from the utility. Predicted harmonic distortion is easier to correct.

If it seems as though the entire plant could be affected, all inductive and capacitive elements must be factored. All these elements must be attributed with some value of impedance. Any power factor correction systems must be accounted for because a power factor capacitor can actually make the power factor worse in the system when affected by resonant oscillating currents. Thus, it should be made known as to what value of power factor must be held. If the capacitors are switched, it must be made known when the switching occurs.

If symptoms of harmonics seem to exist, it is best to first ascertain whether or not any rectifying equipment is in the plant. It is then necessary to list and document all electrical components on a particular circuit. This plant map can be invaluable in totaling loads and impedance values, as well as for indicating where sensitive equipment is in the plant to allow for planned filtering, if required. Concurrently, the utility should be able to provide data relative to the source of power (from it) and the available short-circuit current, along with known impedances for upstream circuitry. After all, the power companies are partially driving the requirements of harmonic filtering and power factor correction.

When testing for harmonic distortion levels, certain test equipment is required. The method of testing must also be determined. A

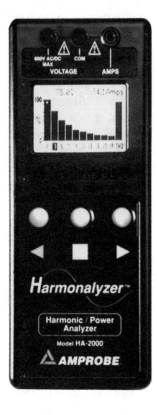

7-24
Harmonic distortion analyzer. Amprobe Instrument, a Division of Core Industries Inc.

current transducer, voltage leads, and some type of metering device such as that shown in Fig. 7-24 are needed. The device in Fig. 7-24 allows the user to measure electrical wiring to determine harmonic distortion levels. This device can display, from the readings, waveform shapes or graphs of a select harmonic. All this data can then be downloaded to a computer for further compilation and can be compared to allowable limits and acceptable values. It can then be determined whether or not corrective action is necessary. Corrective action can be tuned, harmonic filters or placing inductors in the circuit. Corrective action can be expensive, so a good analysis by a specialist is recommended.

Radio frequency interference

Radio frequency interference (RFI) is not harmonics. Harmonics is the distortion of the sine wave due to power rectification in variable speed drives. Radio frequency interference is exactly that, interference to frequencies above the audible by fast-switching devices similar to those used in variable-frequency drives (VFD). Filters can shield the input and output of a VFD to minimize the possibilities of interference.

Radio broadcasting frequencies are typically above 150 kHz and below 100 MHz. Whenever current or voltage waveforms are nonsinusoidal, there is always the possibility of radio frequency interference. Newer switching technology in converter and inverter devices has started to infringe on the high-frequency domain of the radio waves. The Federal Communications Commission (FCC) forbids interference of these signals. Thus, new standards are being discussed every day as frequencies of inverters increase. Switching frequencies, or carrier frequencies in drives, are approaching 15 kHz.

Because drives use the switching capability of different transistors, they have come under scrutiny. When current increases quickly, as is the case with today's power semiconductors, noise can be detected. The distance from the emitting device, shielding, and filtering are all factors relating to the existence of RFI. Other factors impacting RFI are the horsepower and current rating of the drive, the output and switch frequency, the impedance of the ac supply, and how well shielded the power modules are.

Filtering using inductors and capacitors is presently the best way to suppress RFI emissions. The problem regarding radio frequency interference and drive equipment is still ambiguous. Emissions and occurrences are rarely proven and rarely considered. Enforcing the adherence to the laws and standards has not been paramount. As the occurrences increase, so too does the need to police the issue.

Power factor

The topic of power factor is another of those issues often misunderstood. *Power factor* is the ratio of the instantaneous, or active, power and the anticipated power. When power is being analyzed as a pure sine wave, all is predictable and understandable. When we deal with nonsinusoidal power, as in drives, the analysis takes on new meaning. Typically, leading and lagging power factor assumes capacitance and inductance in the system. This assumption is only valid when sine wave power is evident for voltage and current.

Power factor actually consists of a displacement factor and a harmonic factor. As seen in an earlier chapter, power in the equation $P = EI$, is the product of voltage, E, and current, I. Voltage and current waveforms in phase means their respective zero-crossing points are the same. This condition allows for all the available power to be used as productive power. When the current gets out of phase from the voltage, it is termed *lagging*. The net result is that the power is not productive for this half-cycle. This is the displacement power factor. The harmonic power factor is of a lesser degree, as it basically is the effect of the wave distortion in the same phase sequencing, as previously described.

Power factor is expressed in terms of VARs or K-VARs, or 1000 (kilo) VARs. VARs are volt-amps, reactive, which is the reactive power. VAR is a product of the current, RMS (a peak average); and voltage, RMS. The power factor is proportionately equal to the watts divided by the volt-amps. A device used to measure power in both balanced and unbalanced systems is shown in Fig. 7-25. It can measure leading and lagging conditions. Power factor correction occurs when a predetermined value for a system must be met, such as when the utilities demand that certain levels of power be maintained. A standard circuit for power factor correction can use capacitors and inductors to provide a tuned circuit placed between the supply and the converting/inverting device.

Electronic-drive applications

Variable-torque drive applications are why variable-speed ac drives came into being. The need to save energy and reduce the speed of a fan or pump resulted in the variable-speed drive. Typically, variable-

7-25 *Power factor meter.* Amprobe Instrument, a Division of Core Industries, Inc.

torque applications have centrifugal loads, allowing for the drive and motor to increase the torque, or current content of the circuit, as speed increases. The curve for a typical variable-torque system can be adjusted to incorporate different ramping, such as an S ramp or another type of slow start. Most fans and pumps are variable-torque drive applications, but there are exceptions that, when misapplied, give the drive manufacturers and users headaches.

Positive displacement pumps and fans are the culprits. They have high-starting torque requirements because a large load is present upon starting due to the mechanical construction of the fan or pump or just simply because fluid (air or liquid) is present in the system under high pressure. The drive must compensate for these conditions to start. These might even be called constant torque applications.

Constant torque is another misnomer. You might think that torque is required to perform any amount of work. Unless the electric motor is off, it must be performing work; therefore, torque is required. Thus, torque is required constantly, right? Is this what is meant by constant torque—having torque output continuously? Unfortunately, that is not what is meant by constant torque, and this is where the confusion begins. Constant torque would be better named *constant high torque* or *high constant torque*. It is simply a matter of looking at the torque versus speed curves for variable torque and high constant-torque applications. In the constant-torque application, starting torque requirements are high and continue at those levels until at full speed. If someone does not understand the concept of constant torque, how can the drive be applied properly, even in variable-torque applications? Even when constant-torque applications have been qualified, the potential exists that the variable-frequency drive might have trouble starting the load (remember VFDs must overcome slip to start producing higher amounts of torque). The VFD might be undersized, without enough current available, even in an overload condition, to start or run the application.

The issue of completely understanding constant torque is why variable-frequency drives have a bad name. Many an engineer or technician can tell a story of the drive that didn't work, but maybe the drive was misapplied. Or maybe it was a manufacturer's new design and was not field-proven. Maybe the motor was undersized or maybe the drive was. Whatever the case, constant-torque applications must be examined closely.

Certain issues need to be considered when using an ac variable-speed drive for an application. In the industrial environment, if the application is not a pump or a fan, the torque requirements will probably be high and constant. To determine whether a VFD is practical for an application, ask if the shaft that connects the motor to the load can be turned by hand. If so, a VFD can probably work for that application, assuming the torque requirements are not too high. Motor nameplate data is always good to obtain, especially when the horsepower rating on the motor might not be clear. The full-load and no-load current values on the motor are more appropriate for selecting a drive. Matching the drive, which is a current output device, with the current requirements of the motor assures a safe, efficient, and proper installation.

Troubleshooting drive systems

Troubleshooting is a vast subject and varying degrees of troubleshooting exist based on the expertise of the troubleshooter. Drives come in many designs and packages. Some are digital and some are not; some are ac and some are dc. Others are servos or stepper drives. Many drives have onboard diagnostics that can help pinpoint a problem. Sometimes the diagnostics cannot. For instance, a digital drive can indicate that a main input power fuse is blown. The first step in troubleshooting is to ascertain why the fuse blew. If nothing is obvious, another fuse is placed in the fuse holder and power is turned back on. This time, if you are watching when the fuse blows, the reason might become apparent. A drive can only tell what fault occurred and when. Some drives can offer an "onboard" manual that, for a given fault, offers suggestions as to where the cause might be. All an onboard manual requires is extra text and memory in the computer system of the digital drive, but many drive manufacturers do not put this functionality into the drive. This data can also be pulled from a hardcopy of the manual.

The electronic drive's operation and service manual is as important as the drive hardware itself. Without the manual, the plant or factory is in jeopardy of having the drive out of service until service can be called. The manual is important because electronic digital drives have become so powerful they actually contain too much information. The drive can access too many circuits to analyze and monitor them all. With the microprocessor and extra memory, drives can be programmed to not only monitor many faults, but can also keep a log of them, when they occurred (the electronics allow for an onboard clock), what they were, how and if they were reset, and so on. This data can be saved as the drive runs and when a drive fault occurs, the most recent data is retained in battery-backed memory to be retrieved by a technician even with the drive down. This diagnostic and troubleshooting tool is invaluable to not only the personnel but also to the service people from the manufacturer of the drive.

Drive faults common to ac and dc drives are undervoltage or overvoltage conditions, overcurrent or overload, fuse failure, overtemperature, and failure to pass its microprocessor "ready" check if a digital drive. Undervoltage faults generally mean that supply power has either been interrupted or the voltage value has dropped below an acceptable minimum. Some drives have ride-through capability that keeps the drive in control for a small amount of time until the glitch has passed. Conversely, overvoltage faults indicate that the incoming supply has surged to a high level of voltage that exceeds the maximum limit. Overvoltage faults also indicate that the bus voltage has risen to such a high level that components are in danger of burning, and the drive must shut down. These trips can also be caused by trying to stop a motor with load too fast, so the regenerated energy cannot be dissipated fast enough. Braking resistors could be used here.

Overcurrent or overload conditions are for the motor's protection. For a certain limited amount of time, output current to the motor can be above the highest setting. These conditions might exist when starting a load that requires high starting torque. If too much time passes, however, imminent danger of damage to the motor exists, and the drive shuts down. Too much time passing could indicate that the motor has stalled, or another type of problem exists that needs investigation.

Another drive fault that mainly protects the drive is that of overtemperature. Temperature monitors are placed within the heatsinks or bus of the drive. If temperature gets too high, the drive shuts down rather than damaging itself. Increased temperature might simply indicate that a cooling fan has stopped and needs to be checked. Other faults common to most drives, such as fuse failure and microprocessor trip, are necessary to protect against high levels of current going into the drive. If the microprocessor is not ready or has a problem, it should not attempt to control a motor. No control is better than bad control.

Other fault circuits are monitored by drive type. For instance, a dc drive should check for loss of field and loss of tachometer signal. An ac drive does not look at field signals because none exist. Likewise, any drive that is dependent on a feedback device from a motor to control that motor (i.e., ac vector, servo drives, etc.) should always check for that feedback signal. If this signal is lost, the drive assumes it has a major speed correction to perform and sends a drastic increase-speed command to the motor. This situation is called a *runaway motor* and is very dangerous, to say the least. If a feedback device is in the loop, it must be monitored. If the signal is lost, the drive must recognize this fact and either shut down or go to minimum speed.

Some drives check control power; others have a ground fault sensor to handle these errors. Some drives are phase-sensitive, while others are not. Motor thermostats are often interlocked with the drive to provide extra insurance that the motor cannot overheat (even if the drive is monitoring output current of a motor). These faults vary with the type of drive and manufacturer. It is best to check with the manufacturer and find out what happens with the drive and motor when a certain condition exists.

Remember that the electronic drive is both protecting itself and the motor it is controlling. When the drive shuts down, it is not trying to put a company out of business; rather, it has detected a problem and that problem needs correction. The drive trips can be a nuisance, especially if they occur often and other computerized equipment stays online in the plant while the drives go down. Some devices are more sensitive than others, but filters, snubbers, and capacitor networks can correct nuisance tripping.

Troubleshooting a drive system can be extensive and not always straightforward. The operative word is *system* because troubleshooting the drive might be only half the battle. For example, we might find we have an overload fault at the drive, but there is actually a problem at the motor or even in the machine. Tracing the problem is where the fun can begin. A hands-on technician with good mechanical and electrical aptitude is sometimes needed even if the drive is furnishing data about the problem. All in all, drive troubleshooting tends to be application and machine troubleshooting as well.

Because the drive is the least-understood component, it is the easiest and the first component blamed. A drive manufacturer's service person is often called to the job site only to find that the problem was in poor wiring, a motor short, or any number of nondrive equipment failures. The bill for the service trip must still be paid. A trained technician, electrician, or operator could do productive troubleshooting and diagnosis beforehand, saving costly downtime and service bills.

Proper documentation, good records, and proper tools make troubleshooting electronic drives much easier. At least two sets of manuals should be resident at a facility. One set should go in or near the physical drive on the factory floor. The other copy should be kept in an office or company library. If drive parameters must be set or programmed, a copy of these should also be kept in both places. These settings or values might need to be reentered if a dramatic problem occurs with the drive. Only a limited number of personnel should access the drive's program, and the program should be password-protected. If too many people can make changes, confusion will ensue. It is best if only one or two individuals can make changes. A log, if not a standard feature of the drive hardware, should be kept to track when or if a drive faults and what was done to correct the fault. Trends can thus be determined and lost production time fully tracked. This information is also very useful to the drive manufacturer and service person.

Proper tools for troubleshooting the drive can be expensive. Tools include spare parts, and nobody wants to purchase and stock expensive items unless they absolutely must. The decision of whether to have a spare-parts inventory must be based on the quantity of drives and the value of lost production time.

The expertise of the in-house technicians is another factor. Most facilities have one or two technicians who become the resident experts on the electronic drive. Many drives today have a means of making electronic changes. This method might be a standard device or an optional, hand-held device. It is worth having around.

In addition to the standard current probes, volt-ohmmeters, and occasional pot tweaker (yes, some drives out there have many potentiometers that need adjusting), a digital oscilloscope capable of storing images is always useful. Checking waveforms can help pinpoint problems both into and out of the drive. It might be necessary to put a strip chart recorder on a drive circuit to monitor line conditions over a period of time to help pinpoint trouble areas.

Another recommended tool is a digital tachometer. A hand-held unit is shown in Fig. 7-26 and is very handy for measuring the speed of a given motor. As electronic drives control the speed of motor, it is sometimes necessary to physically see if the speed is actually being controlled. The digital hand-held tachometer can be useful to determine speeds of other rotating components or other motors. This data can be used to select motors and gear reduction for a given system.

Becoming intimately familiar with the type of drive, its control and fault scheme, and the machine or application are all important components of any successful drive and is true for both the user and the supplier of the drive equipment. Complete training on the product is necessary. As previously mentioned, good documentation and tools can go a long way to keeping the drive running. Many drive installations, once started, rarely, if at all, fault or cause problems. These are the ones applied and installed with up-front attention. As time passes, the electronic drive is gaining some overdue respect and thus more usage.

7-26
Hand-held digital tachometer.
Amprobe Instrument, a Division of Core Industries, Inc.

Conclusion

Motion control is a facet of factory automation where engineering and technical disciplines meet. They are actually in full use in the motion control industry. Electrics, motors, drives, sensors, power transmission, computers, PLCs, machine vision, communications, system integrators, quality control, and safety are all factors because each discipline is used somehow in a factory somewhere. Industrial electronics is the common thread of all the aforementioned and must be fully understood. The field of motion control is growing in both complexity and opportunities.

Electronic products are constantly being asked to perform more functions, which is possible with digital products, but it also can make the products more complicated and harder to understand. Complexity is not a problem if the technician stays current with the technology and keeps learning. The opportunities are many. There are well over a hundred suppliers of electronic drive products in the

Conclusion

world today. From servo and stepper, to ac or dc, a drive product exists for every application.

Drive installations are happening at record levels. They are being installed everywhere from industrial machines to rooftop air-handling systems. They might work their way right into the home. Price is the only obstacle. Still, there must be an awareness of what these products do both in function and electrically. Technicians must be involved with the design, application, and installation of drive systems if they are going to help keep them running.

In addition, the old question of "Which drive is best for my application?" can still not always be easily answered. Is dc better than ac, or vice versa? Should a servo system be incorporated? When do flux vector drives need to be considered? The answers all lie within the application: what needs to be controlled. The actual answer is not that simple, but the fact is that each application must be explored fully on a case-by-case basis. An application might be well-suited for a dc drive but not an ac drive. Many times, the issue of dc versus ac is simply customer preference; other times it is based on costs.

The dc drives are often used because they are less complex than the ac drive with its inverter section, making the dc drive less expensive for certain levels of horsepower. The motor must always be factored into the analysis for proper economical evaluation. The dc drives have a long tradition in industry, and many factory personnel are familiar with their design and are comfortable servicing the drives by themselves. When properly applied and maintained, brushes and commutators of dc motors are not always maintenance issues. The dc motors can provide large overload ratings and high torques at low speeds, and the drive can regenerate energy back through its SCR bridge onto the supply, which is an attractive and necessary feature for certain applications.

To give the ac drive its due, technology is addressing the issues of regeneration, low speed, and overload performance. Presently, ac drives are capable of giving this type of performance. They might be high-performance or vector drives to provide good, low-speed operation. Braking resistors can handle the regenerative situations, but some regenerative ac drives are on the market. The ac drive, beyond these minor shortcomings, offers several other advantages. Standard, lower-cost, lower-maintenance ac motors can be used. Multiple motors and emergency bypass (backup) schemes are possible with ac drives. High-speed operation is better suited for an ac drive than a dc drive because commutation is easier for the ac drive. Existing ac motors can be reused, and electronic reversing is possible. Obviously, many issues must be covered, and the application and costs prevail as the determining factors.

Industry is leading the drive technology into a single package drive, configurable as ac or dc. The ac drive is simply a dc drive with an extra inverter component. Industry will not pay the premium to have such a unit when only a dc motor will be run. Thus, this scheme will need to be very cost sensitive, but it could happen. The ac drive technology continues to offer solutions and answers to objections from dc drive users. If this trend continues, ac drives might take over.

The argument of ac versus dc is a good one. Every year, ac drives gain more of the share of total drive market in installations. Still, it will be some time before ac drives completely replace dc drives because of the large base of dc motors in existence, which can run for 20 to 30 years. Both drives will be around for many years, even decades, and the electronics technician will have to keep up—the field of motion control is in motion!

Bibliography and suggested further reading

A.E. Fitzgerald, D. Kingsley, Jr. and A. Kusko, *Electric Machinery: The Processes, Devices, and Systems of Electro-Mechanical Energy Conversion*, Third Edition, 1971, IEEE.

Electrocraft Corporation, *DC Motors, Speed Controls, Servo Systems*, Fifth Edition, 1980.

L.E. Bewley, *Travelling Waves on Transmission Systems*, second Edition, 1951, John Wiley and Sons, New York.

1992 Inverter report, Invertec, Boston, MA.

"IEEE Guide for Harmonic Control and Reactive Compensation of Static Power Converters," *IEEE Standards 519-1981 and 519-1992.*

Spitzer, David W., *Variable Speed Drives—Principles and Applications for Energy Cost Savings*, Second Edition, 1990, Instrument Society of America.

8

Sensors and feedback devices

If computers are the brains of industrial automation and communications the glue, the nuts and bolts must be the sensors and feedback devices in the plant. They keep the process together. Compared to computers, electronic drives, and programmable controllers, they are often ignored, until they are suspected of being the cause of a problem. At that point, everyone asks, "How could such an inexpensive component in our factory cause a machine to completely shut down?" These sensing and feedback devices are very important to the uptime and performance of machines. They provide the valuable information that allows any system to correct itself. Without these feedback devices, automating a plant would not be possible.

Look at any piece of automated equipment in the factory. Some type of feedback device is providing position, speed, temperature, pressure, or other process information. This information is vital to the automated plant's ability to keep producing. Many machines and processes use some means of feedback. If there is control, there is feedback. Most electric motors, if controlled, have speed, position, or both types of sensors that keep the motors at speed or in position. High-precision machines, such as numerically controlled (NC) and computerized numerically controlled (CNC) equipment, function properly only with feedback devices. Likewise, any process being monitored or controlled has sensor technology.

Feedback is not always obvious. Many processes occur that we take for granted, such as some loop closing to control a specific process. For instance, consider the industrial oven or dryer. Just like the ones in the home, a device needs to shut the fuel off to the heater to prevent burning. A temperature sensor either cycles the heat on or raises and lowers the heat to satisfy the temperature setting. Likewise, some motors are part of an electrical circuit that includes a motor controller. Voltage from the motor can be monitored to control the motor's speed. Perhaps the best example of behind-the-scenes feedback is the human operator.

How often have we asked, "How is the machine being controlled, now?" And the answer is, "Our operator, Charlie, goes over to the machine and turns a lever a half turn or so and the machine is back to where we like it!" Charlie is the feedback device. He is continually correcting the system. This example is a prime setting for implementing an automatic sensor or feedback device to perform the same function. The result might even be more consistent.

Open-loop control

To understand how and when to use sensing devices, it is first necessary to review open-loop control versus closed-loop control. Open-loop control, in its most basic form, does not lend itself well to fully automating anything. In Fig. 8-1, an input signal, or command, is given, and a direct resultant output achieved. We do not affect the signals, and we do not know if what we requested actually happened. If the process is an open-loop control scheme, no sensors or feedback devices are present in that system. If a valve is opened, liquid flows until someone shuts it off. If a switch is thrown to a motor, the motor runs at full speed until someone opens the switch. Then the motor stops.

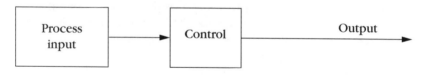

8-1 *Open loop system.*

For example, assume we want to fill a tank with hot and cold liquid while maintaining 100°F. No temperature sensor exists, however, and the tank is out of sight of the shut-off valves. We start the process and simply hope that the correct mix of hot and cold is there, and the tank does not overflow. If we only were using feedback!

These examples all seem like manual systems, not too automatic at all. How does the controller know if or how well the proper function is being performed? It doesn't. Walk through an older factory. Most of the processes are handled via an open-loop system and operator intervention. In these cases, the operator can be likened to the feedback device. Even they make mistakes, get delayed, or just are not nearby when an event must happen.

Closed-loop control

A true, closed-loop system can be seen in Fig. 8-2. Another component in the diagram provides information about the process. This information returns to a summing, or comparison, point for analysis, and new output is provided based on the comparison with the process input signal, or the "what should we be doing" signal. This output is a cause-and-effect relationship between devices. The feedback device notes any actual error in remaining in the prescribed condition while the controller attempts to correct for that error. A closed-loop control scheme constantly corrects itself. A good example of a closed-loop system is that of a cruise-control function in the automobile. A speed-sensing device located near the wheels monitors how fast the car is traveling. That information is fed back to a control point to compare to the desired mph. The more often the error signal is sampled (sent, received, and corrected), the better the overall system's accuracy.

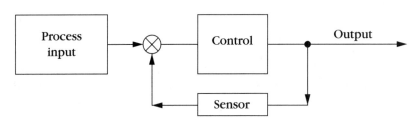

8-2 *Closed loop system.*

The proportional-integral-derivative (PID) control scheme is a common closed-loop system. The PID control loop was discussed earlier in the process control chapter, but has applicable meaning pertaining to feedback devices. Without these transducers or feedback devices, no proportional, integral, or derivative loops are available to control. No closing of any loop whatsoever occurs. Two types of closed-loop feedback exists, negative and positive. Negative feedback reduces the variable in the controller's routine, which changes the process or machine. The direction of the signal, negative in voltage for instance, is interpreted by the controller as reason to change the output accordingly. Conversely, for positive feedback, the feedback signal's direction requests an increase to the variable that the feedback represents.

Closed-loop systems exist in many newer, automated factories and can even be incorporated on many older pieces of equipment. It is sometimes difficult to add to older machines because the feedback devices are precise, responsive, and repeatable, which can be problematic to an older piece of equipment with sloppy mechanics and backlash. It is often better to retrofit much of the machine or use new components when creating a closed-loop system. Other times, installing the feedback device is the sole reason for disturbing the system (by putting a feedback device in, Charlie does not need to keep adjusting the machine). Trying to install one device, however, might mean two or three other components must be changed as well. Ideally, it is more predictable and desirable to have all new components, with known mechanical and electrical values and constants, when automating a system or process. Given the choice, a designer always wants all new components. Economics usually dictates the answer. The term *reengineering* is often used these days, and it applies here. If the project is not fully analyzed and all components evaluated, the automated system might not function as planned. The designers or retrofitters must then go back to the drawing board until the system works. Reengineering should be avoided.

The closed-loop system contains at least one type of sensor or feedback device. Many applications, however, require more than one device. Any of several types exist for different functions. Some sensors can be used simply for counting, while others measure temperature. Some are photoelectric, while another is electrically actuated. Others use fiberoptic cabling to transmit information, while some use two-conductor electrical wire. Whatever the application, an appropriate sensing device is available.

Speed and position sensors

Motion control would not be possible without feedback devices. Because motion typically involves a rotating element, or electric motor, the feedback devices are most often directly coupled to a motor shaft. As the shaft spins, the sensing device gathers speed or position data. Prior to feedback devices in electric motor control, attempts were made to control motors with the electrical data available. The ac motors can provide a form of feedback in the electrical circuit called *counter-electromotive force*, or counter EMF (CEMF). This electrical data can be used by a motor controller to determine how fast a motor is running with some inaccuracies. Depending on how well the speed regulation must be for the given motor and application, this feedback can be adequate. Similarly, a dc motor, when controlled by a rectifier circuit, can provide armature voltage feedback for speed regulation. As the controller sets the output to the motor, the electrical circuit in the ac or dc situation is monitored and maintains the desired speed.

As speed regulation must be held better and better, improved feedback means must be incorporated into our system. Rather than relying on CEMF or armature feedback, we can install a tachometer. This device attaches to the motor and produces a voltage output to the controller in the form of volts per thousand revolutions per minute at the motor. With this information, we have a more accurate way of comparing and correcting the actual speed with the commanded speed. Other devices can provide even more precise information for motor speed, and their use only betters the speed regulation for particular processes.

Encoders

One such device to enhance speed and shaft position is the *encoder*, or digital pulse tachometer. These are shaft-mounted devices that work on the principle of counting electrical light pulses per revolution of a disk. Light is sent through a disk that has coding, or slots, and a light pickup sensing device counts the light pulses. Thus, data regarding location and speed of the disk can be sent to a computer for decisive action. The encoder basically translates mechanical motion into electronic signals. A depiction of this electronic signal is a pulse train, shown in Fig. 8-3. This feedback device is often called an *optical encoder*, tach (short for tachometer), pulse tach, or pulse generator.

There are two types of encoders. The more common and least expensive is the incremental encoder, while the other is the absolute encoder. There are also two versions of each: rotary, the most common, and linear. We refer mainly to the rotary type. The incremental encoder consists of the light source, disk, light receptor, grid, and amplifier. As seen in Fig. 8-4, the circular disk has several slots along its perimeter. The disk can be made from glass material with marks imprinted as the slots or made of metal with machined slots on the outer edge. Light is emitted through the slot, received by the receptor through the grid assembly, amplified, and sent to a host for further use. This entire sequence can be seen in the diagram in Fig. 8-5.

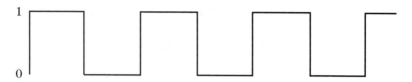

8-3 *Digital pulse train.*

8-4
Incremental encoder disk.

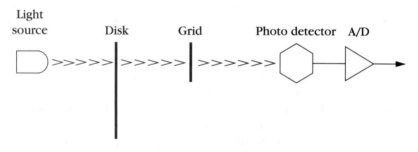

8-5 *The operation and parts of an encoder system.*

The quantity of slots on a disk is important to the application. Encoders are often specified as having 1024 or 2048 pulses per revolution, generally in powers of two, which is consistent with binary math for computers and aids in the absolute encoder disk design. Pulses often dictate the resolution, or how accurate the feedback signal is. If high precision is required, more pulses per revolution are needed in the beginning. The more slots on the disk, the more expensive and larger-diameter disk is necessary. A marker pulse must also be placed somewhere on the disk as a reference.

With an absolute encoder package, the disk is the big difference, as can be seen in Fig. 8-6. Here, the disk has a unique and almost

Speed and position sensors

8-6 *Absolute encoder disk.*

pretty slot design. There is a definite reason for the pattern: these slots are actually concentric, getting larger as they get closer to the center of the disk, and a binary relationship exists to the pattern. By passing light through all the slots at a given instant, the encoder can provide an exact position. This method of absolute decoding solves for the binary patterns on the disk. These types of encoders are mainly used for positioning applications, where position is needed if power is interrupted. The disks do not require a "homing" sequence to find a starting reference point.

Most encoders employ a quadrature form of decoding. Here, a second channel employs a second light source and receptor. The second channel is physically positioned within the encoder a half-slot distance away from the other channel, providing more available data points, as seen in Fig. 8-7. More data points provide better resolution (i.e., more pulses per revolution). The quadrature approach also allows for the detection of rotational direction, forward or reverse.

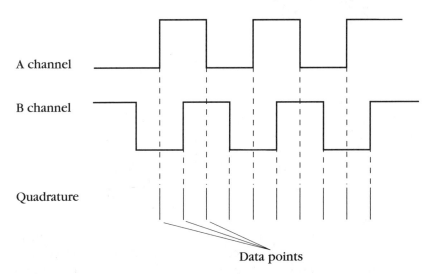

8-7 *Quadrature decoding.*

236 Sensors and feedback devices

These units can be large or small and mounted differently. Thought should be given to applying these devices. Glass disks can sometimes collect a film on their surface, impeding good light transmission. They also can break. Metal disks, although more expensive, might be a better choice. Various housing methods exist to protect the electronics and disks within. A photograph of a mill-duty digital pulse tachometer is shown in Fig. 8-8. The mounting and housing of the encoder is important to its successful and extended use. If mounted loosely or misaligned, premature bearing failure on the rotating element of the encoder can occur. Likewise, adequate enclosure around the electronics and disk keep them working properly for longer periods.

8-8
Mill-duty feedback devices. Carlen Controls

With such a device providing more accurate speed information faster to the controller, motor shaft speeds can be maintained to even 1 or 2 rpm in a steady-state load condition. Of course, loads tend to change, making the encoder/control system work harder to continually make corrections. But that is its purpose as a feedback device. Consider the alternative without the feedback: Charlie? Good, clean, fast feedback is necessary and is the ongoing battle for true motion control in the automated factory. As the desired accuracy of a particular feedback device increases, so too does the cost of the device.

Figure 8-9 shows a sample specification for a digital pulse tachometer. The electrical and mechanical specifications are shown. Both are important to the successful installation of the device. The electrical specifications indicate how and what type of electrical out-

Speed and position sensors 237

```
ELECTRICAL SPECIFICATIONS                              MECHANICAL SPECIFICATIONS
Code            -   Incremental, Optical              Housing       -   NEMA 13
Cycles per                                                          -   Material - 6061
  Revolution    -   on Code Disc:                                       Anodized Aluminum
                    60 - 2540 (others                               -   Seal - O-Ring
                    available)                        Shaft Size    -   0.4998" (+ .0000,
Supply Voltage  -   12-24 VDC +/- 5%                                    - .0005)
Supply Current                                                      -   0.6245" (Optional)
  (88C30)       -   CMOS = 150 mA max,                               -   303 Stainless Steel
                    125 mA, typical                   Shaft Seal    -   Teflon Graphite
Output Signal   -   Square Wave, including:           Moment of
                    - 2 Channels, phased                Inertia     -   550 gm-cm²
                      90° Inverted                    Weight        -   12 lb.
                      Push-Pull Signal
                    - Differentially Driven
                    - in Quadrature, +/-2°
                    - with Marker Pulse
Temperature     -   Operating, -40°C to 100°C
Output
 Protection     -   Power Supply Reversal


ORDERING INFORMATION

Example:    CC450T  -  1024  -  ABZ  -  12  -  FM  -  C
              1        2        —       —     3     4

     1 -  CC450T   =  BASIC (Mill Duty) DESIGN
  _____   450T     =  Single Shaft
  _____   452T     =  Dual Shaft

     2 -  CYCLES PER REVOLUTION
  _____   XXXX     =  CPR needed

     3 -  MOUNTING
  _____   FM       =  Foot Mount
  _____   56C      =  NEMA 56C Flange Mount
  _____   MM       =  NEMA 56C Motor Mount (per GEP-387 Option Item 22.a)
  _____   FC       =  FC Flange Mount (per GEP-387 Option Item 18.e)
  _____   DS       =  Base Plate to replace GE-DS3820
  _____   S        =  Base Plate to replace Selsyn

     4 -  OUTPUT TERMINATION
  _____   MS       =  MS connector, with mate (quick disconnect)
  _____   C        =  Conduit Input (½" NPT) terminal board version
```

8-9 *A sample specification for a pulse tach.* Carlen Controls

put should be expected. The mechanical specifications show how well-suited the particular unit is for the environment and duty cycle. Both specification lists allow us to comment on and compare this device to other feedback devices for motion control.

Electrically, the unit is supplied with a low voltage of 12 to 24 Vdc. This voltage value must be held within plus or minus 5 percent, or the output functionality can suffer. The output is a low-voltage pulse in the form of a square wave. The square wave consists of two channels, 90 electrical degrees out of phase from each other (quadrature). The signal is differentially driven with a marker pulse (home or null). The pulse-to-pulse accuracy is equal to plus or minus 2 arc minutes in one revolution, RMS. It has a workable frequency range of

0 to 100 kHz, which refers to the clock, or crystal, speed for stable velocity determination. Its operating temperature range is 0 to 70°C, or −32 to 178°F.

Mechanically, this particular housing for mill environments is machined anodized aluminum, with the shaft being made from type-416 stainless steel. Both metals are corrosion-resistant. The shaft mounting is 20 millimeters (mm). The shaft rotation contains no backlash. *Zero backlash* is the absence of mechanical play if the shaft were turned. The shaft can turn in both forward and reverse directions without affecting the attached motor. The whole mechanical and physical purpose of the pulse tachometer is to basically ride "freewheeling" with the motor shaft, adding as little inertia and extra torque as feasible. This unit has low torque and inertia values. Its slew speed is rated at 5000 rpm. *Slew speed* is a state in an application where time is made up in a motion profile. In simpler terms, the motor is run at its maximum possible speed for short periods.

Bearings are a very important issue with any feedback device attached and expected to run on a motor. If the bearings are a weak link, the entire process can go down without notice. In this case, the bearings have a life expectancy of 20 billion revolutions. The lifetime in hours depends on the speeds the motor and tachometer run. The bearings are sealed with grease, making them maintenance-free. The cover is extruded aluminum and the connectors are MS, or military style, which are screwed-down for extra protection. The tachometer has good shock and vibration characteristics. All in all, if the tachometer can stand up in a mill environment, it can survive most application environments.

Resolvers

Resolvers are another type of feedback device used in motion-control applications. They are different from the encoder and pulse generator in that there are no electronics at the resolver; it is separately excited from a source external to the physical resolver. The resolver is a rotary transformer with a rotor and stator component. It is a small ac motor, excited by a voltage to itself. Figure 8-10 is a picture of a resolver taken apart, showing the rotor, or armature, and stator components. Figure 8-11 shows a fully assembled resolver with wiring bundle. The reassembled resolver is then attached to a motor with a flexible disk coupling, as is seen in Fig. 8-12. Alignment is crucial. The flexible coupling can provide some cushion, but the feedback device must be mounted as concentric to the motor shaft as possible to minimize premature failures or bad readings.

8-10
A separated resolver showing the armature and stator.

8-11
Fully assembled resolver.

8-12
A resolver mounted to a brushless motor by way of a flexible disk coupling.

The wave shape to the resolver is actually what is observed in the circuit. Because the resolver is excited by a sine-wave, low-voltage signal, the signal can be decoded within the controller (with power source). Decoding is accomplished through a resolver-to-digital (R/D) conversion. The converter observes displacement in the returning sine wave to determine shaft location. Thus, the loop is closed to the motor, or rotating device. Accuracies for resolver systems are presented in arc-minutes and usually are in the 6 to 7 arc-minute range.

The attractive features about a resolver are the lack of electronics, which are located at the motor assembly, thus making the resolver more suitable for nastier environments. The resolver is an absolute measuring device. It can retain the exact position of a motor shaft even through a power outage. The feedback from the resolver can be driven great distances, over 1000 feet if necessary, with good resilience to noise. It is always recommended, however, that signal wiring be routed separately from any power wiring if possible. A single resolver can provide resolution accuracies up to a 14-bit number, or 16,384 counts. Dual-resolver packages can also be used to greatly increase the measurement accuracy in a particular feedback system. In these cases, a fine resolver is geared to a coarse resolver at a certain ratio. Common ratios are 64:1 and 32:1.

Magnetic pickups

Different versions of *magnetic pickups* exist. This class of feedback device can act like a resolver or a gear, or toothed-wheel, which can be installed in conjunction with a magnetic pickup sensing device, as shown in Fig. 8-13. The objective is to get rotating shaft position data accurately and consistently. The magnetic pickup sensor works in unison with the toothed wheel to count the teeth on the gear as they pass. Once set up and properly aligned, this magnetic sensor is fairly reliable. Depending on the environment, initial cost expectations, and motor/controller application, the magnetic pickup device selected is an important component to the process. Its selection must be fully evaluated.

Process control devices

In addition to motion-control sensors, there are many other applications for feedback devices. These devices can be of different types and have dedicated functions, including temperature, pressure, flow, ultrasonic, etc. These other sensors have very similar needs to the motion sensors previously discussed. They will need a good, steady-state power supply. Stable and fast clocks need to be incorporated into the

Process control devices

8-13 *Magnetic pickup device.* Carlen Controls

control loop. They need to network with plant host computers, as well as many other discrete computerized pieces of equipment.

Transducers

A very common feedback device used in gauging, measuring, and detecting displacement is the transducer. A *transducer* is a device that converts mechanical energy into electrical output. A better, more appropriate definition, however, is that variations of this device can convert any input energy, including electrical, into output energy. The output energy will, in turn, be different from the known input energy, thus supplying useful feedback information from which closed-loop control and correction can be achieved. There are hundreds of different types of transducers, and just as many ways to provide the input and output to the transducers. Pneumatic, hydraulic, and electrical types are the most common.

Transducers provide an electrical output, sometimes sending a voltage signal (0 to 10 Vdc) or a current signal (commonly 4 to 20 mA). An off condition would be 4 mA, 20 mA full on, and values in between are the range. This range signal is generated from a stimulus to the transducer and corresponds to a direct action from which the tranducer's output signal is going. For example, a process involving

Sensors and feedback devices

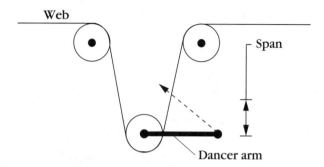

8-14 *Dancer feedback scheme.*

the unwinding and rewinding of a material as shown in Fig. 8-14 uses a linear displacement transducer, sometimes called a dancer. Changes in the dancer position correspond to a voltage drop across a resistor. This voltage output value from the transducer indicates how loose or tight the web material is, and a correction can be made by the motor controllers to increase speeds or torques.

Other types of transducers involve the piezoelectric condition, which produces motion from an electrical stimuli or produces an electric signal from a motion, usually a strain or load. These types include strain gauges and accelerometers. Another device is the rotary variable differential transformer (RVDT). This device employs a rotor and stator relationship to produce voltage. No slip rings are used, as the electrical output is via the electromagnetic relationship of the stator windings and the rotor. It produces a voltage, usually 0 to 10 Vdc, whose range varies linearly with the angular position of the shaft. Figure 8-15 shows the device's performance curve of a typical RVDT for output voltage versus input shaft position.

Temperature control

Process control often also requires temperature control. The temperature can attain extremely high levels; thus, an industrial grade of thermometers has emerged. The class of temperature-sensing devices used that generate electrical output are called *thermoelectric devices*. One common device is the *thermocouple*, which uses two metallic components and conducts an electric signal when their junctions are at different temperatures. It is sometimes referred to as a *thermal-junction device*. Other temperature sensing devices use a direct output linear to the measured temperature whenever feasible. A *thermistor* is an electrical-resistive device whose resistance varies with a change in temper-

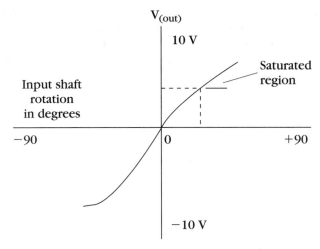

8-15 *A rotary variable differential transformer (RVDT) performance curve. Output voltage versus input shaft position.* Carlen Controls

ature. A *thermostat* is yet another temperature-sensing device that uses a bimetallic scheme to sense when a predetermined temperature has been met. At that point, a single output signal is provided to start or stop another piece of equipment, such as in a furnace. Temperature-sensing equipment in the factory of today must interface with many higher-level systems. The simple thermometer does not suffice, and digital displays, along with microprocessor-based units, are used to communicate with other devices and chart and record temperature data.

Pressure sensing

Present day pressure-sensing devices must not only be rugged for industrial use, but also fast-acting and high-precision to be useful. These devices, much like their temperature-sensing counterparts, are installed virtually inline with the fluids they are measuring or sensing. The integrity of the unit is thus challenged as it physically sits in the wet and hostile environment. Common pressure ranges go up to 500 millibars (mBar). The accuracy of the pressure sensor is rated by its linearity, or error from steady-state input/output; its hysteresis, or the difference in response to an increase or decrease in the input signal; and its repeatability, or the deviation over many readings.

The pressure sensor can convert the pressure signal into an electrical signal, while other times this conversion is performed in an external controller. This facet is sometimes overlooked in the pressure-control loop. This conversion module must physically reside somewhere, and the application and environment typically decide where that location is. Once located, the conversion is simple, usually taking a 3 to 15 psig or 0 to 100 psig pressure signal (scaling is important) and changing it to a 0 to 10 Vdc or 4 to 20 mA electrical signal. The controller then scales the electrical signal further, and a resulting output signal is generated.

Flow and level control

Many traditional mechanical meters and feedback devices have been overwhelmed by electronics, communications, and electromagnetic fields. Flow and level control is now a field of several disciplines. In the magnetic flowmeter, which measures the electrical conductivity of liquids flowing through a pipe, the basis for operation is electromagnetic. Magnetic flowmeter technology was discussed in Chapter 3 in the hydraulics section. The flow of a liquid in a pipe (a generally mechanical occurrence) can be determined much better because of technology. The power source is electrical, and dc excitation must occur. The benefits of ac versus pulsed dc power are debatable. In addition, the device is worthless unless meaningful output is transmitted to a central source.

A digital display is always necessary in the automated plant. Response times, the magnetic field, process noise and noise reduction, calibration, maintenance, configuring the memory and display, self-diagnostics, and communications are all important considerations. Installation, grounding, and hazardous locations must be considered for potential pitfalls. This type of feedback device is representative of most feedback devices in that it now contains much more electronics than its predecessors.

Level control has become more exact via electronics. The principles of physics have remained the same, but now closed-loop control is better achieved through modern-day technology. Housing the electronics at the feedback device has always been difficult because fluids and electricity do not get along. Electronic solutions for faster response and higher accuracy drove this industry to developing ways of housing the electronics to allow devices to reside in the environment in which it is monitoring. Many times level control is just an on or off state—is the desired level reached? Other times, we must know how close we are to the desired level. Whatever the requirement, a suitable device exists.

Photoelectrics

Another complete sect of sensors is light-detecting electronic devices, or *photoelectrics*. These sensors and controls detect the absence or presence of an object. Sensing whether an object is present or not via a light-based device allows plant personnel to control a process and make discrete movements on the production line.

There are three basic elements to any photoelectric system: a light source, a transmitter, and a receiver. There are also advanced control techniques that use logic modules, timers, and amplifiers to amplify output signals so data can be sent to other plant locations. The light source emits a strong beam of light that is distinguishable from ambient light and sunlight. The light is usually in the infrared spectrum and can be a visible beam emitted from a light emitting diode (LED). Figure 8-16 shows the basic photoelectric sensing elements.

8-16 *Components of a photoelectric system.*

Once we have a light source, we need a means to transmit that light signal and a means to receive. The receiving device is a light-sensing diode that recognizes whether light is absent or present from the light transmitter. From here, the output to another relay or device relative to the receiver's sensing comes into play. There is often a focal, control unit that controls both the triggering of the light source and the actual output signal. The receiver can be triggered when it has sensed light, or it can be triggered when it senses no light. This setup is application- and user-dependent.

Types of photoelectrics include reflective or proximity, retroreflective, and through beam. The distances the light must be transmitted often determine which type of photoelectric sensor to use. Photoelectrics issue a contact signal when an object has been sensed (or not sensed). This contact signal is then sent to a relay for either instant use or later use. A time delay can be incorporated from when the sensor first sends the contact signal to when it is issued to its final destination. For example, when an object on a moving conveyor is sensed and continues to move, the time delay allows for the object to arrive at a predetermined position before the sensor output is received. Thus, the object can be loaded, unloaded, or have another function performed on it because it is now in position, and the conveyor can keep running without interruption.

To select a type of photoelectric sensor, one should analyze the application. Distance, as mentioned earlier, is one of the most important issues. How far is the sensor going to be from the object? What is the object's size and does that size change? Is the environment dirty? Is there physical room for the wiring? These questions can help determine the type of photoelectric sensor and control scheme.

The through-beam photoelectric sensor is probably the most common. It can be seen in Fig. 8-17 with its elements and action. The light source transmitter and receiver are located opposite of each other. Light is sent in a straight line to the receiver. When an object passes between the transmitter and receiver, the beam is broken, thus triggering a contact closure. This type of sensor is popular because it can be used in somewhat nasty environments and has the longest range of photoelectric sensors. Alignment is crucial between transmitter and receiver.

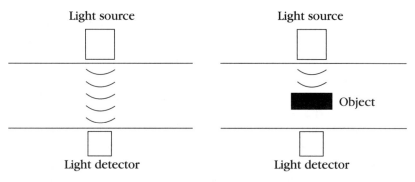

8-17 *Through-beam photoelectric sensor.*

The retroreflective photoelectric sensor is shown in Fig. 8-18. Here the transmitter and receiver are placed on the same side of a process, with a reflector opposite them. The light beam is transmitted, reflected, and received. Once the beam is broken, the object is detected. These sensor systems are relatively inexpensive, but have somewhat shorter ranges than the through-beam types. This sensor, however, is easier to install, and alignment is not as crucial. The object being detected cannot be more reflective than the reflector being used, which can lead to false triggering.

The easiest to install and perhaps the least-expensive photoelectric sensor is the reflective or proximity sensor. It is also the most sensitive and can be affected in many ways. It works on the principle of reflectivity of an object, as is shown in Fig. 8-19. The sensor must be adjusted for the reflectivity of the object being detected, and all other

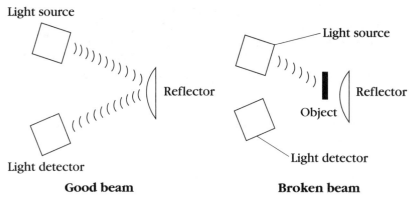

8-18 *Retroreflective photoelectric sensor.*

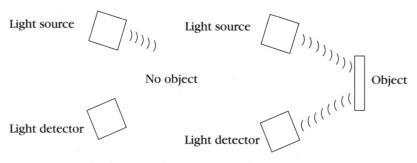

8-19 *Reflective or proximity sensor.*

objects, especially those in the background, must be tuned out. The environment should be relatively clean and the range expectancy not too great.

Applications of photoelectric sensors are many. Photoelectrics in the factory are used for counting, sorting, jam detection, and general inspection. They have many more uses, but they do have some limitations. If they fall short performing in a particular application or installation, other sensing equipment can be the answer.

Other sensors and feedback devices

Ultrasonic, or simply sonic, feedback devices are used quite extensively in many industrial applications. For these devices, the application and material being sensed must be compatible with sound waves or misreadings can occur. For instance, in a metal-stamping press application, sound waves are projected onto the surface of the metal and bounced back to the receiver. The time it takes for the waves to return indicate the distance between the material and the source. This type of device is very good in applications involving solid materials; however, materials with porous surfaces do not work well.

Machine vision is actually a form of feedback device and sensor system. It uses video imaging to monitor a process and provide real-time output to affect the process. Granted, the machine-vision system, either online or offline, is an expensive approach for controlling a process; however, a single sensor solution might not be available, and many sensors and corresponding feedback devices might need to be applied in conjunction with a computerized central station. This system can be cumbersome, and the net cost can be equivalent to a simple machine-vision system.

Likewise, laser technology has allowed for precise measuring and sensing of parts and processes not possible five years ago. The laser beam can work well in nasty environments, and its accuracy is rarely challenged. On the downside, current laser implementation can be expensive. Automatic gauging and weighing systems are also common applications involving sensor technology.

Methods of transmitting sensor data

Sensors can gather data, but unless that data is passed on to another device for analysis or compilation, sensors are not needed. Therefore, one function goes hand in hand with the other. Data transmission is covered in more detail in Chapter 9, "Communications." This section merely broaches the subject from the standpoint of the sensor devices themselves.

The most common forms of sensor output to another electrical receiving device are discrete means using voltage or current. A contact closure is the quickest and simplest form of discrete output. Once a predetermined level is met, the sensor indicates it by the closure of the contact. This scenario is a major part of traditional process control using relays, simple sensors, and relay logic.

A big advance in sensor output has been that of sending a 0- to 10-Vdc output signal, 0- to 5-Vdc, or even a 4- to 20-mA signal to a device that can receive such a signal. The current, or mA form, of the voltage signal can be furnished by incorporating a resistor component into the circuit. More elaborate forms of output from sensors and feedback devices include pulse trains of a certain voltage and frequency and even serial communication of ASCII information. Of course, as with any more complicated system, the more complex, the more potential for problems!

And problems do arise. The signal must sometimes travel great distances and must be amplified. Sometimes, the signal needs conditioning. When installing feedback devices and considering their use, it is helpful to determine whether or not distance and noise will be a factor. If so, appropriate steps can be taken to avoid these problems. An amplifier, which can also double as a signal conditioner, can help get the signal over great distances in the plant with minimal voltage drop. The 4- to 20-mA signal is popular because it can travel greater distances. Electrical noise can be a factor, however, and routing is important.

The future of sensor technology

All in all, sensor technology will advance side by side with the other industries to provide factory automation. Consistent, reliable communications must always be a necessity, for without the reception of good, usable data, the process suffers. Advances in the semiconductor and microprocessor industries will greatly enhance sensor technology. In addition, advancements in laser technology, fiberoptics, and superconductivity will create new methods and tools for sensors. The needs of the process and the application of the feedback devices will dictate future requirements.

One type of sensor that combines a biological component with a traditional sensor component is called the *biosensor*. It has not yet caught on as the sensor of choice but perhaps will find its way into many newer applications. Present industries that use the biosensor are pharmaceutical, food, and beverage companies. The biosensor senses when a process (such as a fermentation process) has come to its end.

The biosensor consists of a biological component, which can be an enzyme, nucleic acid, or receptor; any biological materials with properties from which information can be detected. The information is such that a specific compound, or even a specific element, can be identified or tracked in a process. The sensor portion is usually electrochemical, sonic, strain-gauge, or optic and works in full conjunction with its biological counterpart. With microprocessors and advances in light spectrum techniques, this type of sensor product will gain popularity in the future.

As data acquisition requirements become ever increasingly necessary to appease quality control and ISO-9000 standards, sensors and feedback devices will play a larger role. This situation is already the case in many facilities where advanced automation systems exist. Gathering information helps document that procedures are being adhered to and quality control measures are in place and working. Sensors and feedback devices efficiently gather the information to control a process and provide the necessary documentation. The alternative is to send Charlie out and have him "tweak" and record for us. The first choice is always preferred.

9
Communications

Automation in the factory would not stand a chance if machines could not communicate with one another. Machines must talk to other machines. They must send and receive data vital to production. If this data gets lost or hung up somewhere along the transmission line, or if it arrives with unwanted noise or trash, the machine usually stops running. Sometimes, the machine or process appears to be running, but upon receiving bad data it might just sit there in limbo. When we ask why machines stop running, the answer is faulty data transmissions. A simple machine can often be reset to clear the problem, but the process might not be able to tolerate the reset, and costly downtime is incurred—most undesirable!

Signs of communication schemes are throughout the plant. Not only are they apparent in the machine and process equipment, but they are also part of the plant's host computer, manufacturing systems, and accounting systems. Ideally, these computers should be able to talk to one another. Plant host computers must have current, up-to-the-minute information to perform the multitude of tasks that a day-to-day business requires. Communications is the thread that ties everything together. Shipping, scheduling, receiving, record keeping, and reporting are all dependent on how well data is transmitted throughout the plant.

What constitutes communication in the factory? What wiring is actually crucial to communications? Local area network (LAN) cables are usually coaxial or fiberoptic. Signal wiring used as feedback might be construed as communication between sensor and controller. Still, there are the basic serial and parallel ports that must be considered. Practically every in and out connection to a control device, except the power wiring, can be considered communication wiring. This wiring is illustrated in Fig. 9-1 by the several wires and cables going into and out of the controller. Each scheme must be considered for integrity, and each plays a key role in the control process. A problem with any single wire or terminal can shut down an entire process.

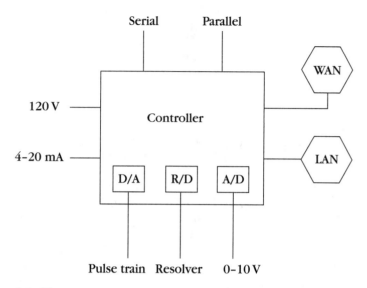

9-1 *The various wires going in and out of a typical controller.*

Communications is more than computerized devices talking to one another. It can be the signal coming back from the feedback device on the motor or valve. It can be the 110-V signal to a relay to stop a process. Communications can take many forms. In its basic format, communications is usually a lower level of electrical energy moving from location to location. It can be pulsed at a frequency and can have a magnitude, or amplitude, of electrical force (typically voltage), which is the difference between analog and digital communication schemes (Fig. 9-2). Communication can also be in the form of higher levels of electrical energy (excluding power wiring). It can be video, sound, or electronic. And it can be an operator yelling to another operator to order that hamburger for lunch! Communications is a vital phase of industrial automation. Without it, electronic data is confined to a local area.

Communications in the plant is more than the telephone call to the plant floor to change production runs. That might have been how changes were made in the past (and even today in nonautomated factories). But today, communications is the electronic information traveling at very high rates of speed over a conductor to and from smart devices. Sounds simple enough, but with so much happening over those conductors, accidents happen. Transmissions get garbled and sometimes lost. The machines are only following instructions. They do not think for themselves—not yet anyway!

Communications

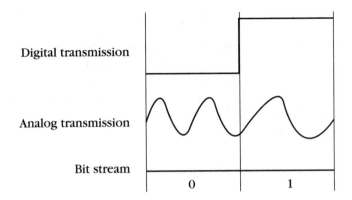

9-2 *Comparing a digital bit stream of data to an analog scheme.*

Instead of that physical telephone call to the production manager, the high-speed communications approach is this way: an order comes into the plant-wide manufacturing system. The order immediately shows purchasing which parts are in and which parts need to be ordered. This system is also available to the production people and they find that, to meet delivery, they must start a new production run immediately. This data is shared with many individuals in the plant and can be referred to as system-to-man interface. There is still the completely separate issue of machine-to-machine interfacing and man-to-machine interfacing. Together, this network of data transmissions automates the plant.

Any time data is transmitted, it must adhere to the rules of that transmission, or the protocol. In any transmission, there are rates of speed at which the data is moving. Some data rates are slower than others. Some are more impervious to noise. Regardless of these factors, set methods and forms must exist for the data to be moved across these conductors. This is the case in a simple computer and printer setup on the plant floor. For the computer to transmit good, usable data to the printer, it needs rules for that transmission. In more elaborate systems, this protocol is extended and more complex; communication networks sometimes have tons of data on a communication bus, all of which is moving at high rates of speed, avoiding collisions. Each bit of data is assigned a register, or address, so that when a piece of equipment needs that particular piece of data, it selects it at the appropriate instant. This type of communications bus is a *network* and is much more complicated than the serial and parallel communication schemes still prevalent today.

Protocol

Protocol governs the format and timing of data communication, and it is appropriate to look at the individual elements. There is a good deal of handshaking going on. Without handshaking, data would not transmit in the proper order or even at the proper time. Each manufacturer of a computerized device, capable of data transmission, defines its own protocol. The communication medium is chosen, the rate of transmission is selected as required by the application, and the actual sequencing of bits determined. The mode of transmission is then decided.

There are three basic modes of transmission: simplex, half-duplex, and full-duplex. A simplex channel allows for one-way, or unidirectional, transmission of data. It can send or receive, but not both. An example is the computer's printer. It is set up to receive data from the computer. It cannot send data to the computer. The second mode of transmission is the half-duplex mode, where the communication can occur in both directions. This communication can occur in only one direction at a time. An example is the serial port from computer to computer. Sometimes, one computer is receiving data, and other times it is only sending. This scheme is also used in telephone systems. The last communication mode is full-duplex. A full-duplex channel can transmit data in both directions simultaneously. This type of transmission is the most versatile and is seen in many network systems in use today.

Other protocol functions describe the breakdown of a string of bits and what each bit means as it travels to another device. For instance, titles for various bits are given in a transmission scheme. There are start bits, check bits, parity bits, stop bits, request-to-send bits, and clear-to-send bits. A start bit signals that data is now being transmitted. Parity is an error-detection scheme whereby an odd number of transmission errors has been detected. Stop bits are added to every character in a data transmission to signal the end of the character. As these strings of bits travel across a conductor, their order is very important. The receiving device contains decoding software that, when accepting this initial string of data bits, prepares the device to receive the entire transmission of data. Communication can get tricky when transmission speeds get faster and microprocessors (at the sending and receiving devices) get busier. Factor in noise and crosstalk, and you've got an interesting situation. That is why it makes sense to carefully choose the communication medium.

Communication media

The physical material that allows the electronic data to pass from device to device has many names. It is called wire, cable, or conductor. Usually made from copper or aluminum, it is the communication media and has evolved over the past decade. The outer casing of the metal is usually a plasticized coating, extruded onto the wire as the wire is drawn. Besides the actual material, the communication media includes the full physical form of the bus, or network. Connection methods are important, as are casing (around the conductor), grounding, and shielding. Several types of media are available. First, there was lighter-gauge, two-conductor wire with another wire used as ground. This scheme evolved into the twisting of the conductors within the casing to minimize eddy currents and reduce noise impregnation. Shields were then added to this version of conductor and are discussed later in this chapter.

The media has seen a rapid movement to coaxial and triaxial cable. The latest trends in communication media are fiberoptics, telephone lines, radio waves, and microwave technology. With these newer means of transmitting electronic data come higher costs and other issues concerning noise, connections, and software. As time marches on, however, costs should come down and experience will teach us the proper methods for implementation.

With each type of media comes added capability. Twisted-pair conductors are typically used for process signals and should not be run more than 2000 feet. Transmission rates are limited with twisted-pair wiring. Rates in the 100-kilobaud range are seen but as the rates get higher other types of media should be considered. In these cases, it is typical to see coaxial and triaxial cable in use. This communication cable has two conducting materials, each sharing the same axial point within the cable (coaxial). The same is true for triaxial cable, except there are three conducting materials. Depending on whether or not the coaxial system is set up for broadband (many simultaneous signals) or baseband (one signal at a time) transmission, the maximum rate of transmission is dictated. Rates of over 10 megabaud are not uncommon with broadband systems.

Fiberoptic cabling has gained acceptance as the best noise-resistant communication medium. If we look at the basic method of isolating electronic signals, optical isolation, we find that this principle is sound and useful with fiberoptic cabling. Electrical noise, having no physical way of conducting into the fiberoptic system, is no longer an issue. This cable, compared to equivalent lengths of coaxial cable, is lighter and diametrically smaller but still not as cost-effective. The more effort to manufacture, or the more complicated the wiring, the higher the costs. As better manufacturing methods are developed, costs of fiberoptics will come down.

Radiowave and microwave technology is also coming on strong. Cableless, wireless, and electromagnetic are the hot new buzzwords. The television industry is driving a lot of this development. For industrial and factory use, it has a niche application in remote areas where physically running cable is impossible. This technology can also make wide area networking more practical. With satellites and telecommunication schemes gaining in capability and acceptance, the future of communication media is approaching a new frontier.

Transmission distances

An often-asked question is how far cables can be run for a given application, and the answer can depend on whether the cables are power or signal cables. Whichever the case, limitations exist, and much of the routing methodology is simple common sense. Ask suppliers of any new equipment what they recommend for safe distances without sacrificing performance. Distances are determined by the type of transmission, the transmitting scheme, and the receiver's ability to accept degraded signals. Obviously, the longer the distance, the more likely that electrical noise of some type will be encountered along the way. The shorter the route the better!

Serial communication has limits. RS-232 lines can typically send and receive safely up to 50 feet. Beyond this, the configuration of the signal must change to a differential form, similar to RS-422 or RS-485.

Networks typically must transmit over long distances. In a large plant, these transmissions can be over a mile long. Good noise-resistant cabling must be used in these instances. Coaxial and fiberoptic runs are the most common. For power, high-voltage cable must be sized according to the amount of current that will be carried. In addition, if equipment is sensitive to voltage drops, this factor must be taken into account by increasing the diameter, or gauge, of the wire. Line drivers, repeaters, and signal boosters can be used for long-distance transmissions. Use these only when required because more components in the circuit create more noise or disturbances. If coupling points and junctions must be made in long runs of cable, ensure that the junction point has good connections and is somewhat sealed to make it as noise-immune as possible. The simpler the system the better.

Transmission speeds

The rate at which data is transmitted across communication lines is called the *speed of transmission*. Bits per second, or baud rate as it is normally called, is the common unit of value placed on transmission rates. The baud rate is illustrated in Fig. 9-3. This terminology is used mainly for serial and parallel communication schemes, but the terminology is appropriate for all data transmission systems (i.e., networks, telecommunications, etc.). Common baud rates are 2400, 4800, 9600, 19,200, but there are slower rates, such as 1200, 600, 300, and 240. With all the emphasis in the factory on getting data to and from devices fast, the slower baud rates are becoming extinct. Give us the faster rates! Sometimes they are referred to as K-baud, with K being equivalent to a thousand bits (19,200 baud is equal to 19.2 kilo or K baud). At 9600 baud, one character can be sent approximately every millisecond. Beyond 19.2 kilobaud, we start to get into the high-speed lines, mainly used in networking. These transmit millions of bits per second.

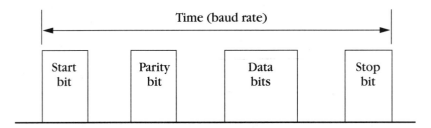

9-3 *Baud diagram.*

Serial and parallel communication

The traditional, most common means of communication in the office, home, and factory is by serial and parallel. Electronic data sent and received from one device to another is eventually broken down into individual bits. All electronic data must begin and end as a binary digit, or bit. Transmission of data, one bit after another, over a single conductor is called *serial communication*, which can be seen in the illustration of serial transmission in Fig. 9-4. A ground wire is also needed in this conductor circuit. There must be at least four wires in a serial communications scheme: the ground wire, a signal ground wire, and separate transmit and receive wires. The ground wire, when grounded properly, helps keep voltage spikes from damaging low-voltage level,

Bit 8 × Bit 7 × Bit 6 × Bit 5 × Bit 4 × Bit 3 × Bit 2 × Bit 1   ⟶

9-4 *Serial communications.*

sensitive communications componentry. Shielded cable can greatly reduce unwanted electrical noise from entering the signal conductor.

Acronym terms for typical serial ports, or communication connections, define the type of signal configuration. RS-232 is one form, where the RS stands for return signal and has a certain protocol that must be adhered to. If one of the crucial bits is missed or if one device tries to send when it is not supposed to, both devices sit there and do not perform any function. Thus, serial communication from machine to machine in real time is not desirable. A microprocessor must have the available time to handle the transmission and decoding while focusing on the machine's operation. It is more typical to find serial transmissions to and from a machine before it starts running (i.e., downloading instructions) or when it stops running (i.e., uploading collected data to a data acquisition system). The connector for a serial communication scheme is typically a 25-pin connector. All 25 pins are available for use, but not all are employed. Most often, only the first few pins are used in the connection.

Variations of serial communication exist. These are the RS-422 and the RS-485 differentiation of RS-232. The basic difference is in the driver. The signal in an RS-422 scheme is sent in a differential mode, thus allowing transmission over longer distances.

Uploading is the sending of electronic data to a host computer, usually for later use in statistical process control and data acquisition systems. *Downloading* is the converse of this. Information to run a machine must be sent to a machine's memory for use in its program before it runs. If sending instructions were attempted while the machine was executing its program, a greater risk for error and potential machine stoppage would occur.

Troubleshooting a serial communication port is straightforward. A documented protocol is hopefully available to allow us to check for certain irregularities. The first step is to compare the bit configuration with what is desired at both the sending and receiving devices. The number of data bits, stop bit, start bit, baud rate, and parity should be the same at each device. A device that can assist in serial communication troubleshooting is the serial analyzer. It actually takes the transmission cable and deciphers what was transmitted. Not every company owns one or can afford one, but if a company's business is built around serial communications, it might be a good investment.

Serial and parallel communication

After checking the serial port configuration, actual line symptoms can tell us what might be wrong. For example, if two characters are displayed when only one is required, this might indicate that the system is set up for half-duplex and should be changed to full-duplex. In addition, switching the transmit and receive wires of a system (host and peripheral, or sending-device and receiving-device) can establish whether or not any communication is present. Some serial ports require handshaking using the request-to-send (RTS) and clear-to-send (CTS) routine. By jumpering these pins on the 25-pin connector, you can establish the presence of this scheme.

Another check is that of the ground. Earth ground should not be used for the signal; dc common or signal ground should be used in a serial communication system. Also check that the distance of cable is not over the customary 50 feet for an RS-232C port. If it is, drivers or optical couplings might need to be incorporated. As for shielding, if any is available with the signal cable, it should be tied to earth ground at one end only. Do not ground both ends. By methodically checking that a serial communication system is set up and installed properly, problems can be greatly reduced, at least for this portion of the project!

Simultaneous transmission of multiple bits over multiple lines is called *parallel transmission*. Parallel transmission requires one wire for each bit of the system in which it is transmitting, as shown in Fig. 9-5. An eight-bit system thus needs eight wires plus a ground wire, a 16-bit system needs 16 wires plus a ground wire, and so on. Parallel transmission is faster than serial but more costly over greater distances. Typically, ribbon cable assemblies are used in parallel data transmission schemes. While this is a fast and effective means of com-

9-5
Parallel communications.

munication, especially for instances using a small, local terminal, it is not as common as serial communication.

Troubleshooting a parallel port is not as extensive as with the serial port. A handshaking procedure might need to be followed and checked thoroughly. The most common problems in a parallel system occur with incorrect addressing. Ensuring the peripheral device is configured properly with respect to the address setting can save troubleshooting time later. Remembering that a parallel port sends data at the rate of a byte at a time rather than a bit at a time can also help when troubleshooting the port. Much else is similar to a serial communication port.

Networks

A network is a communications scheme in which registered, or addressed, data is transmitted over long distances at high rates of speed. Often called a *data highway* or *information highway*, this communication scheme is preferred in most plants and factories these days. A graphical interpretation of an LAN is shown in Fig. 9-6. The need to link multiple computers, programmable logic controllers (PLCs), and machines has led to networking. *Networking* connects many dissimilar, computerized devices in the factory or office. These devices might be dissimilar in some respects but are similar in other ways, thus allowing for the interconnection. Besides being able to transmit over thousands of feet at millions of bits per second, the network also offers good immunity to radio frequency interference and electromagnetic interference. In addition, the network is omnidirectional, in that data can go virtually anywhere at any time. This type of network is commonly known as the LAN.

The LAN has gained popularity because it allows for the sharing of data between individual, computer-based devices within a factory rather than relying on one large high-speed computer. Of course, that large, high-speed computer can attach itself to the LAN and be an active participant. The attractive feature of the LAN is that, within the factory, a programmable controller can have several input/output drops, be connected to a computerized numerical control (CNC) machine, and link itself to an operator console across the factory floor, all via the LAN. The key to this type of network is that all the devices connected are located in the same general vicinity.

The LAN's main components are the LAN driver, the software, and the actual transmission cabling, its hardware. Coaxial cable and fiberoptic cable are the basic choices in an LAN-based communica-

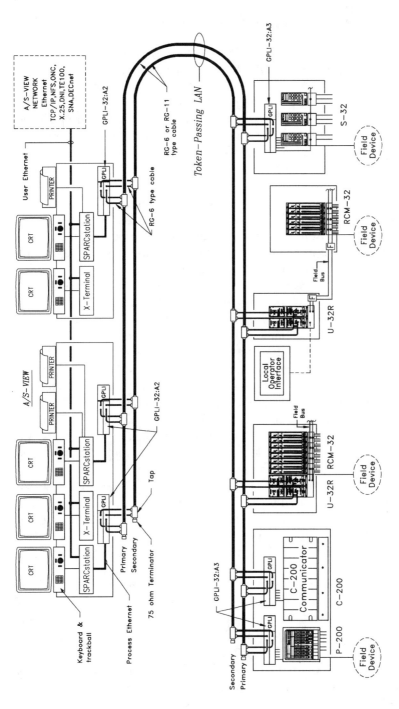

9-6 *Local Area Network (LAN).* Micon

tions arrangement. The television industry has made coaxial cable very popular. Fiberoptic cabling is fast becoming a challenger to coaxial cable. Reducing the cost will make it the conductor of choice. Fiberoptics is more noise-resilient than coax. It works on the principle of sending pulses of light through a conductor, whose inner walls are basically flexible mirrors, allowing most of the light to be transmitted. A light source unit able to translate electronic data into light pulses and a light receiver able to reverse the decoding are the other necessary components in a fiberoptic network. Prior to coaxial cabling and fiberoptic cabling, signal wire was lighter-gauge two-conductor wire with a ground wire, which evolved into twisting the wires to reduce the eddy-current phenomenon. As shielding became the norm and shielding methods of properly tying one end to earth ground, more noise reduction was achieved. The signal wiring is probably the most important in an automation sense, but it is also the most sensitive to noise. It has become necessary to clean up the signal and keep its integrity during transmission.

Following are some typical electrical, software, and mechanical specifications for an LAN system. The LAN interface should have software and drivers capable of several thousand I/O points, each with its own tag or name. The update time for this I/O scheme should be 1 to 2 seconds or better. Several ports should be available, both serial and general-purpose LAN interfaces. Data rates, parity, data bits, and stop bits should be software selectable to facilitate port configuration. Multiple devices, sometimes up to 50 adjunct controllers, should be attached to the LAN (multiple devices might require additional hardware). Regarding the central processing unit (CPU), memory, and clock frequency, these should be the latest, fastest, and most reliable as practical.

Another electrical consideration is that of the token-passing scheme. Some standard method is usually incorporated by each manufacturer. Coaxial cable is usually the 75-ohm type, with flexible and semirigid trunk connection units. Transmitting levels and sensitivity levels, in decibels (dB), should be in the 60 to 70 dB and 14 to 18 dB levels, respectively. Repeaters should be allowed, and several repeaters might be needed. EMI and RFI tolerance for an LAN should comply with federal communication standards. Mechanically and environmentally, the LAN hardware must be able to operate in an ambient temperature range found in normal factory environments, which is usually 0 to 50°C or 32 to 122°F. This rating does not require any additional cooling methods. The relative humidity tolerable levels should not exceed 95 percent, noncondensing (water and electricity do not mix well). In addition, boards and sensitive components should carry protective coatings to guard against corrosion and chemicals.

Two common LAN architectures are Arcnet and Ethernet. Both have been in use for many years, and each has a niche. Ethernet is the communication scheme of choice with many controllers because it offers the greatest transmission speeds. Rates of 10 megabaud are not uncommon with this method. Arcnet, on the other hand, carries with it speeds of 1 to 2 megabaud, which is fast enough for many instances. Limitations exist as to how many devices can be added to each, and this should always be considered before choosing networks.

Troubleshooting tools for analyzing a network exist. They include items such as the LAN "sniffer" and the signal-level meter. These devices can detect the presence of network signals, which can be invaluable when trying to pinpoint problem areas. Test software is also available and can be used at the start-up of a system and later if problems arise.

Network theory and implementation have minimized communication problems from where they were. Software and LAN drivers are now the most limiting factors. Each plant wants to move enormous amounts of data quickly. The challenge now is how do the LAN programmers handle all this data? Noise is not the problem it once was, either, and with newer transmission media becoming available, the network's future looks promising.

Wide area networks

Wide Area Networks (WANs) connect a number of remote plants or facilities, regardless of the plant's physical location. The WAN can be looked at as an extended LAN, as seen in Fig. 9-7. With telephone transmission facilities gaining more power every day, it is very practical to interface complete, remote factories. These networks take tremendous effort in software development, drivers, and implementation. In addition, the hardware for such a system must be capable of fast action. Costs of a WAN are currently high. To install a WAN, the reasons must be many and the commitment to a long, deliberate installation must be made.

Modems (modulators/demodulators) have preceded WANs as the interconnecting means between facilities. Sending data from a computerized device over the telephone line to another device has been happening for years. Although cumbersome, the modem has been a tool that has routinely kept us in touch with other sites. The telephone and facsimile machines are the other methods. The time has

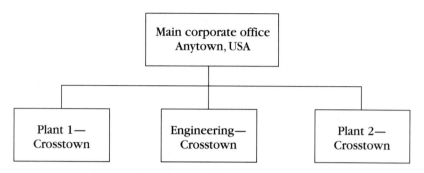

9-7 *Wide Area Network (WAN).*

come for plants to link to each other to share information more readily. The prevalent WAN is coming.

Noise

Electrical noise has become one of the biggest challenges confronting the automated plant's personnel. It often cannot be traced to any one device or source. Many times it is intermittent, which makes finding the problem even more of a chore. Electrical noise, crosstalk, hash or trash, or even sometimes called garbage, noise is the culprit in many electronic plant shutdowns. The key here is that years ago machines and processes had few microprocessor-based components. The microprocessor and the low levels of voltage in the circuitry surrounding it are very susceptible to electrical noise. By now, we are all aware that it is important to save our files when using the computer. We have been trained to save because all data could be lost if a power fluctuation occurs. With electronic data traveling bit by bit across a conductor at low levels of current, it doesn't take a whole lot of higher-level energy to disturb one bit or destroy it, which is basically what electrical noise can do.

When the subject of electrical noise is discussed, there are many ways to describe it, find it, and handle it. Describing electrical noise is interesting. We have all heard of "trash talking," but this trash talk is a different kind. Trash in this case is unwanted disturbances on our electrical lines. When discussing noise, we are mostly concerned with noise emissions and noise immunity; where does it come from and how can the control system work adequately if noise is present. A noise emission is defined as electromagnetic energy emitted from a device, whereas immunity is the ability of a device to withstand electromagnetic disturbances.

When looking at noise emissions, we find that most electrical noise is man-made. There are natural emissions from the voltage coming off all electrical components, which is sometimes referred to as *thermal noise emission*, and the sensitivity is a law of physics. The good news is that this type very rarely affects the electrical system. Another lesser type is the natural, atmospheric noise attributed to lightning storms. The last type is the man-made noise, and because it is created by humans, it can be controlled by humans. This type is made up of conductive, inductive, and radiated noise emissions.

Looking at radiated noise, we are concerned with that type of emission that travels through the air. The other type is that conducted over a wire. Being able to pinpoint where the noise originates from helps eliminate the problem. Identifying all the circuits, isolating them, and locating the point of common coupling is necessary, which is dependent on how two, and sometimes more, electrical circuits have been incorporated into the system. Electrical coupling is the common point where the circuits meet. This type of noise emission is often inductive but sometimes capacitive. Inductive emissions occur when the magnetic field around a live cable affects another cable, also referred to as an eddy-current phenomenon. Capacitive noise usually occurs when two electrical circuits share a common ground.

Many different types of noise-related phenomena exist. Being able to distinguish one from the other and avoiding each is important. These types include harmonic distortion, EMI, ground loops and floating ground situations, power fluctuations, and RFI. Complete power outages and brownouts can also be thrown into the category of noise because interruption of service is an issue. If the interruption is short (a cycle or less) the sensitive equipment might keep running. If the interruption is longer than relays, most computerized equipment, and other electronic devices shut down, thus demanding resets of the equipment throughout the plant, which can be costly. Each has its own cause and each exhibits its own effect.

Harmonic distortion

Harmonic distortion is the phenomenon by which a device in the plant is converting ac power into dc or vice versa. This rectification of electrical power usually disturbs something. These disturbances can be local to the source of the power conversion or can travel upstream or downstream along the circuit. These disturbances are more noticeable in the factory of today because so much equipment is computer-oriented and sensitive to nonsinusoidal waveforms. This phenomenon is relatively new to the past decade and is becoming a scapegoat when noise-related problems, which cannot be attributed to any other issue, do not go away. Quick solutions are filters at the source or at the piece of equipment exhibiting the disturbances. Harmonic distortion is discussed at length in Chapter 8 on motion control, as most often these motor controllers are performing a power conversion function, and the problem must be dealt with from that vantage point.

EMI and RFI

EMI is not easy to pinpoint and usually does not occur frequently. This interference occurs when a sensitive device, such as a computer, receives the EMI and trips, faults, or starts acting odd. Electromagnetic fields can exist around high-voltage, high-current carrying devices such as arc welders, large horsepower motors, and electric furnaces. This type of interference is best addressed when locating equipment within the plant. If not, loss of data, bad data, and false contact closures can be the result.

Likewise, RFI acts much like EMI in problem but not in source. RFI is the presence of high-frequency waves in a general vicinity. These high-frequency signals can interfere with data transmissions and create erroneous data at a microprocessor-based piece of equipment, whether it be a personal computer or a computer-controlled machine. The common sources of RFI are variable-speed drives with high carrier frequencies, wireless phones and microphones, and walkie-talkies.The best solutions are not to have any high-frequency devices near sensitive pieces of equipment, and, if they must be close to one another, the use of a filter at the source can suppress the interference. Both RFI and EMI are looked at more closely in Chapter 8.

Grounding

Ground loops and poor grounding are often sources of electrical noise. A ground circuit is supposed to route higher levels of current to ground rather than destroy valuable electronic components and shock humans. Unfortunately, this is also another route for low-level electrical disturbances to travel. The ground route can send unwanted disturbances out to other sensitive devices, and it can also receive unwanted electrical noise. A ground loop exists when there is potential, or EMF, between pieces of equipment whose grounds are connected in the same system. A typical ground loop condition can be seen in Fig. 9-8. The best solution is to ground each separate piece of equip-

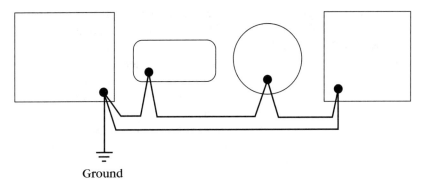

9-8 *An electrical noise problem area is ground loops.*

ment directly to its own ground. The trick is not to create a loop condition. The loop allows unwanted signals from various sources to enter into a completely separate circuit, whose equipment is often very sensitive.

Grounding, when done solidly and properly, is a very good practice. Grounds can become loose over time in environments where vibration and simple activity around the connection occur. Grounds can also deteriorate over time, especially in corrosive atmospheres. Checking and inspecting ground wires (and all other wires) in a circuit is good practice. Ensuring the connections are constantly solid can save damage to devices that normally expect low levels of voltage and can also eliminate disturbances due to ground loops.

Power fluctuations

Because most electronic equipment wants good, clean electrical sinewave energy in a consistent form, we can categorize many power fluctuation phenomena into this section. Nice and convenient as ac is, it is also unforgiving if the supply is disturbed on its way to its recipient. The recipient in the automated factory is usually a computerized piece of equipment. That piece of equipment might have a host of protective devices on the incoming power side, including fusing, noise suppression, filters, and voltage matching transformers. Even with some or all of this protection, problems can still occur. The problems of this section involve complete power outages, brownouts, sags in voltage or current, oscillations (possibly due to harmonics), and power surges. Each has its consequence and each has its protective circuit, both at some cost and risk to the piece of equipment being protected.

A power outage can have many definitions. Did the lights flicker on and off? If so, is this is an outage? Maybe yes and maybe no. If the lights were off for a cycle or two, it is not a severe outage, or complete loss of power. If the lights and many other electrically driven items went off for several seconds, it is an outage. Call the utility and find out what happened. Lightning storms are a good culprit in the summer. Certain sensitive equipment can ride through a couple of cycles without power by incorporating capacitor networks. More crucial and sensitive pieces of equipment must be placed on a circuit with an uninterruptible power source (UPS) system to continue running throughout the outage.

Sometimes the outages occur at the same time over the period of a given day, which can be an excellent clue in determining whether or not the occurrence is external or internal to the plant. A chart recorder placed on one of the supply lines can record when and what happened. With this information, a determination can be made as to what to do to solve or live with the outage.

Sags in voltage to a piece of equipment can also appear to be an outage. The end result is the same—the equipment trips and must be reset. The cause of the sag is just as important. Is there a large motor in the plant that, when started, demands so much current the overall system sags? Is there a similar current-needy device somewhere near the plant? If so, it should be isolated from all other pieces of equipment. Some equipment can handle lower levels of voltage for extended periods of time, but should be prequalified with the supplier of the piece of equipment when these conditions are possible. A brownout is actually a low incoming voltage level over an extended period of time.

Electrical disturbances

Hash, crosstalk, garbage, harmonic distortion, oscillations, ringing, and notches to the electrical system's waveform are the common terms, and they are both unsightly and unwanted. They are unsightly because they are usually picked up on an oscilloscope and, from a visual standpoint, do not want to be seen. From an electrical standpoint, they are unwanted. Sensitive pieces of equipment need clean, smooth low-level waves coming into their circuitry. If not, erroneous data can be produced, software programs can stop executing, or physical damage can occur. There are many causes of oscillations, and there are several solutions. Filters, reactors, and capacitor networks can smooth out the waveform on the incoming line, but the

Electrical disturbances

best effort should be put into finding the cause and actually stopping the oscillations.

Power surges and voltage spikes are the power fluctuations that are probably the most unwanted. They are truly undesirable because they can destroy electrical components without warning. The best protection for these types of fluctuations is fusing or circuit breakers. If fuses are blowing often and randomly, it is important to source the problem rather than to just constantly replace fuses. All it takes is one instance, with all conditions right, for the surge to get past the fuse and take out an important piece of equipment. In addition, if spikes are present in the system, transformers and power supplies are stressed often, leading to premature failure.

The key to dealing with electrical noise is to prevent it. Diagnosing noise and its source later is tough enough. If good wiring practices are followed and attention given to what types of equipment are to be used together, some noise problems can be addressed from the start. With so many retrofit projects happening in the plants today and engineers and designers trying to anticipate problem areas, it is still likely that electrical noise will be present in the system. Testing for it and getting the parties responsible are two important steps to solving the problem. There is always a solution, but at what cost and who should pay?

Following are some steps to take when installing new equipment or retrofitting old machinery in a plant with new. First, know the electrical characteristics of the new pieces of equipment. Find out what types of power conversion devices are used, how fast they switch, and so on. Suppliers of this equipment can answer these questions and must answer honestly or else fix the problem later. Second, make a single line drawing of all the equipment on the electrical system in question. Show both new and old equipment. List all the sensitive pieces of equipment on the circuit and find out from the operators which ones are most sensitive. List any filters already in place. Remember too, when doing this evaluation, that some equipment can be a receiver as well as a sender when it comes to RFI and even EMI.

Next, analyze the plant's incoming ac supply. If we must connect up to this source, we need to know what we are dealing with. Is the supply clean, or is there already a present disturbance? Is the supply stiff, is the short-circuit current level high or low? In other words, can the supply system handle your loading and possible fluctuations? Understanding what's coming in can pinpoint where problems might reside later. Likewise, the plant's grounding system must be measured.

Check if the grounds present are proper and show signs of deterioration. This analysis will also aid in grounding the new equipment. Again, heading off these types of grounding problems is well worth it.

After the analysis, it's time to install the equipment and apply power to it for the first time. Before running the equipment in the production mode, first test at various key locations for the presence of electrical noise. Check in and out of electrical enclosures, control panels, even at the sensitive computer stations. Many electrical noise problems surface right away. Some might not. Be prepared to test the supply and outgoing lines to and from equipment. Have certain test meters and recorders handy; they are necessary to find the source of the problem. After all, there will probably be only a small window of time to get the equipment back online into the production scheme.

Shielding

A common practice of protecting signal wire from electrical noise is *shielding*. As discussed earlier, several techniques can be used when routing signal wire and the type of cable actually used. Coaxial cable provides an ample degree of shielding because of its composition. As seen in Fig. 9-9, the signal wire is protected fairly well. Coaxial cabling is a safe choice when some electrical noise is expected in a system, but not too much. Another common type of signal, or reference, wiring is that of shielded wire, shown in Fig. 9-10. Here, a three-conductor wire is encapsulated with a shield tied to ground at one end to help keep unwanted noise from getting through to the wires. Another method that helps minimize noise is twisted wire. Shielded and twisted wire provide an even better degree of protection.

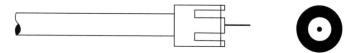

9-9 *Coaxial connector.*

Another form of shielding is installing physical barriers within an enclosure housing multiple electronic components. These barriers only need to be sheets of metal that separate compartments inside the enclosure. It can contain emissions, natural and man-made, from getting to the sensitive devices within the enclosure. As suggested, it is worthwhile to predict this situation ahead of time. Quite possibly, the

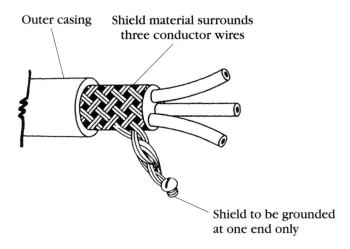

9-10 *Shielding.*

sensitive component or the noise-emitting component can be located from the other external to the enclosure.

Isolation versus nonisolation

This issue creates a lot of confusion when electrical devices are being wired together in the field, but it shouldn't. It is often the scapegoat when things are not working, and it probably should not be. It first must be understood and then, if a manufacturer requires it, it must be implemented. Signal wiring should be isolated, so says the control manufacturer in their manual. What does this really mean? Process signals are extremely important to the success of a process or to a machine operating properly. The signal must be free of electrical noise and linear. By a linear signal, we mean there is a corresponding relationship to an input reference signal and the resulting output. If the signal is bouncing, the process can suffer.

Reference signals entering a controller must be isolated. Isolation keeps line voltage to ground potentials from being present, a problem that could throw off the incoming signal. It is important to know from where a signal is coming. Many times, the source has its dc common tied to earth ground. This source, if a noise emitter, often passes the noise to the signal receiver via the signal, which is why it is common to completely isolate the incoming process signal to a controller. Isolation can be accomplished by solid-state relays or optical isolation.

Because light-emitting diodes are plentiful and practical, it has become common in many low-voltage level control systems to use

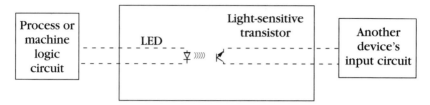

9-11 *Optical isolation.*

optical isolation. Basically, as shown in Fig. 9-11, as the electrical current passes through the diode (LED), light is emitted. A dedicated light receptor located adjacent to that LED receives the light transmission and keeps the circuit flowing. Optical isolation provides built-in circuit protection. No actual electrical current flows through this isolation point. Because control equipment is expensive and board-level components are not easily replaced (it is now customary to discard the entire board rather than replace one or two components), it is extremely important to protect the low-level voltage components from all spikes and surges. Optical isolation is a common method of doing so.

Signal conditioners and filters

Many times the electrical noise cannot be traced to a source and eliminated. Fortunately, dedicated modules can be installed in a circuit to provide the necessary isolation or filtering. Electrical noise and sensitive control/computer equipment do not mix well. Thus, a new business of filters and signal conditioners has arisen. Many filtering packages can be purchased off-the-shelf and installed quickly. These filter networks are often a resistor and capacitor in series parallel to the load being filtered, which can be seen in the equivalent circuit for a typical RC filter in Fig. 9-12. The values for the capacitor and resistor are factored to the amount of filtering desired, the load, and the actual noise predicted. This solution is often the least expensive and works provided no one inadvertently removes the filter. A single capacitor can provide adequate filtering and dampening in a control circuit, although too much capacitance can make a system very nonresponsive and sluggish, which should be avoided. Another solution is to install metal-oxide varistors (MOVs) in the circuit. These solutions are always worth the attempt. At the worst, the filter will not filter enough, and another approach must be considered.

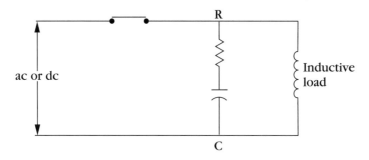

9-12 *RC (resistor-capacitor) filter.*

Throughout this chapter we looked at what communication really is along with some of the basics. We have looked at serial, parallel, and network systems and discussed standards and troubleshooting. Many electrical disturbances and interferences can occur as we implement new electrical and sensitive equipment. Preventing incompatibility and doing up-front investigation can save long hours of troubleshooting later.

Bibliography and suggested further reading

Dulin, Veley, and Gilbert. 1990. *Electronic Communications*. Blue Ridge Summit, PA: TAB Books.

Horn, Delton T. 1992. *Designing and Building Electronic Filters*. Blue Ridge Summit, PA: TAB Books.

Tomal and Widmer. 1993. *Electronic Troubleshooting*. Blue Ridge Summit, PA: TAB/McGraw-Hill.

10

Machine vision

Video graphics has made major strides into the industrial automation sector. Many factory inspection, guidance, and quality control systems have been completely automated via machine-vision technology. Machine vision has made inspections and process control more consistent and, in some cases, just possible where the same tasks were handled by the human eye not long ago. With advancements in computer and communications technology, image processing has emerged as an integral part of factory automation. Machine vision is now a link to quality and process control, as well as between manufacturers and their suppliers. Machine-vision technology actually consists of many technologies.

Machine vision combines high-speed and high-tech photography with the computer processing of tremendous amounts of data to visually tell us what's happening in our process, allowing for instantaneous adjustments to the process. Machine vision also allows for more dynamic activities on the factory floor. Robots can be guided by a vision system as parts can be oriented for the robot. Another vision function is that of pass/fail, or go/no-go situations on the assembly line. Machine-vision systems inspect the part's progress through the production line. The difference between machine inspectors and human inspectors is that these modern-day systems are more consistent, omnipresent, and do not take vacations.

The five disciplines of machine vision

The technologies that constitute a complete vision system are optics, computers, electrical engineering, mechanical engineering, and quality control. As each of these disciplines has advanced technologically, new doors have been opened in automation, as is the case with machine vision. If microprocessor capabilities and data storage media had not matured, there would not be a machine-vision system. A digitized picture could not be processed or used for any function. Machine vision technology created new opportunities. Inspection, machine guidance, and metrology functions were never attempted before. Likewise, technicians who use and work with vision systems had to become well-versed in all the aforementioned disciplines. This theme seems to run true for practically all aspects of factory automation—one cannot simply be a one-discipline employee!

Optics is the science that deals with the effects of light and, from a human sense, sight. Machine vision relies on the basic principles of optics and physics to function properly. A machine-vision system is extremely consistent. It can be consistently good, or it can be consistently bad. It depends on how well it has been set up. Just as the human eye can be fooled, so too can a vision system. When using optics, we are interested in lighting, cameras, fiberoptics, lenses, and the object being inspected.

Let's start with the object being inspected. We need to define what we are looking at and what we can do once we have looked at the part. For instance, if size or shape changes during the process, we must be prepared to handle this change, both with hardware and software. What color is the part? A black edge on a black background is very difficult to see for either machine or human, whereas a black or dark edge on a white or lighter background provides excellent contrast. This factor is the essence of video imaging. Good examples of this are shown in Figs. 10-1 and 10-2.

Figure 10-1 shows a black briefcase, black plastic bag, black notebook, and black paperweight. When stacked upon each other, the edges become hard to see and even harder the farther away the viewing party is. The image is also affected by the amount of ambient light. Imagine these items as integral pieces to an assembled part being moved down an assembly line. If the edges are not well-defined in a still photograph, what should we expect when the production line is at capacity (where line speed and throughput are critical)? Ideally, if machine-vision manufacturers had their way, all parts would have sharp contrast to each other, as can be seen in Fig. 10-2. Now, all the same components from Fig. 10-1 are matched with white items and white backgrounds. These edges are well-defined and would make for almost 100 percent repeatable inspections.

Besides the challenges mentioned, we must provide additional answers. Are we going to inspect, measure, orient, or perform some other task with the vision system? How will the part be presented to the camera? Where in the plant will the vision system be installed?

10-1
Typical black-on-black, hard to identify edges.

10-2
Typical white-on-black or black-on-white contrasting edges.

Can the entire part fit within the field of view of one camera with the proper lens, or must multiple cameras be used? What is the best magnification to use for the features we need to view? Many questions, but every control/automation application has many to answer. These questions all dictate what type of optical solutions are available.

Lighting is probably the most important. After deciding what is to be viewed, a suitable lighting scheme can be considered. Considering the contour on the part, its color, and so on, we determine what type of light to use and from which angle to supply it. Backlit light silhouettes a part and defines edges fairly well. Ambient lighting is usually not very good, especially in a factory. Additional light might be required where the camera and part meet. Sometimes, just a little more fluorescent or incandescent light is all that is needed. Other times, po-

Machine vision

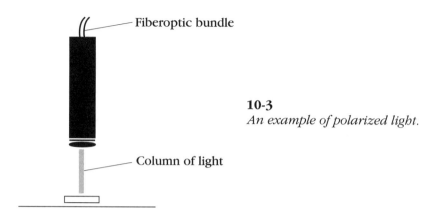

10-3
An example of polarized light.

larized, or columnar, light is required. Columnar light is a beam of light, as shown in Fig. 10-3, that does not break up over a given distance. Polarized light is achieved by directing the light waves through a diffuser. In this way, the light wave direction is changed from a random, or scattered, effect to a unified, controlled direction, allowing higher intensities of the light to be directed onto the object. Certain features on a complex part might need this type of lighting to see the desired feature. Sometimes, the lighting might need to be programmed to come on and off in a certain sequence to see all the features necessary. Filters might also be needed. It all depends on the part, the surroundings, and the feature being viewed.

The next most important facet of the optics package is the reliable, high-speed camera. Tube type cameras were the norm originally, but now solid-state, smaller cameras provide good video images in a much smaller package. The graphic images are converted into digitized pictures. Each digitized frame has an array of picture elements, or pixels. The more pixels, the better the resolution of the picture, but the more expensive the camera. Just as high-resolution monitors have arrays in the 640×480 range (for a total of 307,200 pixels) so, too, do these solid-state cameras. Common pixel arrays for high-resolution cameras are 512×512 and 736×552.

High resolution is not a problem to a high-speed computer. The more pixels the better to a powerful computer system. Image processing is explored completely in this chapter. For further edge definition and better accuracies, each picture element of the digitized picture can be divided into levels of gray. These levels range from white, which might equate to zero, up to a level of black, which might equate to 256, thus giving us 256 levels of gray in between. The reason for so many levels can be understood once we look at what happens to the digitized image in the processing stage, which brings us into the next important component of the machine vision system: the computer.

The processing of any digitized image requires extremely high-throughput microprocessors and significant amounts of storage. Clock speeds must be fast, and data storage immense. For each digitized picture, there are literally thousands of bits of information to analyze. On the production line, parts are constantly moving and being produced. The vision system must take a picture, process it, store relevant data, move to the next part, and take its subsequent picture. The process keeps repeating, over and over.

Figure 10-4 shows a nondigitized image. Figure 10-5 shows that same image digitized and given an array of picture elements in which it is contained. The microprocessor, via a high-level software program, scans the array of pixels vertically and horizontally, as seen in Fig. 10-6, looking for black-to-white and white-to-black transitions. The computer, by being provided with previous reference information (i.e., the part's dimensions, the shape of the part, and physical location), can then provide an almost instantaneous comparison.

10-4
A nondigitized image of a round object.

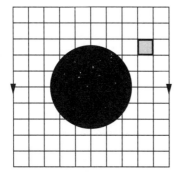

10-5
A fully digitized image of the round, backlit object, complete with a picture element (pixel) array and one highlighted pixel.

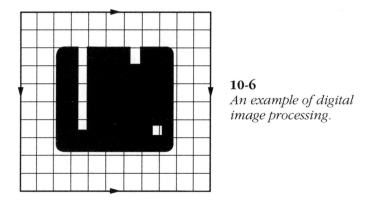

10-6
An example of digital image processing.

In more sophisticated systems and programs, the machine looks for levels of gray and changes in the levels of gray. These systems are required when a hard-to-determine edge on a specific part is encountered. The basic vertical and horizontal pixel scan might not be consistent when looking for just black or white elements within the array. If the element is gray, one scan might consider it white and the next black. This condition of an edge is shown in Fig. 10-7, which results in inconsistent readings. It actually caused the development of gray-level edge detection. The additional data about the edge and its pixel yields further usable information for processing. This information is stored quickly, quickly compared to predefined conditions (pass or fail, for example) or tolerances, and a resultant action is taken in the form of output. Output can be a signal to another device to remove the bad part, provide offset information to a robotic gripper about ready to pick the part, or just simply send data to a printer or statistical data collection system for future use. The whole vision process starts over with the next available part. Thousands of instructions and pieces of data are processed every second.

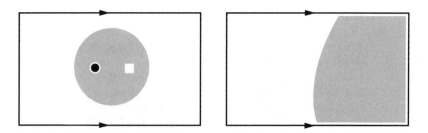

10-7 *Gray-scale edge detection. The image on the right is the same part as on the left but magnified many times.*

The five disciplines of machine vision

An important facet of the computerized portion of the machine-vision system is communications. The image-processed data must go somewhere, usually via a serial communications link to another computer for further use or simply to a printer. The data can also be stored to a local hard drive, but this amount of data can use up available storage quickly. The serial communications ports are generally RS-232 with some systems using certain PLC protocols and parallel ports. When communicating with a robotic controller in a pass/fail, or go/no-go situation regarding an inspected part, the vision system might simply provide a dry contact to the robotic controller, indicating that the part is bad. The robotic controller then instructs the robot to remove the part from the assembly line. In other robotic instances, an actual offset value in some prearranged increments is sent to the robot from the vision system. This data is supplied for an X (horizontal) axis and a Y (vertical) axis. The horizontal and vertical information is a minimum, as parts on the assembly line can arrive misaligned from either direction, or a resultant of both. The height offset, or Z dimension, can also be used.

This system sequence can be seen in Fig. 10-8. The vision system, upstream from the robot, views the part, its location, and its orienta-

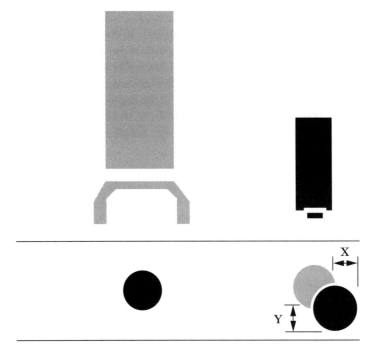

10-8 *An X-Y offset for a robotic guidance system.*

tion in that location and compares the digitized image with a standard or base image. Any pixels out of line are so noted and corrective action is taken by the vision system. This corrective action is offset information quickly transmitted to the robotic controller for injection into its motion program. This offset data allows the robot to compensate before the part arrives. The robot arm can then pick the part, without jamming the system, or dropping or breaking the part. The process is a little more complicated, especially when starting up a new vision system in conjunction with a robot in a parts-orientation scheme. Data must be sent and received properly.

Another facet of the machine-vision system is its electrical engineering content. The whole system operates on some preset voltage and current supply requirements. The computer system relies on stepped-down, lower voltages to process information. The lighting and optics scheme requires a power source. The communications is a form of low-level energy that must be fully engineered to work. Attention must be given to ensuring compatibility exists between devices (PLCs, computers, etc.). Some offline inspection stations incorporate servo-drive systems that need to be coordinated into the whole package. Electrical filters and constant-voltage transformers must sometimes be used to ensure clean, consistent power. All these factors must be taken into account to avoid problems later.

Likewise, one must be concerned with numerous mechanical issues in a machine-vision system. First, it must be determined how the part is to be presented to the camera. The system can take on an online or inline theme if the camera and lighting must be installed on the production floor or assembly line. Fixtures, or devices that hold a part consistently and completely still while the camera takes its picture, must be designed around the particular part being viewed or inspected. Similarly, an offline machine-vision system also requires a fixturing method. An offline system is where the part is taken from the production line area to a lab or station dedicated to more elaborate inspection and analysis.

The mechanical content does not end with fixturing, mounting, and installation. A good understanding of parts handling in general, machine design and tolerances, and a good basic understanding of metrology (the study of weights and measures) is also required. A list of typical metrology symbols and their meanings are shown in Table 10-1. Being able to read a blueprint for a given part is just as important as actually making the part. This familiarity with the inspected part makes for a more complete analysis; how a part is manufactured, how it fits into a subassembly, and how it can be used all enter into the vi-

Table 10-1. Common metrology symbols and definitions

Diameter	⌀
Centerline	℄ — - —
Concentricity	◎
Runout	↗
True position	⊡
Tolerance	+.0002 −.0002
TIR	Total indicator reading
Micron	One thousandth of a millimeter
Coplanarity	Flatness of a surface check
RPY	Roll, pitch, and yaw

sion-inspection equation. The machine-vision system often must be designed around these functions and inspections, which leads us right into the related disciplines of quality and process control.

Quality control is different than process control, and this concept is explored further in Chapter 13. Just as quality control is different than process control, so too is statistical quality control (SQC) different from statistical process control (SPC). The basic similarity between the two is that the company making the product wants zero defects. To achieve this goal, a company must control its process. Statistical quality control and statistical process control are different in that quality control is strictly concerned with the end product, while process control is always trying to maintain or improve on the manufacturing process. With machine-vision systems capable of furnishing tons of real data both inline and offline to the process, dynamic changes can be made immediately to the manufacturing process. Likewise, vision systems can completely inspect incoming parts (components from an outside source for use in the assembly of the end product) and can also inspect final products before shipment. Reports of compliance can be issued and tracked with the particular products for which the reports were generated.

Besides the five main disciplines mentioned, others enter into the world of machine vision. Calibrations to standards for the measurement system itself, software programming using high-level languages, and full-system integration with other machines and processes are also offshoots to the machine-vision industry. A machine-vision system must be carefully evaluated. It can be an expensive investment, but it can do many functions to pay for itself. Deciding just what needs to be accomplished with the part should be done in the traditional question and answer format preached throughout this book. Get involved with every detail. Know what the goals of a system should be. Have vendors supply literature and information. Find out what is out there in the latest technology. The time up front is always worth it!

Digitized video and actual processing

As discussed earlier, the heart and soul of machine-vision systems is its video processing. From a well-lit picture we get a digitized array of elements. These pixels each have a length and a width dimension and are scanned for edges or points, which is the numerical data our computer uses. Today's systems incorporate 16 up to 256 levels of gray (usually in powers of 2) for enhanced edge detection. Coincidentally, the power of two-video processing works best with binary computer systems. Using this imaging process, we eventually select a pixel whose level of gray allows it to be the most consistent for later comparison. It is often specified in machine-vision specifications that the accuracy is plus or minus a value. This value is usually the length or width of the smallest possible pixel produced by the system and is set initially during a calibration routine in which the camera is shown a calibrated image, traceable to the National Bureau of Standards. We then go into a homing routine to zero all the axis motors and set a physical reference. From here, we are now ready to inspect and measure parts.

Online applications

On the production floor of the automated factory exists many high-tech machines. If we equate a machine to a person doing the same work, we must include the vision of that person, or system, as one of the most important elements. Without vision, we cannot tell if a part is present or absent. We cannot tell if the part is bad or good, defective or worth selling to John Q. Public, who has some pretty high standards on quality. The necessity of being able to see what is going on in real-time on the factory floor has created an industry of machine vision in which specialized systems exist for many applications. One such segment is that of the online or inline vision system.

Online applications

The online vision system is physically installed in the assembly line or production area, makes decisions based on what it sees, and informs another piece of equipment, such as a robot, what it sees. Some systems are relatively simple with just one function to perform, while others perform several functions in very small amounts of time. These functions, including taking pictures of parts as they arrive on the assembly line, processing the video, and sending the output in proper form to another device (host computer, robot, etc.). Luckily, high-speed video processing and high-speed computer analysis are available for these complex tasks.

Online machine-vision systems can check for the presence or absence of a component, which is very useful in the assembly of printed circuit boards (PCB). This process tends to be performed by machines in high-volume manufacturing environments, so machine-vision systems play a key role. They are preprogrammed to know the overall layout of the board and know the size and shape of components that should reside in a particular location on the board. A digitized video image of the board, with all the proper components in their respective positions, is the model for comparison. As the newly assembled boards come down the assembly line, a digitized picture is taken. This digitized video image is compared to the model image to determine if the newly assembled board meets the standard. If a component, such as a resistor or capacitor, is missing, the vision system can shut the line down until an operator intervenes to remove the board for manual rework. Obviously, top and side lighting of the board and components in this type of application is crucial. A similar application is checking that the through holes of a PCB have been filled is easy for the machine-vision system. Figure 10-9 shows a machine-vision system dedicated to PCB inspections. Several types of inspections can be made with the same machine. The lighting technique is mostly backlit. Backlighting, or silhouetting, an object and then looking for white and black differences is the best scenario for not only this application but most vision systems. Edges are fairly well-defined even while magnified.

Just as the previous application's ultimate task was notifying an operator after identifying a problem, it could be taken a step further. Suppose the assembly line could insert certain components, maybe those that were notorious for being missed or falling off the board before final inspection. A vision system could be installed to note which component was missing, and send this data to an adjacent robot, which could choose and insert the correct component on the board in the proper location. This task could also involve offset information.

Machine vision

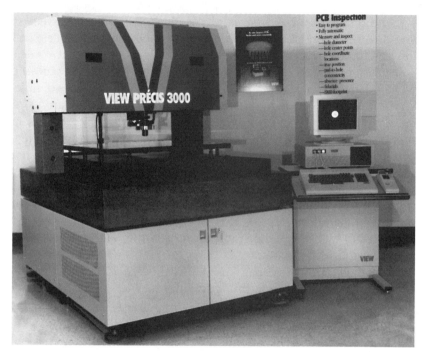

10-9 *A machine vision system dedicated to printed circuit board (PCB) inspection.* View Engineering, Inc.

Offset information is data furnished to the robot in terms of left and right, up and down values that allow the robot to place the part exactly where it belongs. The eyes of the robot are actually a camera, computer, and communications port that sees the offset and passes it to the robot controller. As was shown in Fig. 10-9, the offset data that is passed is actually incorporated into the robotic controller's software routine, and an adjustment is made.

For less-expensive assemblies that are produced at extremely high speeds, it is impossible for human operators to catch all the possible defective products coming down the assembly line. As assembly lines become more and more automated, line speeds are increasing to the point where human inspection is no longer even an option. Machine-vision systems perform pass/fail or go/no-go inspections to products as they whiz by. Many packaged items, for example, can become easily deformed, perhaps in a machine jam, and inside matter can get on the outside of the package, thus rendering that product unsellable. The cost of these packaged products is considerably less than a printed circuit board, so we cannot afford to shut the line down for rework or cleanup. Instead, the machine-vision system sim-

Online applications

ply detects a bad packaged product, sends a fast relay contact to a pneumatic actuator, and the bad product is physically knocked from the assembly line into a bin for later disposition.

An example of a pass/fail system in single-line form with digitized video is shown in Fig. 10-10. Here, ketchup packets are each viewed to quickly determine if any spilled ketchup is on them before being packaged. The saying, "one bad packet can ruin the whole box," applies here. If a packet with ketchup all over it gets into the box, it dries into a mess and sticks to other good packets, resulting in the entire box or even shipment being possibly rejected by the customer. Costly shipping, handling, and rework charges would then be incurred. The machine-vision system helps avoid these situations.

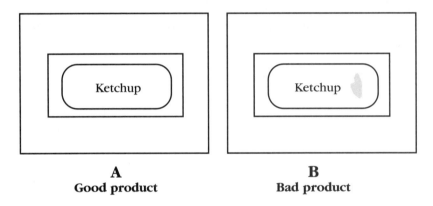

A
Good product

B
Bad product

10-10 *A pass-fail inspection scheme for ketchup packets. "A" is the digitized image to which all packets are compared; "B" is a packet with a spot on it, thus failing the inspection.*

Online vision systems are very good for solving simple inspection and guidance needs because the part is actually brought to the camera, which is completely stationary. The lighting is stationary and consistent. Basically, nothing moves except the part to be presented to the vision system on the conveying system. A single measurement might involve more than one camera; multiple measurements require multiple cameras. The multiple cameras view the same part at different views, which can become expensive, cumbersome to program, and difficult to coordinate. Lighting must be set just right and must be consistent. If the lights get dirty or bumped by an operator, the vision system suffers. These problems have resulted in a new branch of machine-vision systems called *offline vision.*

Offline machine-vision systems

As discussed earlier with online vision systems, it is not practical to do a 100 percent inspection of a part while on the assembly line. Thus, there is a branch of machine vision more concerned with metrology, quality control, and process control called offline machine vision. It is a noncontact system where all the previously mentioned vision components are applied, including computers, cameras, fixtures, servos, and so on. The offline system emphasizes dimensional measurement rather than flaws and defects. In fact, the offline system can provide many more measurements because the environment and lighting are better controlled.

Traditionally, a plant inspector would make his or her rounds periodically, looking at parts on the assembly line and sometimes grabbing one or two to take back to the metrology or quality lab. At the lab, the inspector had various tools to perform many different tests and measurements. Tools used to measure the parts are called calipers and micrometers. The drawback with these tools is the measurer must be capable of using such a device, reading it properly, and then transferring the information to a log book for proper documentation. This process is cumbersome and certainly contains the risk of human error. For closer-tolerance inspections, the part is often placed on a light machine called an *optical comparator*, which basically backlights the part, allowing the operator to measure the silhouette. The process is still cumbersome and still has the potential for human error.

With all the advances in video and computer processing, new machines were introduced that could perform these tasks faster, more repeatedly, and generate the reporting and documentation necessary at the same time. Now, an inspector could have the sample parts brought back to the quality lab, usually a clean-room environment, set the part in the fixture, and simply wait for the results, as seen in Fig. 10-11. Sounds pretty good to all concerned; however, there must be a new level of awareness for the quality control manager and staff. They now had to understand what servo drives actually were, learn new computers and new software packages, and become proficient at fixturing parts. All this knowledge, on top of the statistical and quality control knowledge they should already have.

A photograph of an offline machine-vision measurement system is shown in Fig. 10-12. This machine is a physically smaller offline machine-vision measurement system that could reside in the quality control or research and development laboratories, or be located right next to the assembly line to reduce inspector travel time to and from the

Offline machine-vision systems

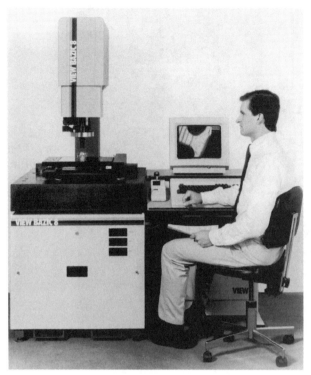

10-11 *An offline machine vision system in a clean, quality lab environment.* View Engineering, Inc.

lab. Figure 10-13 shows a much more substantial noncontact vision system, with the granite table alone weighing several thousand pounds and dedicated to inspections in a controlled laboratory environment. The granite base ensures measurement accuracy via stability and isolation from plant vibrations, but also creates the bigger weight content of the device itself. Using Fig. 10-14, let's walk through the complete process of inspecting a part to see what constitutes the hardware and software content of a typical offline machine-vision system.

First, we determine which features and measurements are the most important to our overall process and our customers. What needs to be measured and why? Once determined, we can decide how to position the part on the moving stage and how to light the part. The fixturing usually involves thought, labor, and cost; therefore it is of value to predict how many other parts could be used with the same fixture or a slight modification of the same. Remember, a fixture is basically a device, sometimes called a jig, that holds a piece firmly in position and without allowing any deformations to occur to the piece.

Machine vision

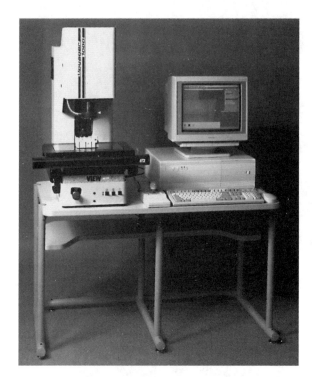

10-12 *An offline machine vision system capable of being located on the production floor.* View Engineering, Inc.

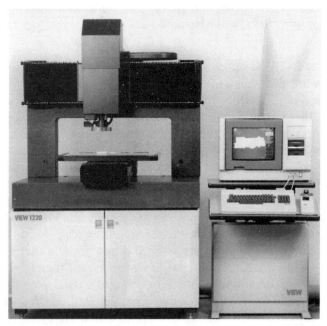

10-13 *An offline machine vision system with a large granite base.* View Engineering, Inc.

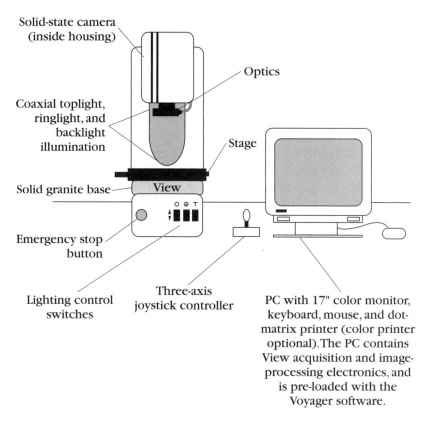

10-14 *Offline machine vision system components.* View Engineering, Inc.

Spring-loaded and screw-down devices are popular. Another concern is how the stage moves, rapidly accelerating and decelerating. Movement is accomplished using high-precision servomotors and servo amplifiers. Speeds of up to 4 inches per second are common, which is how several hundred measurements can be taken in the same time as two or three measurements of a traditional caliper inspection.

Once the part is in place within the fixture, it is now time to choose a lens and calibrate the vision system. Because the vision system uses pixels in its video processing, it is necessary to assign them an appropriate size for our inspection needs, determined by the tolerances that must be held in the process and on the part. If a part must be no longer or no shorter than two microns, plus or minus 0.000080 (a micron is a unit of measure of one thousandth of a millimeter) in a certain direction, our vision system cannot have a pixel three microns in size. Pixel size determines the particular lens and magnification amount required to achieve the proper resolution for the accuracy desired. Once the lens is chosen, the field of view for that particular lens must be calibrated by actually taking a picture of a National Bureau of Standards (NBS) calibration unit to attribute an X and Y dimension for each pixel.

The next step is to "zero" the moving stages. A reference point is needed to start the pixel count and measure from. Each moving axis reads a very high-precision linear scale for servo-loop feedback. This position information, along with pixel size, is the comparison for the offline vision-inspection machine. Like any computer-driven system, the system automatically boots and conducts self-checks upon powering up. At this point, we can inspect our part.

Usually, a blueprint, or CAD drawing, of the part is handy and used to develop the inspection program. The offline vision system is a teach-type system, which means that the operator can jog the servomotors on each axis to the location that best suits viewing an edge or feature on the part. The leading edge of the hole is one of the elements used in a diameter calculation. The tool or edge finder, shown in Fig. 10-15, is used to find the leading edge of the circular object. As the edge finder scans in the direction of the tool's arrows, the first pixel of the edge becomes the leading edge. This same tool is used by the programmer and later the processor to perform the vision operation of the entire feature. These tools are discussed later.

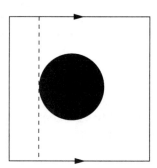

10-15
An example of a vision system "tool" looking at the leading edge of a circular image.

The lighting at this position is backlighting. Backlighting becomes a step in the automatic program and is repeated around the part until all edges and points are defined for video inspection. This process becomes the user program for the given part and is saved to the computer's storage system. The program then executes, taking pictures, turning lights on and off, and moving the servo-driven axes to the next selected location. The images are being digitized, processed and stored, the entire time, and valuable data is generated.

With many computer-driven machines these days, software is the driving force behind the actual hardware. It is therefore important to understand what is available and how it works. Many software packages are often the culprit when a computerized machine stops running. Why? Because the programmer sometimes tries to command the system to perform a task that cannot be accomplished physically or mathematically. Most of the time, the programmer who wrote the original source code installs, prohibits and flags to alert the program user as to when a value or routine is invalid. This knowledge comes with field trials and exposure, which is why we often get revision upon revision of source code.

Common offline machine-vision inspections

The most necessary inspection to be made of a given part is that of the simple measurements against allowable tolerances. The length, or X direction, and the width, or Y direction, provide instantaneous values as to whether or not a given part is good. There are usually many other features on a more complex part, and many of these dimensions need checking, as well. Much of the decision as to what dimensions to check comes from determining that a mating part must be measured to ensure it matches the other parts.

Typically, to check in the Z direction requires a height or vertical inspection to be made, and special lighting requirements might be necessary. This measurement is more of a challenge to machine-vision systems. Often, the surface of the part is hard to focus on, and more light must be shown on the surface. Sometimes, the surface could be glossy or even somewhat transparent. A device called a *Ronchi* grid can be used to project well-defined shadows onto the surface of those parts, allowing the camera to focus on the shadow. Once focused on the image (or actually the images of the shadow projected by the grid), the machine-vision system can decipher from motor feedback position where the Z axis is and store that data relative to a home position or prior Z dimension, to get an actual dimension. The Ronchi grid implementation can be seen in Fig. 10-16.

Another function of the Z axis is to check for surface coplanarity, which involves the individual focusing of several points on the surface, or plane, to which the comparison is to be made. This check is shown in Fig. 10-17. A similar machine-vision check of a part's surface is called *total indicator reading*, or TIR. TIR involves the camera performing several independent focuses at different locations on the surface and placing these values into a mathematical routine, which adds them and furnishes the average. The difference between coplanarity

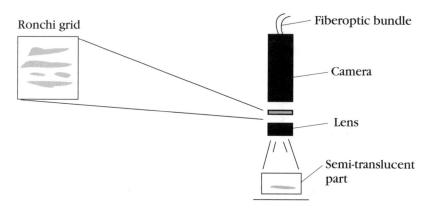

10-16 *The Ronchi Grid concept.* View Engineering, Inc.

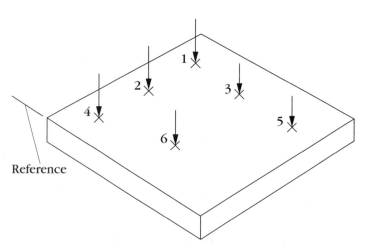

10-17 *Coplanarity of several points on a surface.*

and TIR is that a plane in the Z direction is established first, and all subsequently measured points are compared to the reference plane.

Another set of machine-vision checks, all interrelated, are those called *roll, pitch,* and *yaw*. Sometimes called RPY, this checks for plane offsets in a part in all three axes, X, Y, and Z. Concentricity is a common check for holes, or circular shapes of a part, and their relationships to one another is to a common center point. This machine-vision inspection is usually performed with backlighting. True position is the inspection that looks at a location on a part to verify if the feature is where it is supposed to be with respect to a mating part.

Tools

The term *tools* in the machine-vision world does not refer to hammers and tape measures. Tools are not even the calipers or "mikes" used by the human quality control person to make a measurement. Rather, tools are the graphical icons used in conjunction with the digitized image to define edges in the video-imaging process. They are called *gates* and *edge finders* and can be configured in various ways to make the video-inspection process consistent. The tools chosen can make or break a given inspection in terms of that inspection's validity and repeatability. Figure 10-18 shows common tools available to the vision system programmer. The selection of a given tool to define a point or edge on a given part's feature is crucial to the success of the measurement. Experience and predictability of all possible encounters for that given part or family of parts should be factors.

For example, let's measure a hole in a part using backlighting. The tool will be a window, with the leading edge of the hole being requested via our selection of the scan direction left to right rather than right to left. This tool is shown as example B in Fig. 10-18. This tool ensures that for every part's hole in that location, we should get the best point, or pixel, possible to use in the diameter calculation, predicated that our part is well-fixtured. If we were to use another tool or scan in a different direction, we run the risk of getting a point not representative of the hole's outermost point. It would therefore be a nontrue diameter and, depending on the tolerance, a possible fail dimension when the part's hole was actually good.

One reason to measure and inspect parts with an automated vision system is to collect data. This collected data is used by production, quality, and other individuals to ensure the parts and process are in good working order. This method of collecting data and information for later use in the factory is called statistical process control (SPC) or statistical quality control (SQC). All the aforementioned measurements are carefully chosen and collected to ensure that a family of parts is within tolerance and therefore sellable. The machine-vision system is the means by which the data can be collected and furnished as documentation. SPC and SQC are defined further in Chapter 13, along with methods of charting.

Lighting techniques

Just as the selection of the proper tool is important, so too is the lighting of the part and all the potential configurations of the part. Once

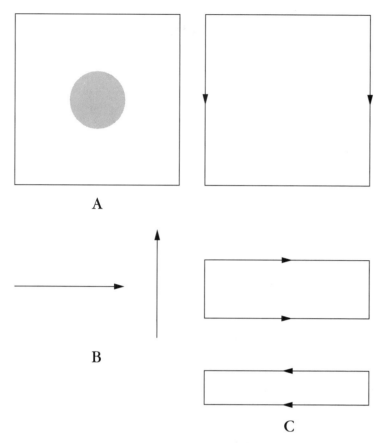

10-18 *Vision system tools and their uses in a program: A is the full field of view tool; B is used to select a single point; and C is the edge-finding window tool whose size and scan direction are selectable.*

lighting is in place, it must be located and fixed in position such that for all possible combinations of parts and ambient conditions, lighting is all encompassing. Thus, the machine-vision system, either online or offline, has a given program for a given part. Within that program are instructions for certain lights and combinations of lights. Some are turned on, and some are turned off, all dependent on what part is being viewed and what feature on that part must be lit. This program can be modified for variations to the part. It is easy to change because it is software, and the parts definitely change (which is why we do these inspections in the first place).

Once our lights are mounted, set, and adjusted, it is imperative that they not be moved, or the program must be adjusted. In addition, the lighting must be frequently inspected itself to see if any cleaning is required. Minimize any and all movement to the lighting when cleaning. Good lighting is a major factor in the camera's getting good, clean images for video processing. Care and maintenance play major roles in a machine-vision system's authenticity and acceptance as a qualified device in parts passing and failing.

Silhouette lighting, sometimes called backlighting, is the most common and most forgiving in terms of position. This light, which puts the part between the camera lens and itself, can be somewhat out of position and even somewhat dirty yet still provide enough light for the camera to get a good picture. This fact is not true with top or side lighting.

Other lighting techniques include through-the-lens top lighting to illuminate a feature that cannot otherwise be lit from below or behind. A ring light around the lens is sometimes a possibility, depending on the part. This projects light from above, but in a somewhat scattered way to catch the sides of a part. One must remember that the vision system's camera can only pick up well-lit images. Just like the human eye, the camera can only do so well in a dark environment, which is why lighting is stressed. Low-lux cameras are becoming more available today, enabling good video imaging in very low levels of light. Limits will still exist, however, and the best approach is to provide consistent lighting.

Fixtures

Fixtures depend on the part, and the parts depend on the fixture. Better said, fixture design is evolved around the size and shape of the part and how the part is lighted. The part's livelihood depends on how well the fixture holds the part in terms of firmness without deformation. Parts must be left in good condition after inspection so they can be sold. The vision system method of inspection is termed *nondestructive*. Parts are not discarded or wasted. When holding a part firmly in place via a fixture, the goal is to leave no detrimental marks on the part after the inspection is complete. This goal is common in the machine-vision inspection process.

Fixture design is a fine art. The fixture must have flexibility to accept a complete family of parts, where possible. Ideally, the perfect fixture could hold all the given parts made by a manufacturer. This goal is not practical because parts come in so many sizes and shapes. In addition, different features must be viewed on different parts. The vision-system fixture must be unobtrusive to the camera and lighting.

The fixture cannot be so large that it hides features that must be viewed or lit. The fixture must be designed such that the parts can be quickly inserted, clamped, and then removed. If this process takes too long, we might as well measure the part by hand! The fixture also should be lightweight so the inspector can physically handle it. More often than not, several different fixtures are in a quality laboratory because of the quantity of parts made at that facility. The fixtures not in use are held in a storage area until needed.

Fixturing is another discipline in the machine-vision application that should not be taken lightly. Good fixture design can make or break an application. It can become a costly part of the quality control process. Because of engineering, materials, and machining costs, fixture design should be given extra thought. If inspections can be achieved with fewer fixtures, time and money will be saved. Fixturing is different from part to part, from designer to designer, and from facility to facility. It must evolve around the part being inspected. A good fixture is a blessing in disguise!

Troubleshooting and specifications

Machine-vision systems differ from manufacturer to manufacturer. They also differ by application, whether they are online or offline, and by how each is programmed. Thus, troubleshooting a machine-vision system can vary. Luckily, there are many common components to any vision system. The camera, lighting, fixturing, and computer hardware are all similar in function. Checking wiring and connections to and from devices is always important. If serial communication is used, this link, both software and hardware, should be checked. Ensuring that components that should not move, do not move (i.e., lights, fixtures, etc.) can save lengthy troubleshooting later. When calibrating a system (yes, a measurement system somewhere along the line must have some calibration), it might be necessary to make adjustments. In some instances adjustments can be made in the software, while other times a physical adjustment might be required. Most of adjusting occurs upon initial startup.

Because the system is electronic, fusing, power supplies, and even electric motors might need checking. Dim lights might indicate wornout bulbs (they don't necessarily fail in an off or out condition), a dirty lens, or broken fiberoptic bundles. Computer-related problems can also occur: disk problems, monitor problems, etc. Consult the manufacturer's manual for these items. Likewise, a joystick, mouse, or keyboard might be used with a given system. Checking cable connections and verifying the item is clean can cut downtime. Probably the most important factor in vision-system uptime is that of cleaning and maintenance. Because it is a system dependent on clear, well-lit imaging, clean lights and lenses are paramount. In a factory, film buildup on these components can happen quickly. When it occurs, the system cannot work properly, and premature failure of hardware occurs. A deliberate, consistent maintenance schedule is recommended.

A machine-vision system, like any other automated factory-grade system installed today, must have a set of specifications that defines, in detail, its componentry and performance. Figure 10-19 shows a sample of specifications typical to an offline machine-vision system.

The future of machine vision

Several years ago, there were over 150 manufacturers of machine-vision systems and related products. Today, there are under 100. The field of machine-vision has taken the same route as the robotics industry. The interest and implementation of these types of products has been greatly reduced because so many other electronic types of equipment exist, today, that aid automation. There are also only so many capital funding dollars to spend each year. There was a flurry of machine vision and robotic sales in the 1980s as many companies strove to gain a competitive edge and automate. These systems hopefully are still in use. It is not to say that the machine vision and robotics markets are small, but a definite reduction in systems implemented has occurred from the thousands installed in the past decade.

Machine-vision systems are also finding niches for areas of expertise. Some vision manufacturers furnish systems in the microprocessor, semiconductor, soldering, and printed circuit board arenas. Others work solely with robotic guidance or pick-and-place systems. Still others can supply systems for virtually any application. Machine-vision systems have become somewhat specialized, and great emphasis is placed on the data collection means and packages that accompany the machine-vision hardware.

The gray-scale technology can only get better. Good contrast makes image processing easy. Many industries, however, do not always have that luxury. Because contrast makes for good edges, the tire and rubber industry is one industry that is a driving force behind elaborate gray-scale imaging and processing, evident in Fig. 10-20 of the good old automobile tire. The black on black edges, although dis-

SYSTEM DESCRIPTION AND SPECIFICATIONS

VIEW 1220

System Configuration
- System software including Simplified User Interface, geometric dimensioning and tolerancing, area centroid, manual mode, and histogram
- High-resolution 13" VGA color monitor (640 × 480)
- 20 MByte or 30 MByte fixed disk drive
- 3½" microfloppy disk drive with 640 Kbytes (formatted)
- Alphanumeric keyboard with dedicated function keys for system control
- Three-axes joy-stick with "ENTER" button
- Granite base and arch
- X, Y, and Z stage assemblies with 0.000040" resolution linear scales
- DC-servo stage positioning system with 0.000040" positioning repeatability
- High resolution CCD solid-state camera
- 256 gray-level image processing
- NIST (NBS) traceable camera calibration standard
- Microprocessor and math coprocessor with 2 MByte RAM
- (3) RS232 communications interfaces
- 6.6X lens with 0.040" field-of-view (additional lenses available as options)
- Edge, surface, and programmable Ronchi Grid autofocusing
- Reflected light, coaxial, and backlight illumination
- Automatic temperature compensation

Specifications

	1220-12	1220-24
Linear Accuracies	12" × 12" × 6" stage X axis ± 0.00016" Y axis ± 0.00016" Z axis ± 0.00020"	24" × 24" × 6" stage X axis ± 0.00032" Y axis ± 0.00032" Z axis ± 0.00032"
Humidity	20% to 80% relative humidity, non-condensing	
Temperature	62° to 90°F Recommended operating temperature is 68°F with no more than ± 1°F temperature change per hour	
Worktable	Maximum velocity – 8.0" per second Workpiece capacity – 12" × 12" × 6" Total weight/load capacity – 200 lbs. Platform glass capacity – 15 lbs. evenly distributed	24" × 24" × 6"
Power	117V, 30A, 50/60 Hz single phase	
Dimensions	Data gathering unit — 48" × 36" × 77" Workstation stand — 30.0" × 33.25" × 54.25"	
Weight	2350 lbs.	2500 lbs.
Shipping Weight	3100 lbs.	3250 lbs.

System Measurement Software

Measurement Features	Part and Feature Alignment
	Calibration
	Comprehensive Measurement Software
	  Point definition
	  Line definition
	  Circle/arc measurement
	  Angle measurement
	  Expression measurement
	  Geometric dimension and tolerancing

Statistical Functions	Basic descriptive statistical analysis package (SPC software available)
Image Processing Tools (256 gray-level processing with 485 × 512 pixel array)	Single Edge Point Minimum/Maximum Point Multiple Point Line Multiple Point Art Multiple Point Circle Outlier Removal Area Centroid Histogram
Manual Mode	Ability to use all image processing and measurement functions

VIEW
VIEW Engineering, Inc.
1650 North Voyager Ave.
Simi Valley, CA 93063-3348
(805) 522-8439
Fax (805) 527-1953
Telex 182-304

© 1990 VIEW Engineering, Inc. All rights reserved.
Descriptions and features subject to change without prior notice.

10-19 *A sample specification for an offline machine vision system.* View Engineering, Inc.

cernible, would not be possible without good gray-scale technology in machine-vision systems. Repeatability and the ability to see other features on a tire (such as lettering) would be difficult to achieve. Any company that manufactures black parts would love to exploit machine vision. This market will continue to grow.

The future of machine vision

10-20
Black automobile tire.

Machine vision has also evolved to complete three-dimensional capability. Laser technology has brought much of this capability to fruition. In the past, it was difficult to get good, consistent measurements and inspection of height, or Z axis, features. Parts had to be rotated or inspected in multiple views, or devices such as the Ronchi grid had to be used. Today, this dimension is not a concern. In addition, higher-throughput computers have made the vision system even better. Extra capability benefits from faster computing speeds and better storage and data manipulation.

Other areas where machine vision will continue to get better are in the camera, communications, and lighting techniques. Better lux cameras, or those that can take good video in low-light conditions, are emerging constantly. As costs (that other factor) continue to become more favorable, implementation of these new cameras will increase. Networks and plantwide communication schemes are becoming the norm today in the factory. Vision systems must have the drivers and full capability to link up or down with many protocols and LANs. Integration into other plant electronic devices is becoming a basic requirement of the vision system. Connectivity to robots, PLCs, other sensors, and computers is expected and is only going to become more and more important as we move into the twenty-first century. As for lighting, longer life expectancy and more industrially hardened devices are emerging. Prism technology, fiberoptics, and lasers are creating better lighting schemes that are more compact, consistent, and flexible.

Obviously, as computer-aided drafting (CAD) systems become standard, more demand emerges for attribute data to be transferred from one device to another. This demand requirement helps to drive more compatibility between systems and makes information available for not only machine-vision systems, but also other automated machines in the plant. Tracking the part and part assemblies through the plant is the desire of all involved in the process because duplication or creation of more work is to be avoided if possible. Care and thought must go into the design and assigning of attribute data to a given part, knowing what is needed through the production line. This concept follows all the way through to the SPC and SQC sectors of the factory.

All and all, machine vision is a major contributor to factory automation. Having received its head start in the automotive industry (like most automation systems), the vision industry has branched out to other markets. A big reason for the expansion is the faith and reliability of those initial systems. When it first came onto the factory scene, machine vision had some bugs that needed to be worked out. Systems were applied where they should not have been. Some parts were impossible to inspect, and certain features were tough with the vision technology available. Computer hardware could not keep up with the burden of all the processing required in video inspection. But hardware and software became better and more reliable. Newer methods for inspecting hard-to-view features were developed, and the vision system gained respect. With this respect came new orders and further implementation. Machine vision has become an integral part of industrial electronics and factory automation. It will be around for many years to come.

Bibliography and suggested further reading

Beiser, Andrew. 1973. *Physics*. Menlo Park, CA. Cummings.
Lenk, John D. 1992. *Lenk's Video Handbook: Operation and Troubleshooting*. Blue Ridge Summit, PA. TAB/McGraw-Hill.
Safford, Edward and McCann, John. 1988. *Fiberoptics and Laser Handbook*. Second Edition. Blue Ridge Summit, PA. TAB/McGaw-Hill.
Shigley, Joseph E. 1977. *Mechanical Engineering Design*. Third Edition. New York, NY: McGraw-Hill.
View Engineering, Inc. 1993. *Voyager 1000 Tutorial*. Simi Valley, CA.

11

Machines and system integration

Throughout this book we have analyzed the pieces that constitute industrial electronics and factory automation. Once we have the pieces of the puzzle, they need to be put together into a usable form. Sometimes, it is in the form of a machine made up of many electrical, computerized, and mechanical components. Other times, the pieces compose several linked machines. This link might be via a communication medium, or merely the fact that one machine interacts with another. Bringing all the pieces together, tying them together, and making them work is the integration phase of automation. This task is becoming more and more difficult as technology changes so rapidly.

Before we march ahead into the automation world, we need to look at how and why we have reached this point. Machines have always existed in some form ever since the invention of the wheel. The difference between the machines of yesteryear and today is in the electronics. The mechanics have changed somewhat but are virtually the same. Steel is steel, and motion is motion. Controlling that steel and motion is the new element in the equation. Of course, that factor of "increased productivity" also creeps into the equation. Engineers and designers are constantly on the verge of defying physics whenever they can. Electronic control provides many possibilities to these inventors.

With electronic control has come a new discipline: the integrator. This individual, or group of individuals, has emerged to provide the specialized expertise necessary to combine, program, and coordinate all the pieces. Because the technology is so diverse, these specialists must make systems out of dissimilar components. Individual components are specified and even purchased for a particular machine or purpose before the integrator is called in. The integrator must then make everything work. The better scenario is of course to involve the integrator from the beginning. Many projects are even awarded to a systems integrator based on the integrator's ability to provide the solution for a fair price, which is termed *turnkey*. Turnkey means the customer simply walks up to the machine and turns the key. The machine then starts and runs—simple as that! Of course, because we are dealing with precious design and procurement dollars, it is not always a completely new installation. Parts must be reused to save money—the classic machine retrofit (which can cost more in the long run). Integration, retrofits, turnkey installations, and why machines run (and sometimes not) are explored throughout this chapter. Using the technology discussed throughout the earlier chapters in this book helps explain how machines, processes, and production come together in the factory.

The steel, computers, motors, sensors, drives—all are the components that eventually become the machine. The integration of these components makes a system. The machine becomes a system when the mechanical parts are merged in harmony with the electronics, which is referred to as the machine system or the integrated machine. It can have many shapes and sizes. It performs a multitude of tasks, many of which have not even been considered to date. The integrated machine is that mechanized compilation of smaller units created to perform tasks previously assigned to human workers. These machines do tasks in environments unsafe for humans. They perform tasks many times faster than the human and without tiring. They can lift greater loads and only need power to keep going. These machines are known to us by many different names.

Robotics

The field of robotics is both interesting and vast. It is interesting because it has a history that goes back several decades, perhaps to the Industrial Revolution. It could be argued that forms of robot mechanisms existed as early as the water clock, which recycled itself, in early Greece. The field of robotics is vast because it can encompass many disciplines—electricity, mechanics, and computers. Most machines could be called a robot. If a robot is defined as an automatic device that performs a task a human might do, we have a machine. All robots are called machines.

A log cutter, grinder, and hydraulic press all could be called robots. But they are not. The line is drawn in the definition of the machine as a robot when elaborate computer control is incorporated that allows for performance unachievable with traditional mechanical means. This performance can be likened to the numerically controlled (NC) and computerized numerically controlled (CNC) machine tools. These machines had independent control of multiple motors via a program. Their physical layout was suitable to adequately machine their parts. There was no need to make them look like a human body, but they actually performed robotic functions and moves. The robot of today has many components that mimic human functions. There are robot arms that can move in many directions like the human version, vision systems that resemble the human eye, the computer (brain), motors (muscle), and sensors (nerves). It's a good thing we were not conditioned to think that a robot looked like an automatic dishwasher! That device performs the task of washing dishes, which was previously done by humans and now even has microprocessor control. The field of robotics is indeed interesting.

The image of a robot has been instilled in us by virtue of the Czechs. The Czech word *robota* is believed to be the root. *Robota* means forced work. Theatrical presentations have portrayed mechanical workers and left certain images with us as to what we conjure up when the word *robot* is spoken. Most of us think of mechanized human-like figures, whereas the reality is that a robot is nothing more than some metal, motors, and control that performs a mundane task over and over. Not only do robots perform routine tasks consistently, but they also perform many dangerous tasks in place of human workers.

True robots are actually robotic systems, consisting of a controller, operator (human) interface, and the machine itself. There must also be an application. A robot is often completely designed around the needs of an application. Where the robot physically rests, how many axes are used, and how it is integrated with other input and output units must be considered. These issues have created different classifications of robots. There are mainly four types: the gantry, the articulated arm or arm robot, the SCARA robot, and the basic Cartesian style. Each is reviewed, but some common issues regarding each type of robot must be discussed first.

Every robot system has some common ground. They each require a source of power, some type of feedback, a set of instructions, and the ability to make a decision while running. The source of power is usually ac and is many times isolated. The feedback methods implemented are used in the decision-making process of the computer, better seen in Fig. 11-1. A feedback device furnishes a signal routed into the controller cabinet. The computer, controls, and even the amplifiers are often physically located away from the actual robot machine hardware. Most feedback devices are resolvers or encoders attached

306 Machines and system integration

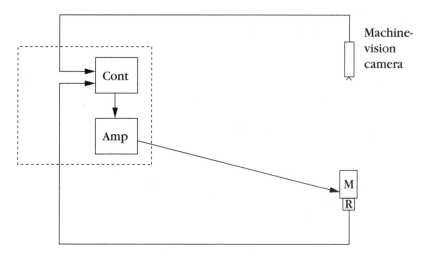

11-1 *Feedback signal from a motor compared to a vision system signal within the controller.*

to brushless motors. This feedback is brought into the controller, compared to commanded position and speed, and an error-corrected, amplified signal is sent back out to the corresponding motors. This sequence of events is subject to the prearranged program, and within that program is the ability to analyze the data and make fast decisions to correct the system.

In a robotic system, each line of motion is called an *axis*. Each axis must have a prime mover, or a mechanism, to initiate motion. This motor is usually a brushless servomotor. The feedback device is connected directly to the motor shaft for the truest feedback available. Each motor needs an amplifier, or electronic drive, ahead of it in the circuit to control the voltage and frequency of power to the brushless motor. Coordinated motion control is necessary whenever the robot has two or more axes, which involves a shared-bus topology. Each controlled axis interacts with another axis via a master instruction set to affect all amplifier signals out to the respective motors. In this way, the servomotors end up in a final location together. This scenario can be seen in Fig. 11-2 as the Cartesian coordinate system is employed for two axes.

Figure 11-3 shows a three-axis system. Here, the final destination has an *x*, *y*, and *z* component. Many robotic systems have the capability to be taught a position. The teach-type approach is simply this: using a hand-held terminal with keys for jogging, or moving, any of

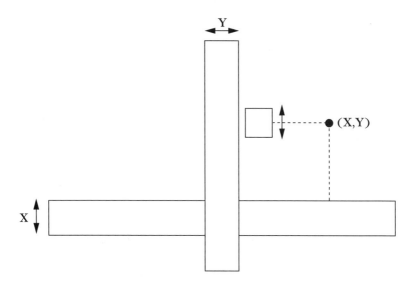

11-2 *A typical X-Y Cartesian move to a point.*

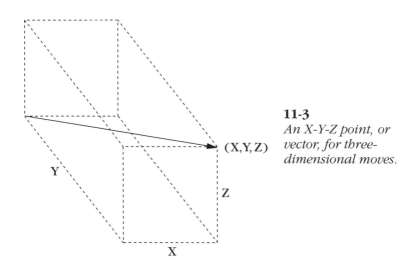

11-3
An X-Y-Z point, or vector, for three-dimensional moves.

the motors, the final X, Y, Z location can be set without needing to move each motor separately, although the move is often accomplished in this fashion to ensure the exact, desired location. The x axis motor is moved to its location according to the chart. Y and z are also moved. At this time, a key is struck on the pendant indicating that this is where, in the program, we want these three motors to move to. Upon hitting the key to save the locations, the controller then reads the resolver position off each corresponding motor, and the position is logged into the program as a point. The same procedure is repeated for every point after that. In this way, the robot system is taught where all its moves are and when to make them.

Depending on how well the walk-through was done during the teach mode, the actual move can look like a vector. The line of motion appears to be a straight line from point to point, rather than first moving the X axis motor, then the Y, and then the Z. Vector movement saves running time when the program executes. In addition, circular interpolation might be needed by the robotic controller, especially for circular moves. It is sometimes necessary to rotate about a part being ground or finished on many machine-tool applications. The rotation must be smooth, or the finish on the part can be laced with chatter marks. These marks are unwanted, and the part must then be reworked. Chattering happens when the servos make discrete X and Y steps when making the move. The desired effect is to have the controller interpolate as the motion is occurring and make the discrete steps appear to be nonexistent by breaking them into even smaller units. This function is challenging for any robot and is shown in Fig. 11-4.

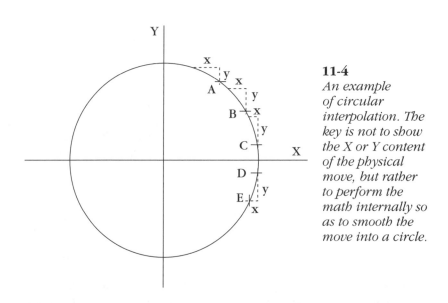

11-4
An example of circular interpolation. The key is not to show the X or Y content of the physical move, but rather to perform the math internally so as to smooth the move into a circle.

Other important functions of the robotic system are move speeds, accuracies, and load-carrying capacity. These can also affect one another. Heavy loading of any motor typically makes it react a bit more sluggishly, which is actually a result of movement speeds being slowed to drive the given load. Going from heavy to light loads and vice versa are tough robotic-control functions. These situations are hard to stabilize. Accuracies are thus affected by loading and speeds. High accuracy must also start with zero backlash mechanics. Gearboxes, drivetrains, and so on are permitted no play if the coordinated robot is to successfully position. Tolerance levels are fairly tight for robotic positioning systems.

Positioning is probably the most important function of the robot. Pick-and-place robots must pick up a part and place it in an exact new location every time. The controls, motors, and sensors must all work in unison. In addition, the end effector, which is the device at the end of a robot arm that actually picks up the part, must be sensitive and gentle yet flexible. The actual component that comes into physical contact with the part is the wrist of the robot. A delicate glass part cannot tolerate a rough-handling end effector—nor will the manufacturers of the parts. They will not accept broken parts. Care must be given to the selection of this portion of the robot. The design of the arm and its wrist must be considered for the part and how the move needs to be achieved. There are different joints at these wrists, much like the human wrist, which can rotate, twist, or revolve. Others can protract or elongate if needed. These joints are the most crucial because they come into contact with the part. After grabbing the part, the other, larger axes take over to move the part to another location. Still, the end effector must be delicate when setting the part into its new resting place.

One last factor to consider when selecting the robot for a given job is its work envelope, which can be seen in Fig. 11-5 along with typical lines of motion for multiple-axis robots. Robots should not be bulky, but they must be substantial enough to handle the loads and

11-5
The work envelope of a robot in one plane.

speeds required of the application. Like a boxer's "tale of the tape," the robot's specifications must include the work envelope, or reach capabilities, which must be considered when all axes are fully extended. A little extra reach capability should be considered when selecting the robotic system for the application. If you lock yourself in, you'll find yourself physically unbolting the robot housing from the floor and moving it. The work envelope can be the most important factor in selecting a type of robot for a given application. Each of the following four types of robots is discussed from that vantage point.

Gantry robot

The gantry robot provides the greatest work envelope by virtue of its design. As can be seen in Fig. 11-6, the gantry robotic system can move as far as the physical track system permits. Many gantry designs use long ball screws as part of each track. Each ball screw has its own servomotor and drive system. It is extremely important that both motors run at exactly the same speed—starting, accelerating, and stopping at the exact same time. If they do not, a condition called *racking* can occur, and the traversing member can bind.

The traversing member houses another set of tracks that run perpendicular to the outside rails. On these tracks still another set of ser-

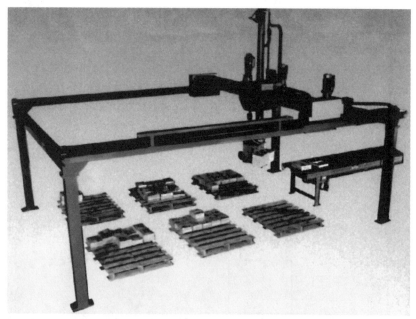

11-6 *Gantry robotic system.* C & D Robotics, Division, of Ohmstead, Inc., Beaumont, TX.

vos are used to gain up and down motion and hold any end effectors. The gantry system's controller is usually housed below, as this type of system often encompasses an entire work area. This system is typical in automobile assembly production lines. A car, being a large unit, moves into the work envelope of the gantry. The gantry can be used to punch holes, rivet, paint, or even weld. The gantry can be taught positions and moves as its program. One important requirement of the gantry is to cover several feet in a few seconds because of its outer rail design.

Articulated arm robot

The other types of robots are smaller in physical size and have floor or base mounts, including the articulated arm. The articulated arm robot is the most complex of the others similar to it: the SCARA and the Cartesian. It has the ability to position in the world coordinate system. In other words, it operates in multiple planes. It can get to practically every X, Y, and Z point, both positive and negative, within its work envelope, as seen in Fig. 11-7. This type of robot often comes as a six-axis unit, capable of three additional axes of motion about the wrist. This added capability is necessary for specialized tool functions. These types of robots offer tremendous flexibility in the loading and unloading of parts into other machines. With added functionality, of course, comes extra expense.

11-7
Articulated arm robot.
Staubli Unimation

SCARA robot

Another type of robot similar to the articulated arm unit and often confused with it is the SCARA robot, shown in Fig. 11-8. It has quite a bit of the work envelope of the articulated arm but with one major

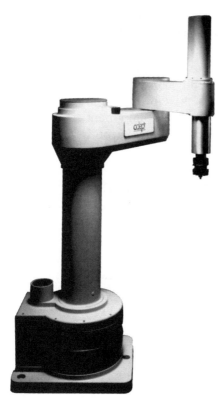

11-8
SCARA robot. Adept Technology, Inc.

difference—it only functions in one plane. A closer look at Fig. 11-8 shows that the rotating elements are only moving in one plane. One plane limits the work envelope, but the cost of such a robot is lower and many more can be implemented in an assembly line. One can be seen in action in Fig. 11-9. Multiple SCARA robots can be seen in action on an assembly line in Fig. 11-10. Applications of this type of robot include assembly-line insertion and removal of components, packaging, and light-duty drilling. Vision systems are often used to orient parts coming down the assembly line.

Cartesian robot

The last robot type in this group is also the least complex. It is the Cartesian type robot and is a coordinated linear actuator. It is shown in Fig. 11-11. It is a low-cost solution to basic robotic control for material handling, rectilinear (simple X and Y moves), and dispensing applications. It has a much lesser work envelope but is modular in design. It varies from the brushless servomotor concept, using linear

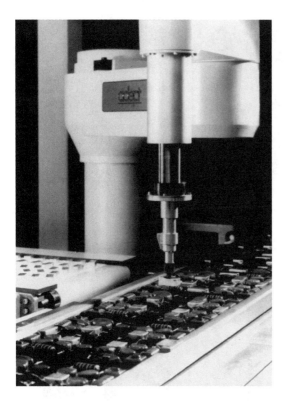

11-9
A four-axis SCARA robot in action on an assembly line. Adept Technology, Inc.

actuators to keep costs down. Simple positioning is possible with this type of unit.

All of the aforementioned robots represent numerous advancements in technology. The robotic industry in general, however, is not where it was 10 years ago. In the mid 1980s, there were well over 100 manufacturers of robots and robotic systems in the United States. Today there are but a handful, and these are more-specialized, computerized machine builders. In addition, many units are imported, and many units installed 10 years ago are still in operation. Users have also become "robot smart" and can implement controls onto machinery to gain flexibility and automation by themselves.

Current robot design

Robot designs have become sleeker, and more capability can be packed into a smaller package. For instance, robots now are expected to have all cabling, conduit, piping, and components housed within the robot itself, adding protection and extended life to the installation. The robot systems are expected to handle interpolation, load changes, and be more or less self-adapting to both normal and un-

11-10 *An automated packaging system with 11 robotic workcells.* Adept Technology, Inc.

11-11 *A linear motion module or Cartesian robot.* Adept Technology, Inc.

usual situations. Fuzzy logic, neural systems, and artificial intelligence are being looked at closely for the robotics industry. Computers, software, and controls continue to get more powerful. As long as the requirement is there to protect workers in hazardous locations, the robot will continue to be implemented. Its use is commonplace on the assembly line, but the workers the robots replaced have actually been displaced. The robotics industry has however created numerous adjunct jobs relating to industrial automation and electronics. The industry is truly a multidiscipline industry. A look at all the individual electronic devices discussed up to this point are virtually all applicable in some way.

Other machine systems and applications

Every chapter of this book has addressed one component or another that could be found in the average factory. From the incoming electricity to the packaging robot, machines and processes are how factories keep producing. Electronic controls and equipment keep machines and processes in line. Without them, the demand would far outweigh the supply, and many times today it seems that way anyhow. This perception could be because of just-in-time manufacturing philosophies and the edict that raw material and stock inventories must be kept to the bare minimum. Delivery cycles of products are thus extended, and it appears to the customer that factories cannot keep up with demand.

Many electronic machines are capable of higher speeds and producing more product than before; however, today's needs come from a world economy and marketplace. There is more demand. One day a machine produces a certain product, and the next day a different product is produced on that same machine. This concept is production planning and is based on new orders and demand. Although electronics has brought with it the ability to minimize retooling times by just changing the software, other factors do not allow instant changeover from product to product. Thus, the production planner runs as much of one certain product as possible, as long as the machine is cranking that product out.

Some machines make base products sold to other manufacturers for further processing prior to becoming an end product. These are called *coordinated web converting* and *processing lines*. Other machines merely perform specific functions, such as grinding, finishing, drilling, and milling parts in the manufacturing process. These machine-tool systems often do not make a final end product, but rather machine a part prior to its use in the final assembly. These machines can perform work on the metal and steel components within a large plant. These components might come off a machine within the facility and need to be repaired or reground. Several machine tool units in a given facility can each have a designated function to perform on a part as it comes down the assembly line. These units are called *machining centers* and often need a master cell controller that governs the control and operation of the several individual machines.

Literally thousands of different machines are in place in industry today. It is safe to say that no two are alike, even if they were originally purchased from the same original equipment manufacturer (OEM). Each machine owner has custom needs for that machine, and if those needs are not implemented by the OEM, the users make those changes themselves. Tying a machine into the supervisory control system and interlocking it with standard plant safety mechanisms definitely makes it unique. Control techniques and applications thus vary from facility to facility, even if the same final product is manufactured. Thus, we can now look at a few individual applications of electronics and controls on machines.

Web-converting line

The first machine is sometimes not called a machine. It is called a *line*. More specifically, it is called a *web-converting line* and is shown in Fig. 11-12. It coordinates the activities of many motors and other machine functions. The line, or machine, is converting a web of material as it unwinds a roll of raw material. The web can be slit into narrower strips, coated and dried, laminated to another web, painted or printed on, and then wound into a final roll.

A line such as this needs high-speed, high-response control equipment to run the motors, electronic drives, and other machine

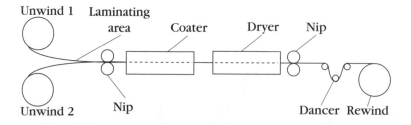

11-12 *A coordinated machine for laminating, coating, and drying paper products.*

controls. If not, material, or webs, break or unroll. The goal is to run the process at the fastest speeds possible. Thus, controllers must coordinate between devices, sometimes down to the millisecond range for electronic and processing response.

Machine tool, NC, and CNC industry

Other machine applications that use electronic controls extensively are the machine tool, NC, and CNC industry. By the mere virtue of their nomenclature, these machines have computerized front-ends and exhibit coordinated control. A good example is that of the grinding system shown in Fig. 11-13. In this system, a part must be presented to a grinding wheel and its surface ground to a predefined finish. The part must turn slowly and in conjunction with the grinding wheel. Another axis must move the part back and forth against the grinding wheel to obtain a consistent grind. Several passes are made until a smooth finish is achieved. Obviously, the three axes of the machine must coordinate to minimize the number of passes, back and forth, that must be made. The fewer the passes, the more pieces can be ground in a given day.

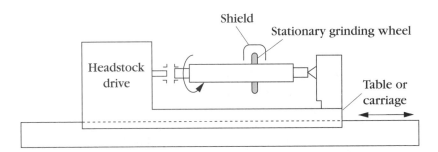

11-13 *Computerized grinding system.*

Sometimes, the machine tool grinder is part of a complete machining center, as found in an automobile engine manufacturing facility. Engine blocks move down the assembly line into this machining center. The control for this work area can be seen in block form in Fig. 11-14. Once the block is in the center, it is routed to different stations to have various work performed on it. Multiple blocks can enter the machining center at one time. A master controller, or cell controller, keeps track of all the engine blocks within the center and what work is being performed on each at any given time. A buffer area is where engine blocks, both coming in and going out, stages, or holds, units until the

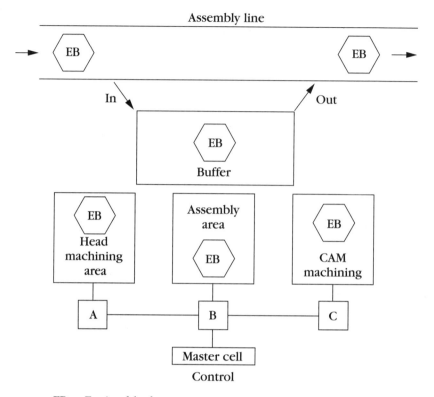

11-14 *A complete machining center with individual workcells.*

next station is ready. The master controller of the entire machining center is in charge. It is the traffic cop. In this way, the machining center is not a bottleneck to the assembly line. At one station, the engine block might be drilled and honed out. At another station, the heads might be ground. In still another part of the machining center, the engine might be fitted with its crankshaft or connecting rods. Further machining and assembly can be incorporated as required. Each machine within the center must be electronically controlled and able to communicate with the host supervisor. Individual machine functions, or programs, run on the part upon the go-ahead from the supervisor. This center is a good use of electronic controls in the factory.

Presses

Another application for electronic control and feedback devices is that involving presses. Presses can mold plasticized parts, or punch, stamp, or bend metal. The function is the same, applying several tons of pressure to a material to form the desired piece. One application uses a hydraulic press to form plastic parts. The controller in this case requires feedback from a positioning sensor to know where the press cylinder is and when to decelerate to get a smooth, even pressure on the plastic to achieve an even disbursement of material for uniform thickness. This application is shown in Fig. 11-15 and involves controlling a single hydraulic servo rather than an electric motor. Several analog inputs and outputs are required for temperature and pressure control to govern the process. Faster cycle times are achieved and scrap reduced with this electronic control.

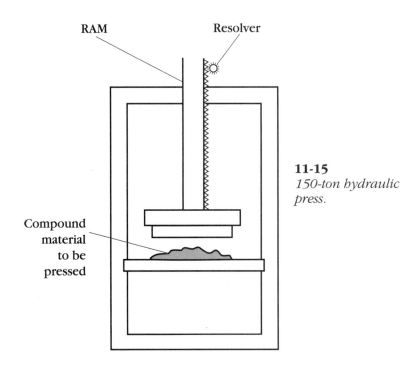

11-15
150-ton hydraulic press.

Electronic controls are only justified if their initial costs can be recovered in better performance, more product, and less scrap, all meaning return on investment. Implementing electronic controls is a challenge, especially if a facility has never used a particular computerized device in the past. Several concerns exist. As with any system, the issues of electronics, mechanics, and software must be addressed. Although many managers think electronic control and automation are very good ideas, they sometimes do not understand what is fully required to implement this equipment, which has spawned a new breed of specialists called system integrators.

System integrators

The system integrator might be resident at a particular factory as an employee. He or she might be part of a design and engineering group within the company. With so many corporations downsizing their engineering staffs and outsourcing these services, however, the system integrator is usually a contracted entity. The integrator is often fully specialized on certain applications or electronic products. Many times, the system integrator is a former employee of the company for which the contracted engineering is to be done. The integrator knows that company's rules and procedures and even knows quite a bit about the machines and controls already in the plant. And this is why companies are downsizing; they do not want to carry the overhead of the employee when they know that talent can be available on an as-needed basis (although that hourly rate might be higher for a shorter period of time).

The integrators and their service is fast becoming a big industry in the 1990s. A manufacturing facility gets new budgets every year. Projects are planned to be completed within a given year's budget. If the facility no longer has the expertise in employees to implement a system such as that required, a system integrator or a supplier of electronic product with system engineering capabilities is called. Each has expertise in specific products, software, and machine applications. This technique comes into play more so for retrofits of old or existing machines, which can be a nightmare. If the wrong integrator is selected or the electronic equipment is misapplied, the project can take forever to finish and might not work the way everyone had hoped. A good case in point is a fancy, high-tech baggage handling system that was to be implemented into a new airport. The baggage handling system did not run properly, and this held up opening the airport. Millions of dollars were lost. This type of situation happens very often on a lesser scale at many plants and factories around the country, which is why retrofits, turnkey systems, and integration must be fully qualified before a company can march forward with an idea.

A system integrator checklist is important when embarking on a project. System integrators often imply they are well-versed in a particular software language when they only have a manual and some source code and have scrolled through some programs. This minimal knowledge does not mean they can quickly and effectively write, debug, and run a machine software program. Experience is difficult to qualify with an integrator unless the integrators are known to the end user or come highly recommended. Software and programming hours are hard to track, and it is hard to see whether any real progress is being made. Interrogating integrators as to their expertise and application experience is always recommended.

As for retrofits, these are always interesting because so many things can be overlooked. Retrofits also carry with them the surprise factor. Because most are older machines, many components being reused do not make electronic controls happy. If a positioning system is dependent on feedback from a mechanized device and that device has backlash, or play, between components, feedback cannot be consistent. The controller only corrects to the signals it receives, and it is so easy for someone to say the controller doesn't work. That answer is the easy excuse, and further investigation by someone more knowledgeable can provide the proper solution.

The systems integrator is typically a person or group of persons capable of acting as a contracted third party to implement a system. These individuals are engineers or programmers who have certain specialized skills for the application. Because so many disciplines interact in machines and systems work these days, integrators can and do have many specialized talents. For instance, motion control applications involve knowledge and expertise in many areas. The American Institute of Motion Engineers (AIME) offers a certification for individuals in this industry. The certification process does not consider education and experience with electric motors only; computer controllers, PLCs, drives, sensors, and many other technologies are part of the scope. Remember the term automation engineer? They are the new breed of specialist needed in industry today.

Systems integrators provide solutions. They do not promote certain types of hardware unless the end user wants it or the integrator feels it is the only acceptable hardware. The integrator promotes the hardware that can do an effective job and that they are confident they can implement. Integrators can offer a turnkey solution or work together with plant electricians and technicians to install electronic equipment. A turnkey system is often preferred by the end user, but is harder to implement by the integrator, which is often a subject of misunderstanding. The end user expects to go to the machine, turn the key, and have the machine make its parts. The integrator might have a different interpretation. A complete, written specification of what is required should be agreed to by both parties.

Why do machines stop running?

Once a machine is started up and running, everyone is happy. If there were problems with late delivery or the actual startup, all is forgotten (for the moment) as parts come off the machine and out the door. There usually comes a time, however, when the machine stops running. There is never a good time for a machine to fail, especially in the production environment. When a machine quits running, the technicians spring into action. It is their job to find out why it stopped and get it back on line, making parts again.

Unfortunately, the troubleshooting process is not always clear. To further complicate the picture, many times the problems and reasons why the machine goes down are unexplainable. Try telling your boss, after losing 4 hours of production time at a thousand dollars per lost hour that the machine just came up and ran after you hit machine reset. You might lose credibility, but it might not be your fault. Electrical noise, intermittent disturbances, and other factors come into play with sensitive electronic devices. The Ishikawa diagram, a simplified troubleshooting tree for machines, is shown in Chapter 13, Fig. 13-8. It provides an overview of where to look for problems and what types of problems can occur. It is not meant, however, to be the full, complete source for troubleshooting. That source should be the service manual, which should have been supplied with the machine and controls. One copy should be left inside the electrical enclosure near the machine along with any pertinent drawings. Another copy should reside in a file room or manager's office.

Sometimes, the unexplained remains unexplained. Until the next occurrence. Eventually, if records are kept, a trend may develop and steps taken to correct the trend. Many electronic controllers now have battery-backed historical log functions for handling faults and diagnostics. When a fault is incurred, the machine's controller logs the time, what the fault is, and how the machine was reset, providing history, or trending, in lieu of keeping a written record at the machine that usually gets dirty, lost, or forgotten. If the plant technicians can become knowledgeable about a machine and its controller, the high-priced manufacturer's field service personnel are not summoned. These trips can cost thousands of dollars, and simple solutions are often found.

Downtime analysis

Good downtime analysis and time studies are only usable and reliable if good records are kept. Bad data in equals bad data out.

Records on the plant floor are often incomplete and misleading. Outside personnel can be hired to come in and conduct analyses, but it can be expensive and unpredictable to have someone watch a machine. Disciplined machine operators and technicians can log good, pertinent data for later use when a problem occurs. A sample downtime analysis log is shown in Fig. 11-16. This format can be used, but it can be customized for a particular machine or purpose. Once data is collected, one can summon the manufacturer of the machine or controller and productive troubleshooting can commence. When a user calls a manufacturer with incomplete information about what re-

Downtime Analysis Log

Company Name:

Address:

Machine Name:

Powered by:

Motor Data:

Incident	Date	Time	Problem or fault	Solution	Initials	Comments

11-16 *A sample downtime analysis chart for any machine.*

ally happened, the manufacturer has no clue as to what could be wrong. This scenario frustrates both parties, and the fuses get shorter. Lots of good data and good, preliminary troubleshooting (using the manual provided) can go a long way in getting the machine up and running and keeping it running.

Training, spares, warranty, and field service

Training, spares, warranty, and field service are fairly straightforward and a fact of life in the factory. Employees need training, retraining, and more training. Spare parts must be inventoried to minimize downtime. Warranties of electronic products should be for longer periods. Field service should not cost as much as it does from a manufacturer. All these statements are easy to make and sound the feelings of many people in the plant, from management to the operators. All are sensitive, expensive realities of industry today. Continually evaluating each to know where one stands with personnel and hardware, and not neglecting any issue, helps keep a user of specialized, electronic equipment current. Know where you stand!

Training has never been more important in the manufacturing facility. With so much computerized equipment, there is good news and bad news. The computer offers better diagnostics and troubleshooting and makes the control scheme cleaner. On the other hand, computers need a program, are sensitive to electrical disturbances, and if the main CPU fails, how does one troubleshoot? Employees change jobs and functions within the industry. An employee who is trained on a particular system might not be available later, so other employees must be trained and cross-trained—technicians need to be trained on many different products and applications. Training does not end with a few days here and there; it must be ongoing. Technology is constantly changing and manufacturers change their designs to suit *their* needs, not necessarily the end user's. The technician must keep up with these changes. Yes, training is expensive to a company, but it is justified. How important is it to keep these machines running? If even the machine operator can perform minimal troubleshooting, the machine might run for an additional couple of hours. At $1000 per hour in downtime, a lot of training can be bought!

Spare parts inventories are one of those black holes that companies wish they did not have. Unfortunately, when a machine goes down, the first question asked is, "Do we have a spare board?" That question is asked by the technician doing the troubleshooting, by the manager, and by the manufacturer of the controller. The manufacturer says that you, the user, should be stocking these boards. Or, the manufacturer says that the boards you have in stock are obsolete. And so you are shipped two boards, one to use in the machine now and one to put in the storeroom as a spare. The probabilities are that the spare board will never be used. Thus, the spares inventory grows and grows. Technology changes and new electronic equipment is installed in the plant. More and different spares are placed into the storeroom. The inventory continues to grow. Manufacturers do not take back unused spare parts. That's just the way it is. One solution to the end user, if 24 hours of downtime can be tolerated, is to rely on the manufacturer of the machine to overnight delivery spares to you. Let them be your storeroom, thus saving thousands of dollars in spares inventory that might never get used.

Warranty issues always seem to arise. First, they arise when negotiating the purchase of an electronic product. One year from date of shipment is standard and covers parts and labor. Thus, when a board fails within the first year, it should be fully exchanged for a new one. Some manufacturers offer two-, three-, and even five-year warranties. And they should. Many electronic products have become fairly reliable, and there is no reason why longer warranties should not be the norm. Also, many times with systems projects, an electronic product ships and then sits in a receiving area for several weeks before use. The warranty period should start at usage time. Warranties can be negotiated, and longer warranty periods can be requested.

When all attempts fail to get the machine running with available spares and personnel, it is time to bring in the heavy hitters. These are the factory-authorized field service engineers who get paid from the moment they leave their house until they return to it. Their cost per hour can range from $100 to $200 per hour, depending on the complexity and rarity of the product. A captive audience with a down machine will pay big bucks to get back online. It is a hard pill to swallow, but it is the better of the two scenarios. Getting your own personnel trained and keeping them well compensated can head off these expensive service billings!

Conclusion

System integration is here to stay. If anything, the electronics, controls, and automation industries will continue to grow. New equipment will be developed and a company needs to keep up with the technology to grow. Technological change is occurring at an exponential rate. One or two important developments can affect the entire industry. New products will be introduced, quickly obsoleting that which is currently in the plant. Good systems integrators are hard to find. Once a good one is working with a particular end user, that integrator will almost appear to be a full-time employee of the end user because of the many hours spent there.

Old mechanical machines will be replaced with high-tech, faster models. Energy efficiency and reliability will be factors. Robots will continue to be implemented, especially in hazardous applications. Controllers will become more powerful, and technicians will constantly have to train, be trained, and educate themselves to keep pace with technology. Like it or not, we are on this course and must learn to accept the technology rather than fight it. We are driving the technology, or is it driving us?

Bibliography and suggested further reading

For further information on the subject try either the Robotics Institute or the American Institute of Motion Engineers. Their addresses are in Chapter 2.

12

Industrial safety

With electronics helping processes and machines run with very little human intervention, there needs to be a dramatic emphasis on safety. The contrary might be assumed because, with machines doing more, fewer humans should be required to make the product. This point is certainly arguable, but the bottom line is that if there were but only one human still working in the plant and all else was automated, that one human life must be fully protected. Safety is an issue that cannot go away. If anything, as we automate and use more electronics, safety is getting a lot more attention.

As we automate, one might think that minute items are overlooked and sometimes probably are, but attention is given to protecting the human worker. From the design of the lesser components in the system to the actual operation of the machine itself, safety issues abound. Attention to detail regarding safety is at almost every phase of the automation process. Many agencies, such as Factory Mutual (FM), the National Fire Protection Agency (NFPA), Occupational Safety and Health Administration (OSHA), and National Electric Code (NEC) are in existence solely to monitor industrial safety and procedures, set the standards, and ensure that installations are safe to all who work in them.

As new ways are developed to run a particular process, the safety of the worker must be kept in mind. A machine or process can be made to run faster to produce more, but we sometimes push to dangerous limits when attempting increased production. The "safety as we go" attitude should always prevail. As we design, install, and desire to produce more, it is much easier to design safety in at each step rather than deal with a human catastrophe later. Industry is presently doing a good job. The suppliers of the electronic equipment are addressing safety because of liability reasons. The industrial users themselves want to protect their friends and fellow workers as they design and procure automated equipment. Several agencies exist to aid and consult businesses when safety issues arise.

These agencies form a network of experts that set the standards to which companies all over the world consult. The agencies can conduct field and laboratory tests, while others are formed from corporate representatives who have safety or manufacturing backgrounds. These groups set the standards and form the codes and regulations by which safety issues are governed. In addition, degree programs focusing on industrial safety are available for the individual. One of these degrees is known as safety engineering and provides for integrating engineering safety into the product, process, or machine design. Its the old adage at work: An ounce of prevention is worth a pound of cure! Beyond the philosophy of building safety into a process, it is also necessary to implement products that have been built with safe components and have been fully tested. This is part of designing safety into the entire project. If a safe piece of equipment is selected, it is one less item to worry about.

Safety labeling

Safety labeling, safety certification, safety listing, or safety approved all refer to the same thing. A nationally recognized testing facility has determined through extensive testing and evaluation that a product is in compliance. In the case of electronic or electrical products, compliancy means that the product meets or exceeds documented standards for that industry. This step is just one in the acceptance cycle. The standards must be fully accepted by those purchasing the product and by those inspecting product installations locally.

When selecting an electronic product for a given application, it is important to know that the product meets certain safety standards for a particular use, usually displayed in the form of a third-party label (a label with the ensignia of a nationally recognized certifying body). This label means that the particular product or design has been extensively tested. It represents the experience, background, and reputation of a certifying agency. It can be difficult to sell a product in certain areas without this label. Inspectors seem to have an affinity for red tags, and they use them on a particular piece of equipment if they do not see a safety label on it.

The nationally recognized testing laboratories that fall into the category are determined by OSHA. OSHA has in place a program called the Nationally Recognized Testing Laboratory (NRTL) program that sets up agencies around the country to legitimately provide safety and performance testing on industrial and consumer products to acceptable standards. These agencies include Underwriters Labo-

Safety labeling

ratories (UL), ETL Testing Laboratories, NFPA, the Environmental Protection Agency (EPA), and many others. In Canada, it is the Canadian Standards Association, or CSA. Internationally, it is the International Electrotechnical Commission (IEC). These are all renowned agencies specializing in product safety, particularly for electrical and electronic products. As for internationally certified agencies, many United States laboratory testing agencies have agreements with companies in Asia and Europe.

Should one of the aforementioned agencies affix their label to a particular electrical product, elaborate testing has occurred. This testing might have taken months, cost a good deal of money, and, more than likely, many products were destroyed in the process. This process is the only way an agency can truly gain the faith of the consumer. It is often frustrating to the manufacturer of the product to endure such an ordeal, but laboratory testing of a product is a requirement. Testing is not always a big ordeal; it depends on the product being tested, the actual testing required, and if more than one type of model exists for the given product.

For instance, if a manufacturer wants to gain third-party safety certification on small 5-V power supplies, several samples might be required, as some units will be destroyed as part of the rigorous testing. For a more-complex item, such as a variable-frequency drive, one or two might be required. The testing of such a complex product is still rigorous, but the testing laboratory is sensitive to the cost of such a product and takes extra steps to test without destroying the product. The time it takes for certification is also dependent on the complexity of the products, the number of models, and extent of the testing. Some testing laboratories consider the need to expedite testing for a manufacturer. Sometimes, several months of testing can actually outdate the product's marketing edge that existed when the product was first developed. It is suggested that when soliciting independent laboratory testing agencies, request a detailed quotation for a scope of tests to be performed, and find an agency willing to work with an accelerated testing schedule. They are out there.

As a matter of fact, from the manufacturer's standpoint, it is very important to find a laboratory testing agency willing to work together with the manufacturer. This relationship is important for a variety of reasons. One reason is obviously the testing time and costs (lumped together because they interrelate). Another is that of noncompliance. What happens if a product being tested fails? It would be helpful and time-saving if the testing agency showed where and why a component or product failed. Some offer this help. Still another working-to-

330 Industrial safety

gether factor is that of future changes in product design. If a manufacturer comes out with a change in the product (which occurs quite frequently), does the upgraded product require full retesting? Some agencies analyze the extent of the changes and, if practical, merely change the original report, thus saving time and costs. Obviously, if the changes are significant, retesting might be the only option.

In today's industrial and manufacturing environments, if no safety label is on an electronic product, the manufacturer of that product must find one, or that product will see minimal implementation. Figures 12-1 and 12-2 show labels with serial numbers and different label designations. A *listed* label is used for products participating in normal listing and follow-up programs. Every unit bearing a label has its own serial number. The *classified* label provides for conformity to an NFPA standard. It also carries an individual serial number. These labels are often put on a product in the field. The label is the norm today for new equipment installations, especially in hospitals, schools, and other public places. Project specifications now require that only products bearing nationally recognized marks or labels be used in many installations. With so much in the way of insurance costs and liabilities, it is simpler and safer to require the label for one's own protection, present and future.

ETL LISTED
CONFORMS TO
ANSI/UL–508
XOXOX

ETL TESTING LABORATORIES, INC.
CORTLAND, NEW YORK 13045

12-1
ETL's listed label. ETL Testing Laboratories, an Inchcape Testing Services Company

ETL CLASSIFIED
SPRAY PAINT
BOOTH
XOXOX CONFORMS TO NFPA–33

ETL TESTING LABORATORIES, INC.
CORTLAND, NEW YORK 13045

12-2
ETL's classified label. ETL Testing Laboratories, an Inchcape Testing Services Company

Laboratory testing

For an electrical product to carry a label, it must go through a series of tests in a controlled testing facility. Figure 12-3 shows a laboratory testing facility. At facilities such as this, manufacturers of electrical

Laboratory testing 331

12-3 *ETL's testing laboratory.* ETL Testing Laboratories, an Inchcape Testing Services Company

products who need safety and performance testing must use the services and latest testing methods of these facilities. These are the agencies who can provide the testing and certification needed to meet industry standards. These tests are performed to documented standards. Extensive data acquisition and nondestructive and destructive testing at these facilities is their business.

A standard for industrial control equipment has been developed by ANSI and UL and is called Standard Number 508. This particular standard covers many of the items relative to most factory automation products discussed throughout this book. Electric motors and motion control products such as electronic drives, starters, programmable controllers, relays, industrial computers, and even the control panels assembled from the individual components are specified. Any device on the factory floor that is part of the process electronically has been through some safety testing during its development.

The testing procedure Standard 508 is lengthy and fairly complete. It is over 300 pages with addenda (and there are new addenda all the time). It has existed for years and has been amended accordingly, as new electronic products have been developed. While this standard is the most prevalent testing standard, others cover equipment dedicated to hazardous locations; temperature and refrigeration products; and analog, panelboard system products. To catch all other products that might have ambiguous functionality, the standard clearly states that if risk of injury or fire exists from a product, it shall be tested somehow and somewhere.

All equipment subcomponents must undergo certain basic tests for compliancy. Further, more specific testing is performed on the actual products. For instance, most electrical products have some enclosure and markings that must hold up in an industrial environment over time and are safe to the user from the start. What good is safe componentry on the inside if the outside of the product can harm a person? The subcomponentry then must go through many performance tests because the subcomponents are used in many different products. A transistor might be used in one manufacturer's starter, and the same transistor might be used in another's electronic drive. It must first be determined that the transistor subcomponent meets minimum standards of safety. These tests are performed on all product subcomponent devices. The tests include temperature, over- and under-voltage as well as full overload, short-circuit withstand testing, endurance and accelerated aging tests, wire flexing, and several others. Beyond these basic tests, further test requirements and procedures for built-up products from the subcomponents exist.

Also included with the performance testing of these subcomponents is a section on the rating and sizing of components. A device is said to be rated for 10 A of current. Does this mean the device can accept 10 A continuously, intermittently, or as a maximum? What happens after 10 A is exceeded? Does the device explode, overheat, or is there built-in overload protection or capacity? The standard seeks to clarify these questions by dedicating a section to rating the devices.

The device must also carry markings. Some markings are for ratings, while others are used to identify wire terminals. Other required markings are for caution when a condition can exist with a certain device. Parts of the standard must address markings and ratings, as it is necessary for a reference point in the industrial and testing sectors.

Specific tests and procedures are fully described, in detail, as part of the standard. The more complicated the product, the more testing is performed. For example, a programmable controller, besides having all its subcomponents fully tested, is checked for spacing, separation of circuit barriers, and bonding. It then goes through a series of performance tests that include temperature and overvoltage, endurance, dielectric voltage withstand, breakdown of components, and impedance testing. These controllers are then given ratings and checked for proper markings.

The standard is fairly comprehensive, and it has to be. Besides doing the testing, the laboratory provides listings, labeling, and periodic follow-up inspections. These follow-ups are usually every 3 months. Directories are kept of all the products that have been certified by the testing agencies and are usually available free of charge. The term

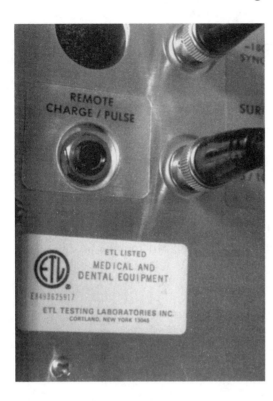

12-4
An ETL-certified product. ETL Testing Laboratories, an Inchcape Testing Services Company

listed goes with this type of labeling. An example of an electronic product with a laboratory tested label is shown in Fig. 12-4.

New electronic products are developed daily, and unless controls are placed on the safety and design of these products, the potential exists for many hazardous installations. Unfortunately, some hazards still exist and a method for dealing with these situations exists, also. This is field testing.

Field labeling

How safe is safe?
  Is that procedure safe?
  "That machine doesn't look too safe to me!"
  "Where's the safety label?"
  How often have these and other discrepancies arisen over whether or not something in the field is safe. Some issues can be resolved by simply consulting a reference or standard. Other times, it is necessary to bring in the experts. Code enforcement officials inspect large equipment installations in their geographic jurisdiction. They look for com-

pliance of product to local codes, specifications, and industry standards. They are another link in the safety chain. They red tag pieces of noncompliant equipment. Red tagging can mean that the installation of that product is prohibited, and operation of that process or machine is curtailed until the dispute or discrepancy is settled.

Several major building code enforcement concerns exist in the United States. They are regional and include the Building Officials and Code Administrators International (BOCA), seen mainly in the northeastern part of the country; the International Conference of Building Officials (ICBO) seen mainly in the western part of the country; and the Southern Building Code Congress International, Incorporated (SBCCI), seen mainly in the southeast. The code enforcement officials who go onsite to inspect large and important equipment installations belong to one of these organizations. They strictly look for safety and specification compliance. If the specification calls for a safety label on a particular piece of equipment and it is not there, they red tag that piece until it is shown to comply.

To settle safety-related disputes, companies go into the field to audit for compliance, perform onsite testing if necessary, rule on the nonclear-cut concerns, and apply what is called a *field label*. It is often simply the settling of disputes on whether or not a machine or process meets a specification. Often, the specification item causing the discrepancy was written into the documents because it was convenient or sounded good at the time. Specification items should be written into the documents to minimize risk, failure, and catastrophe. Specifications are becoming a vast and diverse challenge in the electronics industry when electronic equipment is applied and installed. Electricity is not only the root of all electronic devices but also a potential hazard to life and peripheral equipment.

Other instances of noncompliance might require onsite testing. Though certainly not preferred, it might be the only alternative. The manufacturers bear the costs, which are probably higher now and could have been avoided had they sought certification earlier. Nonetheless, onsite testing is a common occurrence with much of today's electrical equipment. An evaluation can be made by a recognized certifying agency and, if the product meets accepted standards, can be given a field-applied safety label, as seen in Fig. 12-5.

Municipalities, school, and public installations are the strictest. Although the product probably meets the safety standards, if it does not carry the label, the inspector will not yield or evaluate the product. The inspector simply inspects. If a noncompliance exists, a noncompliance exists. The problem is now borne by another party. Let the manufacturer beware!

12-5 An ETL field-evaluated label. ETL Testing Laboratories, an Inchcape Testing Services Company

Performance testing

In today's marketplace, there is much product competition. Decision makers buying products must make fair comparisons. There must be some method to provide other data about a product in addition to its safety content. This method is called *performance testing* and can usually be provided by the laboratory testing agencies who provide safety testing. The performance data can be gathered during safety testing. Performance testing includes assuring that a product is manufactured with high-quality materials and subcomponents. This testing is a byproduct of component testing done in the laboratory. With as much emphasis today on quality, it is almost necessary to have this type of documented information. In addition to quality assurance, laboratory performance testing can also yield facts relative to a product's energy consumption, reliability, durability, and what environments it can withstand. These issues can be paramount in determining whether or not a product is suited for a particular use.

Ironically, devices used in the workplace as safety items are also tested. Horns, sensors, lights, and other items integral to keeping the worker alerted and safe are routinely tested. Other workplace safety items such as helmets, fire blankets, and so on, have rigorous, destructive tests performed. These tests include flammability, shock, and photometric testing. Safety testing agencies for electrical products also provide performance testing of nonelectrical products. They have the facilities and equipment to do both and are usually recognized by industrial, governmental, and building code organizations.

Hazardous locations

In many industrial facilities, it is almost impossible to avoid some type of dangerous gas or liquid. Many processes simply must use these substances to make the product. The solution is to recognize where hazardous areas of the plant are, isolate them, and install equipment suited to that environment. Recognizing a hazardous area and placing it in a remote area of the plant is fairly straightforward. Classifying equipment for various environments is not. Environments can exist in the plant that are potentially dangerous because flammable gases exist. Any spark could ignite them and cause a fire.

The NEC, affiliated with the NFPA, has listed some standard classifications for hazardous areas. They are broken down by class, division, and group as pertaining to the type of fuel and so on. This breakdown can be seen in Table 12-1. Using this chart as a guideline, the designer can properly select electric motors and other electronic equipment for the application. When designing or writing specifications for a new project or retrofit, safety issues relative to environment can be incorporated by referring to these classes, groups, and divisions.

**Table 12-1.
Hazardous area definitions of the electrical code**

Class (or type of fuel)
Class I: Gases and vapor
Class II: Combustible dust
Class III: Fibers

Division (possibility of fuel being present)
Division 1: Present or likely to be present in normal operation
Division 2: Not present in normal operation

Group (specific type of fuel and gases/vapors of equivalent hazards)
Group A: Acetylene
Group B: Hydrogen
Group C: Acetaldehyde, ethylene, methyl ether
Group D: Acetone, gasoline, methanol, propane
Group E: Metal dust
Group F: Carbon dust
Group G Grain dust

The issue of something being explosion-proof is always worth discussing. This term is a misnomer of sorts. Spark-resistant might be the better term, and here's why. If an item is explosion-proof, one might surmise that, should an explosion occur, the device will still be standing and fully intact. The device was truly explosion-proof, right? Well, not exactly so. We want to eliminate all possibilities of explosive conditions. Unfortunately, industry must continue to use hazardous and volatile chemicals in its processes. The alternative therefore is to find ways to safely integrate machinery and electronics with these chemicals. Explosion-proofing should be that condition where no explosion can occur whatsoever. In reality, these events can happen, but industry has adopted the "explosion-proof" terminology, and we must live with it.

Another term relative to explosive atmospheres and making equipment safe for that environment is *intrinsically safe*. Intrinsic means to be completely contained within an area. Thus, an intrinsically safe installation factors into its design a scheme whereby all electrical components are made with proper housings, connections, and materials to deter fire and explosions. Housings of electrical equipment in hazardous areas should be substantial and spark-resistant. An intrinsically safe connection actually becomes a break, or barrier, in the electrical wiring circuit. An analog output barrier is shown in Fig. 12-6. Along with the analog-output intrinsically safe barrier, Fig. 12-7 shows a similar barrier for a typical discrete input. For controller installations, these are acceptable barriers for intrinsically safe environments.

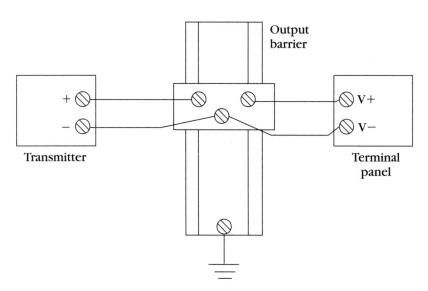

12-6 *Wiring scheme for an analog output intrinsic barrier.*

By minimizing sparking, overload potentials and keeping electricity from direct contact with hazardous gases, the potential for catastrophe can be greatly reduced. These types of environments in the factory do exist. Meeting local codes and standards for these regula-

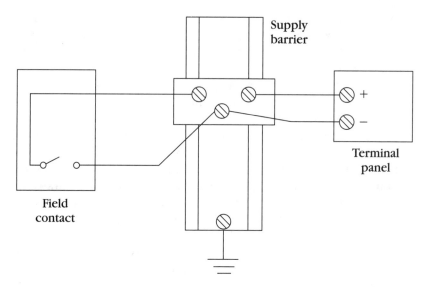

12-7 *Wiring scheme for a discrete input intrinsic barrier.*

tions allows companies to continue producing following periodic inspections. Obviously, the condition of intrinsically safe carries with it interpretation and can create disputes on its validity as a safety procedure. Each case is individual, which is why safety labeling and inspections come into play. The rules and regulations are there to save lives.

Emergency and fail-safe

Accidents do happen and so do emergency situations. When the loss of human life or potential for injury exists, we have an emergency situation. An accident is the result of the emergency situation. The solution is to avoid the possibility of the emergency situation, or, when one exists, give the workers the ability to avert it. Many times this means completely killing power to all electrical components. Sometimes, however, stored energy is present both in the form of stored electricity (as in capacitors) and in the form of motion (as with a machine that is at full speed). Bleed-off circuits can be incorporated to dissipate excess electrical energy, and mechanical means, such as brakes, can be used to stop machine motion. The real challenge is predicting these conditions and implementing the solutions before an accident happens, not after.

It is often argued whether or not to have dynamic braking as part of a motor circuit, especially if the motor controller is capable of regeneration. The regeneration of energy back to the mains allows for controlled braking of the controlled motor. With a loss of power, however, the controller is no longer in control of the motor. Power to run the control circuits is missing, and the motor keeps rotating. No regenerative control is maintained. This condition should be avoided if the continuation of that motor could harm someone or something. A means of avoidance is by implementing dynamic braking, which is a circuit with a resistor bank that has a contactor ready to close and divert energy to the resistors in the event of a power loss. Thus, it does not matter if the motor controller has control or not. Because of this method it is also a common way of braking even if a regenerative controller is not used, it is fully discussed in an earlier chapter. Sizing and circuit design are also included in that chapter.

An "emergency stop" pushbutton should exist on every machine in every manufacturing facility. If not, an unsafe condition exists. This E-stop pushbutton, as it is known, is usually red and larger than the other pushbuttons, but it can look different; it can be a pull cord or a foot pedal. The net result is that it is interlocked within the electrical power circuit in such a way that power to a motor or prime mover is terminated, thus stopping the machine as fast as possible. The E-stop should never go through the logic of a PLC or computer because this route might require more time to shut the machine down. That extra time, maybe even only a few milliseconds, might be the difference between an arm being lost and saved.

Another reason is that if a power failure occurs, the computerized devices will not be processing at all, so an emergency stop will not even happen. The best scenario is to strategically wire it into contactors that stop the power where best suited (to an electric motor, for instance).

Other safety features should include alarms and horns, not so much for added nuisance and noise, but for worker notification that a potentially dangerous stage of the process is about to happen, or that machinery is moving nearby in a loud and noisy area which might not be heard with all the background noise. These horns and alarms are sometimes accompanied with flashing lights. The horn is usually interlocked with a run or start command for a machine. Actual motion is delayed or prohibited until the horn has sounded and, in some instances, another secondary run pushbutton is depressed.

Fail-safe devices are those that operate in a somewhat reverse manner from what might be suspected. For instance, fail-safe electric brakes, when energized, do not perform braking. Rather, they apply braking force when power is absent. They are normally off until power is removed. Many contactors and relays are set up in this manner to shut off valves and other devices in the event of a power failure to ensure that dangerous conditions do not get worse.

Enclosures

Much of today's electronic equipment used in the factory needs a resting place. It is not acceptable to mount electrical devices such as relays and starters on machines and the walls of the factory. This procedure is not safe. It is also unsightly and probably lessens the life of the electrical product. Many industrial users, especially those without much electrical equipment in their plant, mount electrical products in the most convenient place in the building, which is usually on the wall in an out-of-the-way location. This mounting drives building code inspectors crazy. Several alternatives do exist.

Control rooms with restricted entry are common in facilities that have a substantial amount of electrical product in use. Electrical rooms, mechanical rooms, vaults, and outdoor buildings can all serve as ideal locations for most electrical and electronic equipment. In the plant itself, however, it might not be practical or economical to dedicate an entire room to a few electrical components. Thus, there exists the need for an enclosure, which is some type of box, floor- or wall-mountable, made of a metal or plastic that can withstand the local environment and provide a degree of protection to the workers and product within. Enclosures have become the accepted solution for housing most electrical products in today's factory.

The benefit of electronic equipment is that it does not necessarily need to be located next to an associated piece of equipment. Wire and cable can connect the two pieces electrically. Enclosures can be located away from the traffic and actual production areas in the plant. It still costs extra to run cable longer distances, but logical locations usually exist to place enclosures. Once a suitable location is chosen, it is important to select the appropriate enclosure for the application and actual environment. To aid in this process, standard ratings have been established by NEMA for degrees of protection, for both indoor and outdoor enclosures, shown in Table 12-2.

These ratings state the degree of protection required but do not always imply the recommended means of achieving that protection. For instance, an enclosure must be completely sealed and placed indoors. There might be heat-producing devices within the enclosure that do not work well (or at all) when the temperature increase becomes too severe. Thus, an air conditioner might be added to the enclosure for

Table 12-2. Environmental protection classifications

Type	Enclosure
1	Intended to provide a degree of protection against contact with the enclosed equipment.
2	Intended for indoor use to provide a degree of protection against limited amounts of falling water and dirt.
3	Intended for outdoor use to provide a degree of protection against windblown dust, rain, sleet, and external ice formation.
3R	Intended for outdoor use to provide a degree of protection against falling rain, sleet, and external ice formation.
3S	Intended for outdoor use to provide a degree of protection against windblown dust, rain, sleet, and provide for operation of external mechanisms when covered with ice.
4	Intended for indoor or outdoor use to provide a degree of protection against windblown dust and rain, splashing water, and hose-directed water.
4X	Intended for indoor or outdoor use to provide a degree of protection against corrosion, windblown dust and rain, splashing water, and hose-directed water.
5	Intended for indoor use to provide a degree of protection against dust and falling dirt.
6	Intended for indoor or outdoor use to provide a degree of protection against the entry of water during occasional temporary submersion at a limited depth.
6P	Intended for indoor or outdoor use to provide a degree of protection against the entry of water during prolonged submersion at a limited depth.
7	Class 1, Group A, B, C, or D hazardous locations, air-break—indoor.
8	Class 1, Group A, B, C, or D hazardous locations, oil-immersed—indoor.
9	Class 11, Group E, F, or G hazardous locations, air-break—indoor.
10	Bureau of Mines.
11	Intended for indoor use to provide a degree of protection against dust, falling dirt, and dripping noncorrosive liquids.
12	Intended for indoor use to provide a degree of protection against dust, falling dirt, and dripping noncorrosive liquids other than at knockouts.
13	Intended for indoor use to provide a degree of protection against lint, dust, seepage, external condensation, and spraying of water, oil, and noncorrosive liquids.

local cooling of the enclosure contents. The integrity of the sealed, indoor enclosure is not diminished with the air conditioner. Likewise, the same rating enclosure could also tolerate simple muffin fans pulling air in, over the devices, and back out in lieu of an air conditioner. Costs, interpretation of standards, and common sense prevail.

When discussing enclosures, an entire workspace can be enclosed. Many times a machine is designed, by necessity, with dangerous, moving parts. There must be a means of shutting the machine down if someone is near, which is the function of a light curtain. The light curtain is a standalone product that incorporates photoelectric technology with industrial safety. It is a curtain of light, or photoelectric sensors, positioned to enclose the area around a machine. Once the beam is broken, the machine stops, based on the interlocking of the machine control and sensors. Obviously, there should be minimal reasons to be near the machine, or these stoppages can become a nuisance.

Industrial noise

Throughout this book, we have mentioned noise in the factory. Noise can be almost deafening, can cause employees to wear ear plugs, and can cause accidents. Unfortunately, it is necessary to use machinery that creates noise to produce certain products. Sometimes, the end product itself is noisy as it is handled in the plant, such as a metal producer who not only must have big, noisy machines to form and handle the metal, but who make a noisy product. Noise is a real issue and has received more and more attention in the past decade.

Suppose a new piece of equipment is planned for installation into a noisy mill. This new machine has a noise attributed to it. One might think its noise can blend in with all the other noisy machinery and be virtually unnoticed, but this is far from the truth. In actuality, this noise adds to an already bad and unsafe situation. The decibel level in the plant, or in the area where the equipment is going, actually increases. To see how the decibel levels associate with one another and the human ear, review Table 12-8. From the threshold of sound to the physically damaging sound of a jet liner, noise is measurable in units called *decibels*. The decibel scale has been devised to measure the loudness of sounds. It helps safety inspectors provide accurate data to substantiate when and where a loud area exists. Decibels (dB) increase in the scale by a factor of 10, as can be seen in Table 12-8, because the sound-level equation for decibels contains a logarithmic function of 10 in the formula.

Noise:	Intensity (watts/meter squared)	Sound level (db)
The hearing threshold	10^{-12}	0
Whisper	10^{-10}	20
Average office	10^{-8}	40
Lunch before a holiday	10^{-6}	60
A machine running	10^{-4}	80
Many machines running	10^{-2}	100
Assembly line, forklifts, and metal forming machines all at once	10^{-1}	110
Jet taking off	10^{2}	140

12-8 *Decibel noise scale.*

The decibel scale assigns a value of zero to the audible sound considered the absolute minimum level of sound. Levels increase to where a very noisy metal manufacturing facility might have a level of 100 to 110 dB. This volume is deafening, literally, and extremely unsafe. Large forklifts can be moving large loads of metal from place to place, and although they might have horns, it might simply not be possible to hear them because of the background noise. In addition, employees who work in these environments can experience loss of hearing over time (even if they wear ear plugs every day). Today, much attention is given to new equipment noise levels and replacing high-decibel equipment with less noisy machines.

On-the-job safety

In addition to electrical product testing and designing safety into a project, there are many other indications that safety is a major concern in the factory of the 1990s. Hardhats, safety glasses, ear plugs, yellow painting, alarms, horns, mirrors, warning labels, and such all represent a change from years ago when the Industrial Revolution was in full swing. Many people lost their lives before industry realized that production is nice but not worth the cost of a single human life. By today's insurance standards and the rights of the average worker, safety must be factored into the business equation. So often, employees file lawsuits against corporations for physical damage, which in many instances could have been avoided had proper measures been taken. A person sometimes must get hurt or even lose his life to make the changes required for a safer workplace.

The factory billboard can be seen today toting the fact that there have been so many days without a lost-time accident. Workers are given awards for safety suggestions and avoiding accidents. An almost pseudo-paranoia exists with engineers and designers that if they are not always thinking safety first, they might lose their jobs or end up in court. Both of these outcomes are very possible but not likely. It is a definite positive trend to see so much emphasis placed on safety. If just one more life is saved, it is worth it.

Bibliography and suggested further reading

National Fire Protection Association. 1990. *The National Electrical Code Handbook.* Quincy, MA.

Underwriters Laboratories. *UL/ANSI Standard #508: Industrial Control Equipment.* May 1994. Northbrook, IL.

13

Total quality management, statistical process control, and ISO-9000

One might ask how quality control (QC), statistical process control (SPC), and the ISO-9000 standard are relative to industrial electronics. There are multiple answers. First, as factories strive to become more competitive via automation, they also must become just as competitive in the integrity and quality of the product they produce. The automation approach allows for a consistent product, manufactured quicker. If the product is made consistently bad, however, this situation needs corrective action! Quality control comes into play here and in other facets of the automated workplace. So, too, does real-time process control that uses statistical information gathered about the process. Electronic inspection equipment might be used in the process to ensure parts are correct. Statistical data and documentation might be prepared automatically by the electronic equipment as the product is produced. Throughout this book, we have established that electronics plays a major role in process control, inspection systems using machine-vision and feedback devices, and the collection of statistical data via computer and communications systems in the plant.

Electronic equipment must be serviced, calibrated, and maintained, which involves technicians who are trained on the equipment's hardware and software becoming involved with the purchases of new equipment, initiation of new projects, and day-to-day plant operations. Everyone plays a role in the factory today. Documenting procedures, maintenance schedules, and record keeping become paramount if companies want to do business worldwide. Relative to this electronic calibration, test, and inspection equipment has emerged an entirely new base of high-tech product. New businesses have been spawned from the QC and SPC processes and those who service these constantly changing parts of the manufacturing cycle.

One of the more important reasons for maintaining a strong quality program is that the market of today is truly global. There are new rules to play by. A plethora of technologies are being developed quickly. A manufacturer in the United States might get one component from Germany, another from Japan, and assemble them into a subassembly for later mating with another subassembly from Mexico. There must be some order and control to mixing and matching worldwide components. The world marketplace is an extremely competitive one, and the industrial technicians, maintenance personnel, and machine operators must be involved, trained, and conscientious.

It also seems as though the majority of businesses today are electronic. For example, look at all the computer software and hardware companies in existence. Then count the support businesses for those industries. It is evident that the electrical content in industry is much greater than initially thought. Many new industries have also been derived from the quality control, inspection, and test requirements of the manufacturing sector.

With the dependence on machines and controllers to inform us how well a part is being made, it is important to separate quality control from statistical process control. In both cases, data is collected and analyzed. The key difference is that the data is used at different segments of the industrial process and for different reasons. As a product is manufactured, data is gathered in many ways. Machine-vision inspection systems, feedback devices, and programmable controllers gather their respective pieces of information. Parts are measured and inspected for flaws. For statistical process control, data gathered is used to modify the actual process. Data gathered in a quality control format is used to prove that the final product meets predefined specifications and provides documentation for later submittal to customers. The use of data for these purposes is sometimes referred to as statistical quality control (SQC). In either case, some overlap exists and, in many instances, it is not immediately clear whether the data collected is for quality control, process control, or both.

Because we are concerned with controlling the process first, then the quality, and finally the certifications, this chapter starts with an analysis of SPC. Understanding how data is collected, what charts are used, and how this information is analyzed is a good starting point. We then move on to QC, and the inevitable overlap between it and SPC is the next step of understanding. Finally, gaining an in-depth understanding of the ISO-9000 procedure is necessary, as companies all around the world are seeking certification and need to know about this process.

Statistical process control

As mentioned in an earlier chapter, Eli Whitney is credited with one of the first true industrial automation schemes, that of interchangeable parts for firearms. It must be brought to light that he failed miserably at the first production run. He was awarded a contract for several hundred muskets by the U.S. Government. His theory of interchangeable parts and subassemblies being mass produced by different people was tremendous. What it lacked was the ability to change the process, midstream if necessary, to make modifications to the process and have a 100 percent fit of all parts. He did not achieve 100 percent; his first run only produced 14 usable muskets. (This author surely would not want to fire one of the good muskets, either!)

Eli survived, however, and so did his theory. Processes and mass production, over time, became more elaborate and complicated. It was a cause and effect scenario, in that as electronics were introduced to assist in the production cycle, they were also used to monitor and control it. Sometimes, the electronic control in the process could also furnish the data while performing its main function. Microprocessors have allowed for this, and nowadays, it is the standard expected of electronic hardware.

Statistical process control can be defined as running a particular process and making corrections as we go. In other words, using data collected relative to the process allows continuous correction and improvement to the process. The measured data analysis must be compared to a benchmark or standard requirement before taking any action. In addition, a process under statistical control should only exhibit random variations, as all other process functions must be fully corrected or improved as necessary.

348   Total quality management

Common process functions monitored are temperatures, line speeds (which usually involve motor speeds), pressures, and other functions that, when changed, affect the overall product's outcome. This change might be because of raw materials that require different temperatures or pressures to make a conforming product. The means of determining whether or not a change is needed is gathering system data that can be compared to data runs of good, final products. Data can be evaluated in many forms. One is by charts.

Often called data management, or data acquisition, it is the statistician's dream and can be the quality control manager's nightmare. A

DATE: 12-03-1995
NAME: Lot #23
FILE: Sample

F-LABEL: YO1
U-LABEL: DIM1

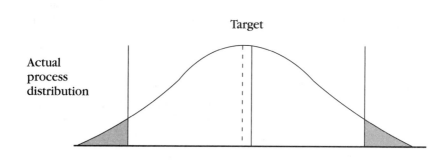

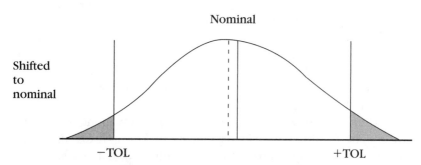

ACTUAL DISTRIBUTION: % under min TOL = 7.6% over max TOL =
SHIFTED DISTRIBUTION: % under min TOL = 6.5% over max TOL =

PROCESS CAPABILITY ANALYSIS

13-1 *A process capability chart.* View Engineering, Inc.

lot of data is collected these days with high-speed, high-memory capable computers. This data must be provided to upper management in a form that is readable, concise, and clear. Decisions about the process must be made and made quickly. These decisions can cost somebody time and money, so the data had better be right. We do not need to spend any more money than necessary. One major tool of the people who compile the data is that of control charts.

There are many of these types of graphs, including process capability charts and regression/correlation charts. Examples of each are shown in Figs. 13-1 and 13-2. The process capability chart can reveal

DATE: 12-03-1995
NAME: Lot #23
FILE: Sample

XF-LABEL: YO1
XU-LABEL: DIM1

YF-LABEL: ZO1
YU-LABEL: DIM2

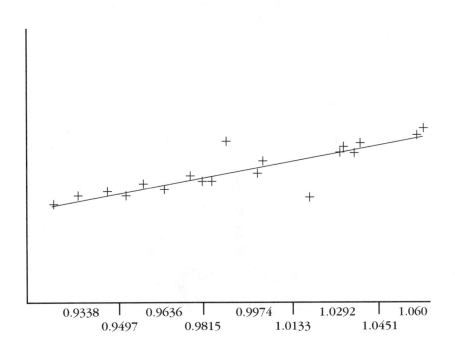

Regression equation: Y = 1.052907 + 0.943001 X
Correlation coefficient: R = 0.90

REGRESSION ANALYSIS

13-2 *A regression/correlation chart.* View Engineering, Inc.

causes of nonconformances. It shows the actual and predictive distribution of the sample. It can display the out-of-tolerance frequency of the process. The regression/correlation charts plot the relationships of process variables, which can be a valuable tool in analyzing the process.

There are several easy-to-calculate, easy-to-plot charts that can indicate more random occurrences happening than allowed. These charts provide the results of online sampling to head off a persistent process problem before too much final product is scrap. One commonly used chart is the *histogram* shown in Fig. 13-3. This chart is basically a frequency distribution shown graphically. It is sometimes called a bar graph. Another common SPC/SQC chart is the $\overline{x}$ *and R chart*; a sample is shown in Fig. 13-4. It is simply the average $(\overline{x})$ and range (R) of groups of data or most-likely measurements and is one of the more common charts used. Sometimes it is necessary to perform a scatter plot when the data collected does not show any coherent relationship. This procedure plots points of data on a graph to see if any relationship exists at all. A scatter plot example is shown in Fig. 13-5.

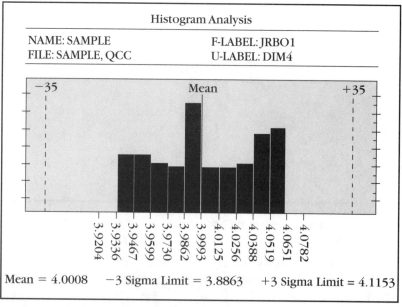

13-3 *A typical histogram.* View Engineering, Inc.

Statistical process control

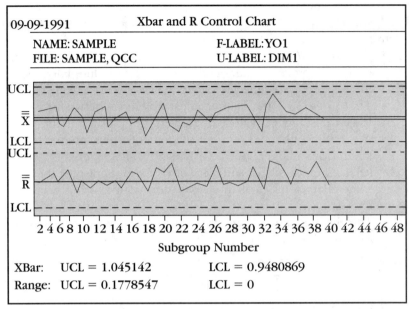

13-4 The $\bar{x}$ and R chart. View Engineering, Inc.

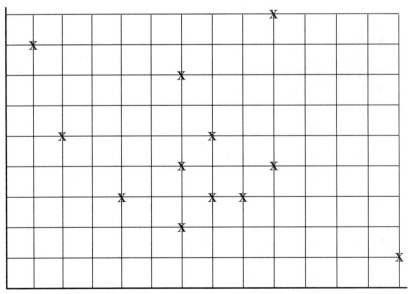

13-5 The scatter plot.

To better understand these charts, some basic terms must be defined, some of which relate to statistics and some to QC and process control:

- The *control limit* is that region within an upper and lower level, or limit. A process is out of control when the values fall outside the upper control limit (UCL) and the lower control limit (LCL).
- The *range* (R) is the mathematical difference between the lowest and highest values in a given group of numbers.
- A *sample* is a limited number of elements (measurements) taken from a larger source (population). If a sample is selected randomly of sufficient size, it should display the same characteristics as the population from which it was chosen.
- The *mean* is the average value of a collection of values. It is expressed as $\bar{x}$ (x-bar) or $\bar{x}$. Its equation is the following:

$$\bar{x} = \frac{x(1) + x(2) + x(3) \ldots x(num)}{num}$$

For example, the mean of the numbers 2, 3, 4, 5, 6, and 10 is 30/6, or 5.

- The *median* is the middle number or value in a set. For example, the median of 2, 3, 5, 9, and 11 is 5.
- The *mode* is the number or value that appears most frequently. For example, the mode is 2 of the following set of numbers 2, 3, 2, 4, 5, 2, 3, 4, 5, 2.
- The term *metrology*, the study or science of measurements, is used very often in statistical quality and process control. A quality control expert is a metrologist.
- *Percent defective* is the amount, or percentage, of units in a given lot that do not conform to specification.
- The *probability* is a prediction of all the possible outcomes for a given situation based on the initial data given. Some might call it the "best guess!"
- A *deviation* is the algebraic difference between one of a set of observed values and their mean value.
- The *distribution* function describes the probability that a system or set of points will take on a specific value or set of values.
- *Kurtosis* is a means of determining how well a normal curve fits a particular distribution of data points on a curve. A sample leptokurtic curve is shown in Fig. 13-6 and represents negative kurtosis, or a distribution curve with smaller tails than a normal curve. Figure 13-7 shows a positive kurtosis curve, which is also called a *platykurtic* curve. The tails are larger than the normal curve.

Statistical process control

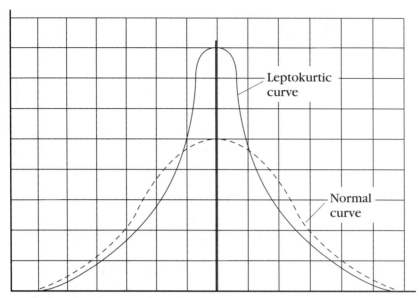

13-6 *Negative kurtosis (Leptokurtic).*

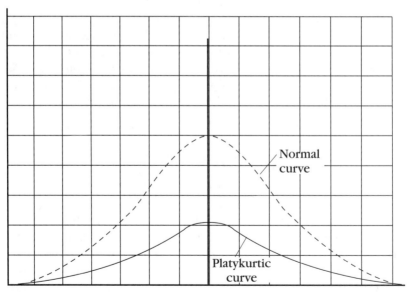

13-7 *Positive kurtosis (Platykurtic).*

Obviously, graphs cannot be charted unless useful data is gathered, crunched, and recorded. In more appropriate words, measurement, inspection, or interrogation data must be collected at the specific point in the process. The data is usually transferred over a communications line to a computerized host device. This transfer can be in ASCII or in dedicated register form so the receiving device can decode and "crunch" it. *Crunching* is mainly repetitively updating and adding to a matrix so that the final numerical values have been properly calculated. Data can then be recorded or saved for later use in plotting or further uploading. Once the data has been output properly, that all-important decision, usually made by the human manager or operator, is made regarding process correction or modification.

Quality control and statistical quality control

Everyone in the corporation must be involved when it comes to quality control, from the president of the company to the individuals having little to do with the manufacturing process. At some level and some point everyone has a role. Previously, there was just a quality control department and a handful of employees who administered the plan. These individuals were all too often not welcomed by their own fellow workers and were treated as more of a hindrance than an asset to the company. Not so today. Quality control is everybody's business. This attitude is the new *Total Quality Management*, or TQM philosophy that is being practiced in countless organizations around the country and around the world. From documentation to design to maintenance, quality control involvement is not discriminatory. The new quality philosophy now incorporates teams of individuals to perform tasks. Employees are now routinely trained so that more knowledge is shared within an organization.

The history of quality

Quality issues date back farther than we think. The late 1700s with Eli Whitney's gun contract is an early example. But surely one or two Roman emperors rejected a chariot, sword, or gift simply because it had a flaw or defect and did not meet their standards. Had a quality program been in place, the acceptance rates would have been higher. (And maybe a few citizens might not have been beheaded!) Quality programs have been in place for years because customers determined if a particular product was acceptable or not.

The consequences would be predictable if the customer were not satisfied. Today, products are more complex and sometimes made from many subassemblies, which are made by other concerns in the production cycle. Tracking a part or subassembly from location to location and providing proper documentation is part of today's quality programs.

Statistical quality control and statistical process control are often confused, and a definite overlap exists. The confusion is because a process engineer might collect data to use in changing the actual process, and the quality control engineer might collect similar data to prove the product is being made to specification. Both need the data, both work for the same company, yet both have different needs for the data. The fact is they could probably share the data, and many times they do. The overlap does not end here. Many charts can be a quality control function, but can also be shown as a process function with different attributes and data. A good example is that of the Pareto charts, used to plot the number of defective products in a manufacturing process in one instance or the causes of a particular problem. The problem could be organizational in nature. Pareto diagrams and charts are discussed later in this chapter.

Early in manufacturing, quality control meant that routine tests and inspections were administered at different points along the production line. Many of these tests and inspections were by hand or eye, definitely allowing for human error. Acceptance samplings were performed at critical junctures in the manufacturing process. The pass-or-fail result of the acceptance test dictated whether or not the part could move on to the next step. Sometimes, all the parts in a particular process were sampled, and this is called a *detailed inspection*. If less than 100 percent of the available parts are checked, this is a *sample inspection*. It must be determined what numbers, or inspected values (measurements, sizes, etc.), are crucial to product quality and then to what standard they are measured against to be deemed acceptable. These numbers determine whether or not the lot is acceptable. The results also have a bearing on whether a 100 percent sampling is to be performed all the time or not.

Usually, a manufacturing facility's quality program is set up as a set of written regulations or standards to which all manufacturing functions must adhere. The plan is administered by a QC manager who usually reports to upper management and interfaces with shipping, engineering, test, and so on. Within the last decade, there has been a great movement in the manufacturing world to enhance and further document a facility's quality control program, paving the way for new reporting and creating many new jobs, all relating to quality.

Ironically, quality control did not get into full swing until the automotive industry began touting it. Automation grew out of the automotive industry, and so has the need to ensure quality. With so much automated equipment used in the manufacturing processes, more data could be collected. In addition, once an automated piece of machinery starts cranking out parts, it does not change its methodology until someone forces the machine to change. A machine can make a bad part consistently bad for a long time! Also, as the large automobile manufacturers started to outsource parts and subassemblies, it became necessary to place controls and regulations on the standards for these parts.

During this same era, two individuals, W. Edwards Deming and Malcolm Baldridge, gave us the basis for quality control programs, from the automotive to the nuclear industry. The focus with Deming's philosophy was to start with management and change the whole organization from top to bottom, not just redoing a portion of the manufacturing process. Deming-based quality guidelines have some flexibility, whereas the Baldridge plan was more rigid. Documentation is more necessary under Baldridge's plan, and statistical understanding is important to that documentation. Obviously, comparing the two philosophies is more a matter of interpretation. The intent is still the same—implement a program, satisfy the customer, and continue to improve products and customer relationships.

For many years, companies entered into and out of quality programs. Escalating process and material costs eliminated quality concerns until the return customer disappeared. The last decade, however, has presented a renewed emphasis on quality control. With the emergence of ISO-9000 as a worldwide standard, quality control programs are now a necessity to do business. Virtually gone, we hope, are the temporary fixes to keep somewhat acceptable (and sometimes marginal) products moving out the door. Also, we do not hear the phrase, "That's close enough," used anymore, as exact standards are used as the barometers in the manufacturing process.

What is quality control and quality assurance?

QC versus QA. Quality control versus quality assurance. Sounds much like a battle, and sometimes it seems like it is. Quality control is that effort in the manufacturing facility to maintain a product to a defined set of standards. QC is accomplished by first establishing the set of standards, implementing them, and periodically testing the product to ver-

What is quality control and quality assurance? 357

ify it is meeting the standards. These standards are the specifications by which the product will be manufactured and, because manufacturers must make a profit, quality control is provided at the lowest practical cost. Quality assurance, on the other hand, is more the pledge by the company and its management that a quality control program is in place.

For a quality program to be in place, a manufacturing facility needs a mission statement, which simply states that the company puts its customers first, desires to make a fair profit, and continues to strive for the perfect product output. From the mission statement must be a documented plan for achieving and maintaining quality for every product made. The degree to which a plan is followed is what sometimes makes or breaks another customer's application of that product. Company quality plans can be turned on or off, depending on the present clientele, financial situation, and current philosophy of existing management. It is so easy and tempting to ship an electronic product without a complete functional test to get a billable shipment. Electronic testing is a very time-consuming portion of the manufacturing process and is skimped on many, many times. The real burn-in occurs during field start-up, which is truly the wrong time for problems.

True quality control consists of several subcomponents. First, the quality program must be fully described. What constitutes a conforming product to the customer's requirements? After manufacturing to a specification, the performance standard of the product must be met: does the product conform to the product specifications and the customer's expectations or actual requirements. Another subcomponent becomes preventative maintenance: if bad product occurrences are frequent, a way must be found to prevent them. The system must be constantly monitored to minimize if not completely eliminate all failures. This portion of a quality program is the performance gauge. The goal in any quality system is to strive for an error-free state. Last, a means of statistically measuring the conformance or nonconformance is needed. How do we stack up?

To clearly see if the goals and guidelines are being met, it is often necessary to put the data gathered into charts and graphical representations. These charts can be statistical quality control, or cause-and-effect relationships for various processes, or illustrations of plant actions to gain a better perspective. One such cause-and-effect chart is the *Ishakawa diagram*, shown in Fig. 13-8. The problem is the mainline piece, and the causes are shown as limbs that lead to it. It looks like the old way of diagramming a sentence in grammar class.

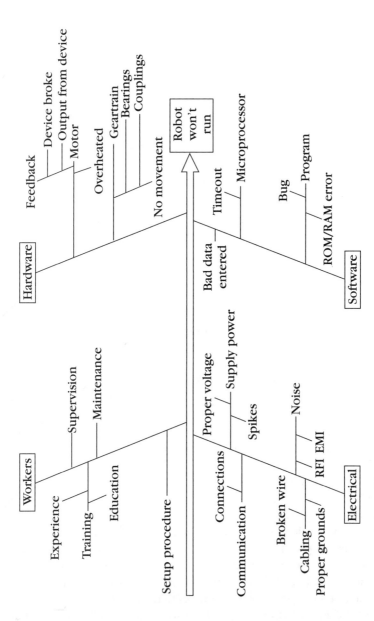

13-8 *An Ishikawa diagram on "why a robot won't run."*

What is quality control and quality assurance?

Another commonly used charting method is by *Pareto analysis*, which is an analysis of the frequency of an occurrence. It is named after an Italian economist to help arrange problems and causes in order of importance. It is helpful in establishing priorities for quality control. The Pareto diagram, shown in Fig. 13-9, is often referred to as a bar graph that shows the frequency of a given occurrence. Electronics manufacturers often use these types of quality control techniques because with modern-day electronics manufacturing and true, effective quality programs, the mere existence and implementation of the product occurred during the same time period.

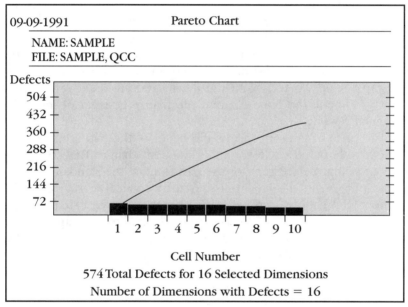

13-9 *The Pareto Analysis.* View Engineering, Inc.

Another important facet of quality management deals with conformance and nonconformance. A gauge, or measurement, is used to measure direct and indirect costs. These costs can be in actual dollars, time, or labor. The *price of nonconformance* (PONC) and the *price of conformance* (POC) are two important concepts to help achieve true quality control. Conformance can be defined as a product adhering to a documented standard. Nonconformance is the failure to meet the standard. Costs are attributed to both conditions. The price of nonconformance is what the rework, repair, or replacement time and money values are because of not meeting the standard. An example of PONC might be when a printed circuit board is placed in a fixture for testing, or burn in, and is suspected of having a defective component that requires repair. The board must be tagged bad (time), troubleshooting conducted to find which component is bad (more extra time), bad component removed (time), new component purchased (money and time), remounted and soldered (more rework time), and finally retested (a full repeat). Nonconformance also interrupts the assembly line and typical manufacturing cycle. It is not favored by many individuals.

Other contributing costs to PONC include, but certainly are not limited to, warranty costs when the product fails in the field, premium freight charges when the product must ship overnight or parts are needed in a hurry, extra paperwork and handling, unplanned overtime, service, or downtime, and returned product.

In addition, customer service personnel must listen to complaints (lost time), situations must be explained (more time lost), and someone must make amends with the dissatisfied customer (embarrassment). All in all, the price of nonconformance goes far deeper than the initial rework.

The price of conformance is the better of the two scenarios because it is, hopefully, a lessor and infrequent charge. This cost is the time and money necessary for the part to meet the standard. An example of POC in the electronics industry might be having all the workers who handle a printed circuit board wear wrist straps to dissipate any static electricity so no stray volts zap the board and destroy a component. This cost ensures the integrity of the board during this phase of the manufacturing process.

TQM

In *Total Quality Management*, everyone is included, from the president down to the cleaning people. The plant electricians, mechanics, machine operators, and even the outside contractors are included with today's quality scheme. This is the *team* concept. The keys are involvement, concern, knowledge, and shared responsibility. Listening, communicating, evaluating, and keeping good records from every meeting make the action work. Having an agenda and sticking to it provides for quick, efficient, and productive meetings. After all, we all dread lengthy meetings, anyway.

A team in a particular plant often comprises members from many different departments. Manufacturing, engineering, maintenance, shipping, the front office, and sometimes outside personnel are placed on a team. This team is assigned the task of solving a particular problem. Brainstorming, scheduling events, and following up are some of the results of a given meeting. Everyone has a say and everyone has a vote. This concept makes for the exchange of knowledge within the different groups and disciplines. Sometimes, cross training is a net result of the team philosophy. Multiple teams can be devised to handle a variety of problems, with some individuals residing on overlapping teams. Teams are an effective way for many individuals to share the responsibility, glory, and satisfaction of solving a particular problem.

ISO-9000

When challenged, a company maintains it has a full-fledged quality program and that it tests every piece of electronic equipment before it leaves the factory. When an angry customer who is in possession of a faulty or defective component disagrees, who is right? Maybe both parties are right. The customer is right because he does not have a good product, and the company maintains it followed quality guidelines. A solution for any problem situation always exists, and this one would get solved, but what about a set of standards so every company is on the same page?

The supplier might indeed have adhered to a quality program. To what degree, however, were tests and inspections performed and to whose standards? The term *quality* is easy to use but frequently hard to substantiate. In the world economy of today, suppliers and customers want proof that a complete quality program is in place, and they want to know exactly how it is implemented. That standard has become ISO-9000, which evolved from BS-S7SO in the United Kingdom.

What exactly is ISO-9000? The initials ISO stand for the International Organization for Standardization, and it has been adopted in many countries around the world as the basis for implementing, controlling, and documenting quality programs. It is not a conformity assessment standard, but rather a quality management system standard. The ISO organization is based in Geneva, Switzerland. In the United States, the American standard has been published by the American Society for Quality Control (ASQC), a member of the American National Standards Institute (ANSI). ANSI, in turn, is the United States member body of ISO. Many important bodies are associated for the establishment and full responsibility of these standards. Another important relationship is that of the International Electrotechnical Commission (IEC) and ISO. The two organizations work closely on all matters concerning electrotechnical standardization, a relationship that is important because of the heavy content of electrical and electronic equipment in use today in factories all over the world.

ISO has established a set of standards that are part of the ISO-9000 series. The Geneva standard is approximately 25 pages in length, while the ASQC's American version is available in a five-part set, with the ISO 9001-1994 standard being approximately 10 pages in length (it is equivalent to ANSI Q91). Varying degrees of the standard exist, each requiring a different level of assessment: ISO-9001, -9002, -9003, and -9004. The highest attainable level is ISO-9001. There are presently 20 key functions certified under ISO-9001. They are the following:

- management responsibility
- quality system
- contract review
- design control
- document and data control
- purchasing
- control of purchased product
- process control
- product identification/ traceability
- inspection and testing
- servicing
- statistical techniques
- control of inspection/ testing equipment
- inspection and test status
- control of nonconforming product
- quality records control
- corrective and preventive action
- internal quality audits
- training
- handling, storage, packaging, preservation and delivery

The major differences between the standards are that design control and servicing are not requirements of ISO-9002 and design control, servicing, contract review, purchasing, internal quality audits, and corrective/preventive action are not required components of ISO-9003. ISO-9001 requires all the aforementioned items plus the main requirements described in the next section.

Each of these standards has a set of rules and standards that must be followed. These rules form the basis for compliancy. Actually, it is the act of not adhering to the rules and standards that causes the commotion, and this is a state of noncompliance. There are two types of noncompliance, major and minor.

A *major noncompliance* is the absence of a documented procedure to address a requirement of the applicable standard. It is an extensive breakdown or the absence of evidence of a documented procedure required by the applicable standard. It is an inability to demonstrate compliance with a technical claim relative to matters affecting produce quality. It represents a material risk to produce quality.

A *minor noncompliance* is a failure to fully satisfy a requirement of the standard with a documented procedure. It is a breakdown in the implementation of a documented procedure in isolated incidents, and does not represent a material risk to product quality.

Requirements

Following is an overview of the requirements as they appear in ANSI/ASQC Q9001-1994, and, unless otherwise noted, represent those requirements for certification to ISO-9001, -9002, and -9003.

Management responsibility, 4.1 Any quality program starts and ends here. Management shall define a policy relevant to its goals and needs. Responsibilities, authority, and a management representative are to be established throughout the organization. Overall, management must be fully committed, make the resources available, and perform periodic reviews of the program.

Quality system, 4.2 A quality system cannot exist unless it is documented. A quality manual must be controlled and maintained. The manual consists of the procedures, work instructions, and basic quality plan. This section weighs heavily on the implementation of the quality plan itself.

Contract review, 4.3 Before entering into any agreement to manufacture or perform a service, it is important (and logical) to agree on what is to be provided. Written requirements should be agreed to by both customer and provider. Changes to the contract and written records should be kept. To meet ISO-9003, this requirement is not necessary.

Design control, 4.4 This requirement is not necessary for certification to ISO-9003 and ISO-9002 because many manufacturers do not design, but rather build to another's design. Those who do design must keep a complete set of plans and specifications for all products. In addition, design input and output must be well-documented. Where and how compliance with regulations and well-defined requirements is established, along with proving the design is safe and can function properly. Design input can also involve other sources, such as technical consultation. Design review, verification, and validation occur periodically and are documented. Any and all design changes must be fully recorded.

Document and data control, 4.5 This is the heart of ISO-9000 certification. Document, document, and document. It is the most common source of noncompliance. This requirement states that all procedures shall be documented and controlled and made available. Either hardcopy or electronic media can serve as the form by which documentation is kept. Documentation shall be reviewed, approved, and controlled. Issues are furnished as they are amended. Any and all changes must be recorded. Keeping up with and following a conforming documentation program throughout the whole quality program and especially the ISO-9000 process practically ensures certification and recertification at minimal costs.

Purchasing, 4.6 With the complexity of many products these days, many components are outsourced. The integrity of the finished product is only as good as the sum of the individual pieces. Because many parts are bought from other sources, there must be a specific way to evaluate suppliers and subcontractors. This requirement addresses both. Again, good record keeping is crucial. Heavy emphasis is placed on documentation, which flows with the component from supplier through assembly. Purchased product verification is also discussed. For ISO-9003 certification, this is not a requirement.

Control of customer-supplied product, 4.7 Similar to purchasing but customer serves as supplier also. This requirement provides for the handling of customer-supplied product. Storage, lost, or damaged product is discussed here. Again, recording and documenting the condition and location of product is the key with this requirement. For ISO-9003 certification, this is not a requirement.

Product identification and traceability, 4.8 It would seem futile to document data relating to a specific part but not be able to locate the part or know which one was used where. This section requirement calls for the tagging and tracing of parts or products through all stages of manufacturing, delivery, and installation. This requirement is extremely helpful in determining where a problem could occur.

Process control, 4.9 This requirement is that segment of the program that actually furnishes the product—the process itself. Control of the process must exist to ensure quality and integrity. The conditions by which control is achieved must themselves be regulated. Most facets of the manufacturing process are covered under this requirement. Production, installation, servicing, and maintaining equipment are all covered, along with the need for written standards regarding each. Workmanship and records on personnel are also addressed in this requirement. For ISO-9003 certification, this is not a requirement.

Inspection and testing, 4.10 With a well-documented quality control plan comes the need to prove that the product is manufactured to those standards. Proof is the actual inspection and testing performed constantly to ensure standards are being met. These procedures must be well-documented. Inspection and testing are performed at three different stages, as required: upon receiving out-source materials, to the product in process, and final inspection. Records are kept of all inspections and tests performed. Often, this documentation and other technical data and reports generated are used as submittals and follow the part or product.

Control of inspection, measuring, and test equipment, 4.11 The inspection and testing phase is as accurate and valid as the personnel and equipment used to perform the tasks. As much of today's testing and inspection equipment is electronic and many times inline with the actual process, it is often necessary to perform routine maintenance and calibration to such equipment. These procedures must be fully defined and documented. With any quality program, consistency and accuracy are always two factors that can get better. The intervals for calibration and maintenance are based on the individual production scheme, environment, and past history. Once the actual control procedure is fully defined as to the actual measurements made, the best-suited piece of inspection equipment is selected. This piece of equipment is therein fully maintained and calibrated, and operators are fully trained. Calibration records are kept. An ISO standard exists for metrological confirmation that can be very useful in setting up a functioning control procedure (ISO-10012).

Inspection and test status, 4.12 Once any testing or inspections have been made to a particular product, it must be quickly determined if the product is good or bad. The better terminology is whether or not the product is conforming to the standard. Once determined, the status of that test or inspection must be made available, which is accomplished by tagging or adding to an existing tag of the product in question. Again, full documentation is required.

Control of nonconforming product, 4.13 The main point in this section is to prevent a nonconforming product from being used any further in the manufacturing process. It involves review by authorities who can provide adequate disposition of the nonconforming product. A nonconforming product can be reworked, accepted with approval or concession, scrapped, or downgraded for use in another application. Sometimes, the disposition is spelled out in the contract, but complete records of the nonconforming product are very important. Any reworked product must be 100 percent reinspected.

Corrective and preventive action, 4.14 This requirement is not necessary for ISO-9003 certification. The goal with this requirement is to eliminate, if possible, the causes of nonconformities. Corrective action includes investigation into how nonconformities happen, the handling of complaints, and determining what corrective action is needed. Another component of this requirement is that of preventive action and reviewing with management the documentation pertaining to this action.

Handling, storage, packaging, preservation, and delivery, 4.15 This section covers much, and these functions represent the larger portion of customer complaints. Documenting the control and procedures of these functions is good, but implementing their successful application is another matter. All handling, storage, packaging, preservation, and delivery of any product is the manufacturer's responsibility. Clear and consistent procedures, when followed, can enhance the success rate. Each of these functions is examined closely by the ISO standard. Good record keeping is a must.

Control of quality records, 4.16 Obviously, because so much of the ISO-9000 requirement involves staunch record keeping and extensive documentation, it is no wonder that one key section is devoted to its control. Maintaining control of quality program records ensures integrity and proves that documents exist. The quality records must be orderly, legible, and easily accessed by authorized personnel. Today, quality records can be on many forms of media, with electronic multimedia a convenient means.

Internal quality audits, 4.17 This requirement is optional for ISO-9003 certification. Much like a full-scale quality audit by customers or third parties, an internal quality audit is sometimes necessary. Periodically, it might be necessary to perform one just to rekindle the importance and need for the plan itself. Areas of weakness and noncompliance can surface, with corrective action being the result. ISO requires documentation of periodic internal audits, results recorded, and corrective action so noted. A good bit of information exists on quality auditing, some of which is available through ASQC.

Training, 4.18 This requirement is mentioned near the end of the ISO-9000 standard but, along with service, extremely important. These are critical because unless people know and understand their jobs, their role in the quality plan, and the equipment around them, quality control is hopeless. After all, quality starts and ends with people—trained people. This requirement calls for the supplier to determine the areas most in need of training and then implement a training program, maintain it, and keep records on who is being trained. Levels of experience also serve as gauges of proficiency and where training needs are. With so much electronic, high-technology equipment in the production cycle (and even local to the quality program for testing and inspection), it is a given that training and retraining must occur.

Servicing, 4.19 The intent of this section is to provide a service plan, implement it, and document it when the circumstances for service in the contract exist. Many times, servicing is not a contract requirement and therefore not necessary for ISO-9003 and ISO-9002 certification. Skilled, trained, and knowledgeable service personnel can salvage the worst products and projects. After all, a product is purchased to work, and many times it's the service personnel who make it work.

Statistical techniques, 4.20 This requirement is mandatory, but each supplier has options as to which forms and methods are implemented regarding statistical information. Trending, charting, graphing, or just simple gathering of test data must be analyzed and presented in some form so as to get the point across. Clear procedures for handling these statistics must be documented, and the needs should be identified.

The requirements of ISO-9000 are spelled out but are not fully black and white, which is why an external audit is conducted. Auditors are looking for compliancy. They are not looking for noncompliancy! There are ambiguous points and there are situations that can involve explanation during the audit. The aforementioned standards are a basis by which companies around the world can compete on equal terms and common ground. In a world economy, this standard is needed.

Where to turn

Several companies are accredited to EN-45012 and provide the service of ISO-9000 certification to other companies. These organizations are called *registrars*. Many of these companies specialize in certain product type areas, such as electronics or electromechanical manufacturing processes. It would be best to find a certifying entity who understands the manufacturing and quality process of the product made by your electronics company rather than by one who has been certifying pretzel factories for many years. One such company that specializes in quality control and ISO-9000 certification is Intertek Services Corporation. It also has expertise in the electronics and electromechanical fields.

Figure 13-10 is an example of a typical noncompliance report used by an auditor to record individual noncompliances observed during the audit of a supplier. Note that the form indicates the standard (ISO-9001, -9002, or -9003) to which the supplier is being as-

Inchcape Testing Services
Intertek Services Corporation

NONCOMPLIANCE REPORT

PAGE _1_ OF _1_

CLIENT NAME: XYZ ELECTRONICS, INC.	CLIENT I.D.#: 1 2 3 4
	REPORT #: 3
	N.C. #: 6

STANDARD: 9001	PARAGRAPH: 4.11	AREA OF AUDIT: INSPECTION EQUIPMENT

NONCOMPLIANCE CATEGORY: ☒ MAJOR ☐ MINOR

OBJECTIVE EVIDENCE:
Vision system used to measure connector lengths has not been calibrated. NBS traceable grid is broken.

SIGNATURE: Robt Smith   DATE: 9/9/94   STAMP:

CLIENT ACKNOWLEDGEMENT:

PRINTED NAME: Joe Jones   SIGNATURE: Joe Jones   DATE: 9/10/94

☒ CORRECTIVE ACTION PLAN STATED BELOW. IMPLEMENTATION DEADLINE: 10/2/94
☐ VERIFICATION OF CORRECTIVE ACTION STATED BELOW.

Request and purchase NBS (National Bureau of Standards) traceable grid used to calibrate vision system. Calibrate!

CHANGED TO MINOR FOR REASONS STATED ABOVE:
SIGNATURE: DATE: STAMP:

CORRECTIVE ACTION PLAN STATED ABOVE. VERIFICATION OF CORRECTIVE ACTION TO OCCUR DURING NEXT SURVEILLANCE (MINOR ONLY).
SIGNATURE: DATE: STAMP:

CORRECTIVE ACTION COMPLETED:
SIGNATURE: DATE: STAMP:

FORM NO. SP 006, REV. 5   REL. DATE 09/94

13-10 *A sample noncompliance report.* Intertek, an Inchcape Testing Services Company

sessed, the paragraph of the standard to which the noncompliance applies, the area (department) where the noncompliance was observed, and specific details as to the nature of the noncompliance. The report further indicates the potential impact of the noncompliance (see definition of major versus minor noncompliance). As indicated at the bottom of the form, the supplier being assessed is given a defined period of time to take corrective action. Generally, corrective action acceptable to the auditor is required before certification can be approved.

Figure 13-11 is an example of an ISO-9000 assessment summary report used by the auditor during the certification assessment. This report assists the auditor in compiling the noncompliances observed

Inchcape Testing Services
Intertek Services Corporation

ASSESSMENT SUMMARY REPORT - 1994 versions

CLIENT NAME AND ADDRESS:

CLIENT ID #

REPORT #:

DATE:

ISO 9001	ISO 9002	ISO 9003	REQUIREMENTS	NONCOMPLIANCE IDENTITY MAJOR	MINOR
4.1	4.1	4.1	MANAGEMENT RESPONSIBILITY		
4.1.1	4.1.1	4.1.1	Quality Policy		
4.1.2	4.1.2	4.1.2	Organization		
4.1.3	4.1.3	4.1.3	Management Review		
4.2	4.2	4.2	QUALITY SYSTEM		
4.2.1	4.2.1	4.2.1	General		
4.2.2	4.2.2	4.2.2	Quality-system Procedures		
4.2.3	4.2.3	4.2.3	Quality Planning		
4.3	4.3	N/A	CONTRACT REVIEW		
4.3.1	4.3.1	N/A	General		
4.3.2	4.3.2	N/A	Review		
4.3.3	4.3.3	N/A	Ammendment to contract		
4.3.4	4.3.4	N/A	Records		
4.4	N/A	N/A	DESIGN CONTROL		
4.4.1	N/A	N/A	General		
4.4.2	N/A	N/A	Design and Development Planning		
4.4.3	N/A	N/A	Organizational and Technical Interfaces		
4.4.4	N/A	N/A	Design Input		
4.4.5	N/A	N/A	Design Output		
4.4.6	N/A	N/A	Design Review		
4.4.7	N/A	N/A	Design Verification		
4.4.8	N/A	N/A	Design Validation		
4.4.9	N/A	N/A	Design Changes		
4.5	4.5	4.5	DOCUMENT AND DATA CONTROL		

FORM NO. SP 007A, REV. 0

REL. DATE 09/94

13-11 *Assessment summary report.* Intertek, an Inchcape Testing Services Company

Inchcape Testing Services
Intertek Services Corporation

ASSESSMENT SUMMARY REPORT - 1994 versions

ISO 9001	ISO 9002	ISO 9003	REQUIREMENTS	NONCOMPLIANCE IDENTITY MAJOR	MINOR
4.5.1	4.5.1	4.5.1	General		
4.5.2	4.5.2	4.5.2	Document and Data Approval and Issue		
4.5.3	4.5.3	4.5.3	Document and Data Changes		
4.6	4.6	N/A	PURCHASING		
4.6.1	4.6.1	N/A	General		
4.6.2	4.6.2	N/A	Evaluation of Subcontractors		
4.6.3	4.6.3	N/A	Purchasing Data		
4.6.4	4.6.4	N/A	Verification of Purchased Product		
4.7	4.7	N/A	CONTROL OF CUSTOMER-SUPPLIED PRODUCT		
4.8	4.8	4.8	PRODUCT IDENTIFICATION AND TRACEABILITY		
4.9	4.9	N/A	PROCESS CONTROL		
4.10	4.10	4.10	INSPECTION AND TESTING		
4.10.1	4.10.1	4.10.1	General		
4.10.2	4.10.2	4.10.2	Receiving Inspection and Testing		
4.10.3	4.10.3	4.10.3	In-process Inspection and Testing		
4.10.4	4.10.4	4.10.4	Final Inspection and Testing		
4.10.5	4.10.5	4.10.5	Inspection and Test Records		
4.11	4.11	4.11	CONTROL OF INSPECTION, MEASURING AND TEST EQUIPMENT		
4.11.1	4.11.1	4.11.1	General		
4.11.2	4.11.2	4.11.2	Control Procedure		
4.12	4.12	4.12	INSPECTION AND TEST STATUS		
4.13	4.13	4.13	CONTROL OF NONCONFORMING PRODUCT		
4.13.1	4.13.1	4.13.1	General		
4.13.2	4.13.2	4.13.2	Review and Disposition of Nonconforming Product		
4.14	4.14	N/A	CORRECTIVE AND PREVENTIVE ACTION		

FORM NO. SP 007A, REV. 0   REL. DATE 09/94

13-11 *Continued.*

and recorded on the various noncompliance reports so that he or she can render an opinion (recommendation for or against certification).

Another factor facing a supplier in the selection of an ISO-9000 registrar is the national scheme or schemes by which the registrar is accredited. Despite the fact the ISO-9000 series of standards is an international series, individual member states have tailored the stan-

Inchcape Testing Services
Intertek Services Corporation

ASSESSMENT SUMMARY REPORT - 1994 versions

ISO 9001	ISO 9002	ISO 9003	REQUIREMENTS	NONCOMPLIANCE IDENTITY MAJOR	MINOR
4.14.1	4.14.1	N/A	General		
4.14.2	4.14.2	N/A	Corrective Action		
4.14.3	4.14.3	N/A	Preventive Action		
4.15	4.15	4.15	HANDLING, STORAGE, PACKAGING, PRESERVATION AND DELIVERY		
4.15.1	4.15.1	4.15.1	General		
4.15.2	4.15.2	4.15.2	Handling		
4.15.3	4.15.3	4.15.3	Storage		
4.15.4	4.15.4	4.15.4	Packaging		
4.15.5	4.15.5	4.15.5	Preservation		
4.15.6	4.15.6	4.15.6	Delivery		
4.16	4.16	4.16	CONTROL OF QUALITY RECORDS		
4.17	4.17	N/A	INTERNAL QUALITY AUDITS		
4.18	4.18	4.18	TRAINING		
4.19	N/A	N/A	SERVICING		
4.20	4.20	4.20	STATISTICAL TECHNIQUES		
4.20.1	4.20.1	4.20.1	Identification and Need		
4.20.2	4.20.2	4.20.2	Procedures		

COMMENTS:

FORM NO. SP 007A, REV. 0   REL. DATE 09/94

13-11 *Continued.*

dards to the peculiarities of specific industrial markets. A supplier who operated solely within the United States or one who exports primarily to the United States should seek out a registrar who is accredited by the U.S. American National Standards Institute (ANSI)—Registrar Accreditation Board (RAB). This rule would also apply to a supplier providing products to the Netherlands, where the Raad voor de Certificate (RvC) accreditation would be most desirable. Certain accreditation schemes (marks) have attained true international recognition, and this fact should also be considered by any certification applicant.

Figure 13-12 illustrates different certification marks. These are of interest to customers and suppliers. Ensure the certifying agency has at least one and that it is appropriate to your business. The marks indicate the source of the accreditation entity to which the registrar is accredited. If the accreditation entity is an EC (European Commonwealth) member, the mark will be well-recognized. These marks also appear on the actual physical certificate, furnished once the certification process is over (Fig. 13-13). The marks shown in Fig. 13-12 represent the Raad voor de Certificate (RvC) and the ANSI-RAB.

13-12 *Examples of certification marks.* Intertek, an Inchcape Testing Services Company

Why get ISO certified, anyway?

This question has many answers, most of which are logical and, let's face it, if the question is being asked the issue has been brought up. It is an eventuality for companies wishing to do business overseas or with customers who require certification of their suppliers. Some of the more common answers to the question are as follows:
- It's a major marketing edge. We have the certification and company xyz doesn't!
- A company's customers might only do business with those companies certified either in the United States or abroad.
- Many project specifications require adherence to the standard.

Why get ISO certified, anyway?

13-13 *The goal: An ISO-9000 certificate.*

Having the certification eliminates taking exception to the specification and negotiating.
- A significant reduction in time can occur during supplier and customer audits.
- It might be found that a particular process has been done incorrectly for many years.
- A particular process might be found to have a much quicker and more efficient approach.
- Discipline is reinstilled in the workplace.
- Employees work in teams and train on multiple functions, minimizing the "we thought that was being done this way" syndrome.
- Customers who require ISO-9000 from their suppliers do not have to test incoming parts as often, thus saving inspection time.
- Product is most likely made consistently with infrequent defects!
- If a customer base is overseas, or projecting to be overseas, ISO certification might reduce international import/export obstacles.

- It is now a must to have a well-documented quality control program in place. Most customers demand it because it is demanded of them.
- Corporate peer pressure is mounting.
- ISO-9000 philosophy is finding its way into nonmanufacturing markets all the time. Schools, services, and others are adopting the policies that underlie ISO-9000.
- The certification process is neither costly nor painful if approached properly.

Obviously, a company does not need to seek certification. It can be a long, arduous process. It is not mandatory, yet (not to imply that it is going to be). But, in the long run, it makes good, ethical business sense to have an exceptional quality program, and ISO-9000 certification is merely a byproduct of that program. It is market driven; not regulatory.

Who is certified and who should be

Lists exist, and once certified, a company wants everybody else to know it! No longer do the salespeople have to say, "We are in the process of being certified," which always sounds like a line, just like "We have a quality program" (when nothing could be presented to document such a program). Those companies who opted for ISO-9000 certification and got it can now hold their heads high and solicit more business. Monthly technical trade magazines run announcements for those companies achieving ISO registration. It makes for good print.

The companies steering certification are not only manufacturing firms but service, support, system integrators, and those wanting to do business in Europe—the list goes on. ISO defines a product in many ways: it can be tangible as in hardware, assembled items, or processed materials. It can also be intangible, such as a concept or idea. Services, knowledge, and software are all products. ISO-9000 carries with it a concept that can be distributed to many sectors of industry, not just manufacturing and service. Conscientious and responsible work, or product for fair pay, has been in existence for thousands of years. Those who performed get repeat business. ISO-9000 is simply normalizing the quality management system requirements and attributes.

How to get certified

First, one must ascertain whether or not a full-fledged quality control program at the company considering certification really exists and to what degree. It is important to ask what condition the quality manual is in. Is it current? Has it been faithfully updated? Is it controlled? Once these questions have answers, it can be determined how much time and effort must go into the quality program before a telephone call is made to a certifying agency. Many certification firms can also come in and provide a preassessment. It is a dry-run audit of the system to find compliance without the pressure of potential certification denial. Keep in mind that their job is to look for compliancy in all aspects of quality control with an emphasis on actual documentation. If your quality control manager is an expert in ISO and if he or she can maintain objectivity, you might be able to do this yourself.

Secondly, how much does this certification cost? Surely, there are varying degrees of costs depending on the current condition of the quality program. Many factors enter into the overall cost equation. There is an application fee. Many times, an initial visit can be made by the certifying body to give a cursory look at what might be required, and from this visit a proposal can be furnished. Depending on how much of a quality program a company has, this visit is a major determining factor. In addition, how well are all phases of the quality program documented? The quality manual review and the onsite assessment are usually quoted together, with travel and living expenses additional. There are four main stages to the certification process:

- the document review to determine if the documented system addresses all elements of the standard
- the assessment to determine that the documented system has been implemented
- periodic surveillance
- periodic reassessment

The initial document review is typically done at the customer site and usually takes one or two days for a company of approximately 100 employees. The formal audit takes about four work-days. Usually, more than one auditor is used at this time. Two auditors might take four days. Of course, the time of audit can vary, depending on the size of the company. After the document review, audits, and corrective action is covered, it is time for certification. But the program does not end there.

Built into any certification is the provision that compliancy will be maintained. This rule is ensured by the registrar by performing various surveillance audits during the life of the current certification. Yes, the certification has a duration. It is not forever. The registration is generally valid for three years. After that, there is reassessment. If all documents are in order and compliance is the norm, the cost and effort of reassessment are minimal. During those three years, a number of surveillance audits will occur at regular intervals.

The bigger portion of the cost burden lies with the quality control program itself. This program truly makes or breaks a manufacturing facility eventually. A company must continually reinvest into the quality program. This reinvestment might be for new inspection and test equipment, retraining of employees, or an overall recommitment by management that quality is just as important as getting a paycheck. It is sometimes difficult to spread the capital and expense dollars everywhere, but the quality control portion of the company deserves special attention. If not, those dollars will have to be spent on the program when forced to (and the amount might be much higher, later).

Conclusion

ISO-9000 certification has an erroneous enigma associated with it that it is both very expensive to attain and there is a high probability that many "dirty laundry" issues will surface. The process of the audit is to find those areas of standards compliancy. If a certain deficiency does not pertain, it's not an issue. Likewise, if a function is found to not be in compliance, there is probably good reason to correct the situation. Certainly, if a company does not have much of a quality program and is seeking certification, the proposition will be an expensive one. With documentation in place and compliancy the norm, however, ISO certification can be relatively inexpensive with few hidden or unexpected costs.

The key to a successful and lasting quality program is constant focus. Sure, a company can get certified and feel like the pressure is off, but the pressure is not off! Customers demand the same or better quality from an ISO-9000 certified organization. More appropriately, the surveillance period is the mechanism by which ISO-9000 certified companies can be checked for compliance. That documentation better be in order!

Documentation and document handling is perhaps the most significant overall ISO-9000 requirement. It must be addressed, and it must be done! Keeping up with the documentation and record keeping as you go is the best policy. Letting documentation get behind just compounds the problem of getting it done and documenting true and actual data as it happens. After the initial investment of time and money, it would be a shame to lose certification because of a lack of diligence in procedure and documentation. Most probably, this letdown might also be seen in the product quality.

To order a copy of the ISO 9001-1994 standards (approved in August 1994) or to inquire, call or write to the address below. There are also other adjunct ISO-related documents that are useful if they pertain to your program. ISO-8402 gives definitions, and ISO-10013 provides quality manual instruction. A complete set can be purchased from the following:

American Society for Quality Control
611 East Wisconsin Ave.
Milwaukee, WI 53202

In closing this chapter, it is rather obvious that we have entered into a new world market that is both competitive and demanding. A company can no longer "stop and smell the roses," because the thornbushes are engulfing them. In more industrial and quality-like terminology: companies must continually strive for newer and better ways to produce or provide a service. If they do not, another company from the other side of the earth will! Do they have ISO-9000 on Mars? We had better find out!

Bibliography and suggested further reading

Burr, John T., *SPC Tools for the Operator*, Second Edition, 1993, ASQC Press.
Deming, W. Edwards, *Out of the Crisis*, 1986, MIT CAES.
Gabor, Andrea, *The Man Who Discovered Quality (W. E. Deming)*, 1990, Times Books.
Quality Control Systems Procedures for Planning Quality Programs, 1989, McGraw.
Tyczak, Lynn, *Get Competitive: Cut Costs and Improve Quality*, 1990, TAB Books.

14

Where we are headed

Automation and electronics really do affect everyone. If not directly, then indirectly. Some might be part of the process, while others might simply see automation in action as they use the electronic checkout at the local grocery store. Likewise, many products purchased at that grocery store were made, inspected, and packaged electronically. Like it or not, the age of high technology is upon us. Change in the field of electronics is occurring at an astounding pace. Therefore, we must ask the question, "Is industrial automation good or bad for us?"

Many individuals believe that eventually machines will perform all the work that humans once did. In this case, there will be no jobs, just products to buy and possibly service. But this simply cannot happen. There must be industry to have growth. Not all professions can be service-related. Somebody somewhere must make a product for the public's use. Granted, some might be imports while others are exported. But first there must be a product manufactured prior to any servicing of that product, and jobs are necessary to make those products. Along the way, more and more electronics will constitute the manufacturing and servicing aspects of the process.

The jobs relative to automation are higher-skilled. To operate in the industrial environment of the 1990s, one must be willing to accept change and constantly learn. Because the technology is changing so rapidly, only those willing to adapt will survive. Corporations also see this trend. Companies are willing to retrain the workers to gain optimum performance from the workforce. Gone will be the mundane-task, no-thought positions. Automation has brought with it the requirement to think quickly and in detail.

True, automation has displaced many workers, but it has created many more new jobs. Automation specialists, electronic technicians, and specialty machine designers are just a few of the emerging new professions in industry. Even in the automated office, electronics has made most workers more productive rather than displacing them. Word processors, facsimile machines, and desktop publishing have thrown the office of the 1990s right into the future, and many people are taking advantage by sharpening their creative skills. The theme here is the same, relearn and retrain. The new high-tech skills are those in demand. And electronics professionals and technicians will be in demand for as long as there is electricity!

This first generation of industrial electronics over the past decade can be called a young science. We are still working many of the bugs out of those first-generation systems in the plants. One industrial philosophy is the "throw it in the dumpster if you can't make it work" syndrome. High-tech manufacturers claim to have all the solutions and sell that fact. When it comes time to implement the new system, it had better work. The start up should not take very long, and performance must be superb. Some production facilities are not very tolerant of new technology. They only want to make products and more of them. Thus, if a particular electronic device is not working, causes considerable downtime, or does not meet expectations, it ends up in the dumpster. Going back to the old way of making the product or going with a more solid system integrator are more attractive choices at this point.

As times goes on, the electronics industry learns from its mistakes and misapplications. Newer and better, more reliable products emerge. Each generation of electronic product gets faster, more powerful, and more reliable. This phenomenon of newer high-tech equipment and its comparative life to a machine is a relatively new one. The cost and usefulness of a new computerized system, when compared to the life of the actual machinery, is high (cost) and short (useful life). The bottom line is that computer and controller technology is moving much faster than all other facets of industry. These systems are obsolete quicker than systems of yesteryear. This phenomenon keeps the technicians and other plant people in a constant state of learning. Many new opportunities are created every day as new automation frontiers are encountered. As one job is lost to a new way of performing that task, another is gained in the form of a programmer or technically skilled individual required to support the automated machine.

Service and support

These are the hot issues with today's industrial electronics. Producers want to keep producing. Electronics manufacturers want to keep

Service and support 381

supplying high-tech equipment. Everyone is extremely busy, and deadlines are everywhere. Nothing is more frustrating than not being able to talk to a knowledgeable person about an electronic product. Actually, talking to a voice-mail system when an answer is needed immediately is probably more frustrating, but this is the way of conducting business in the 1990s. Complex products are sold and installed throughout industry. Suppliers of these products want to grow, and fast. The support and service portion of their business lags behind because economically the sales and installed base must be there to justify the customer service personnel.

As the installed base grows, having the technical people on-hand to handle problem telephone calls is difficult. The additional dilemma is simply having enough good, technical people on-hand. It's hard to predict how many and what types of problems customers will have. But as the sales of installed electronics increase, the support required follows proportionately, if not at a higher rate. The suppliers who can live up to the promise of "the support will be there when needed" will get the repeat business.

It takes time and experience with the product to have the technical support personnel up to speed. A desperate customer with a down machine in a production environment cannot live with answers such as, "I don't know," or "I'll get back to you on that." The customer needs immediate solutions because management is "breathing down the factory's neck." The customer, hopefully, has been trained on the product, stocked spares, and has full documentation in the plant. These steps can expedite the situation. The norm, however, is that customers have electronic equipment installed, it works immediately, and they never get trained, do not buy spares, and lose the manuals. In these cases, the customer's lack of planning should not constitute an emergency when problems occur. It's hard to feel sorry for the end users who do not try to help themselves. Still, the issue of having good technical support when needed is a major one.

Servicing the electronic products in the field has become a large business itself. Many suppliers of electronic products have their spares and service departments set up as separate cost centers, which means to the customer that the "free service or spare" will be hard to come by. When high-tech products are purchased, the end user often expects extra attention and special treatment. With service and spare groups responsible for maintaining levels of profit, it is very difficult to get these extras, let alone get them covered under warranty. Obviously, if a product fails within the warranty period, the supplier must honor the warranty. Disputes do arise, however, especially when determining causes of electronic failures. Who's to blame?

Going into the field to start up electronic equipment or service it after installed is more extensive than one might think. More extensive in the sense that so many electronic products today must be worked on by factory-trained specialists or the warranty becomes void. The frequency of these service calls is also higher than one might think. Not because a particular product or installation requires many repeat field service calls (remember the "make it work or it goes in the dumpster" syndrome), but so many different products are being implemented in industry. One day, XYZ supplier is in for service, and the next day ABC company is in. These calls can be for real breakdowns or problems, but can also be part of a service and maintenance agreement. Whatever the case, many thousands of dollars are spent each year for service by industrial users of electronic products.

These plants must spend the money. They must because they just invested millions of dollars to modernize their facility. They must spend the extra few thousand to keep the equipment running. A whole new profession of industrial technician has been born, both on the industrial end and the supplier's end. Technical individuals are needed to keep this equipment running, and there must be many who are knowledgeable. Knowledge involves training and retraining. It also involves schools. Schools must keep up with the technological trends that industry is promoting. Getting good, substantial training on electronic products is essential. If a plant can fully train and keep these technicians (these people have to be paid well), they can bypass the $200 per hour field service person from the supplier.

Keeping the talented technicians who keep the machines running is paramount. It is hard to find individuals who are both capable and willing to fix a machine at 3:00 AM on a Sunday. These individuals are just as valuable as the staff engineers. Likewise, suppliers of electronic equipment must compensate their field service personnel accordingly. These people often travel and perform under pressure. Customers have production deadlines to meet, and a machine out of service puts that deadline in jeopardy. A field service person works under these conditions almost all the time. Unless compensated properly, the good service people will migrate to less stressful jobs.

Where are we now?

It is the classic tale of old versus new. Smokestack technology versus high-tech—it is merely change, and it has happened since the beginning of time. All concerned must be willing to accept that change and move forward, but it is hard to do, especially when we finally think we have mastered a machine or software and then the manufacturer suddenly changes to a new standard. Or maybe a technological breakthrough occurs that changes everything. We must learn anew all over. Learning is part of accepting change.

Besides the plethora of electronic equipment installations in industry and factories around the world, many other nonindustrial changes are going on today. These changes are rooted in industrial electronics in many ways. Some of the technology evolved out of industry. Other products are simply made or designed electronically. Whatever the electronic content, activity is all around us. It is happening in the home, stores, offices, and in your automobile. If you are not using something powered by electricity, you're safe, but your luxuries are very limited!

Many homes are automated. Many household appliances have had computers in them for years. These items cannot be serviced by the homeowner, and are either repaired by the supplier or thrown away (the dumpster theory on the home level). The skills we are sharpening in the electronics industry sometimes do get used at the home. This fact might be more evident down the road. A main home controller controls lights, electrical outlets, audio/visual equipment, security systems, and the heating/cooling systems. Look out Jetsons, the space age is now.

Commercially, we have seen automation for quite some time. Banking, travel, and retail business can no longer exist without computers and electronic automation. Automated tellers, electronic wire transfers, and account information are all signs of the times. Travel agents must be networked and linked with all the major airlines' databases. Retailers depend on point-of-sale transactions for inventories and stock. Interestingly, it is sometimes hard to tell who and what industries are driving the electronic development. Many implementations of new technology emerge in factories as well as in the grocery store. Telecommunications is now a trendsetter for high-speed switching, fiberoptics, and communication highways. All in all, these advancements have made business and life more convenient—right?

In the transportation industry, electronics can be seen in many areas. Our automobiles have microprocessors controlling everything from the engine operation to overall diagnostics of the car. Today, we do not fix a car until after we plug into the computerized diagnostic center for analysis. Other transportation-related electronic needs are evident with the airlines and their need to serve millions each week with travel plans, reservations, ticketing, and even aircraft navigation. These computer systems are some of the most powerful on earth.

The defense industry and weaponry are perennial leaders in using technology for advancement. Many high-tech industries have emerged from the defense industry, especially with machine-vision systems. These systems evolved originally from the missile guidance systems technology, which were in full effect in the Gulf War. This war was high-tech and electronic. Computer-guided missiles, night-vision glasses, and advanced communications (on the ground, in the air, and by satellite) were instrumental in that war. The military has always been on the leading edge of technology when defense budgets are well-funded, allowing the private sector to then take the technology and enhance it.

Every day is a new technological breakthrough. At times, it's as though technology is running away, and we are not in full control. These developments must be refined and accepted by the populace. As time goes by, many new technologies become accepted, such as electronic voice mail. Many people hate answering machines and thus cannot stand voice mail systems. When properly used by both parties, however, (one has to leave a message and another has to return a call or at least respond), it is a great tool. Acceptance does occur. Sometimes, electronic items are forced upon us until we accept them. Even Grandma and Grandpa are screening their calls with answering machines today.

Future developments

Future developments hold the key to how far automation and electronics will evolve. There might even be another industrial revolution or two. New breakthroughs in science and technology dictate how fast we evolve. The issues of competitiveness and acceptance by both industry and individuals prevail. Superconductivity, solar power, microprocessor advancements, and laser technology are just a few technologies that can move development in one direction or another. Technologies are presently emerging that might have an eventual impact but for now are virtual unknowns to industry. Throughout this book, we have analyzed what constitutes different sectors of industrial automation and the associated electronics. Each of those separate industries or product areas will change in the future. Following are some predictions:

The computer industry Obviously, this industry has not stopped evolving and will not. It cannot because too many other industries are dependent on it. Look for smaller microchips, faster processors, more powerful and flexible memory, and higher-bit technology in practical use.

Future developments

Programmable controllers PLCs as we know them in the traditional sense are being displaced by other controllers. Motion controllers, computers, and smart workstations are getting much more I/O capability. With faster processing and more throughput at these other types of devices, the future of PLCs might be in jeopardy. Stay tuned.

Electric motors Superconductivity is a key. Strict energy savings standards are another. These two issues will make motor manufacturers change their present-day offerings. Motors are the prime source of motion in the factory. Unless another major breakthrough occurs, such as Faraday's Law of Induction, we will continue to use electric motors.

Electronic drives Ac or dc, dc or ac? This question's answer is coming. Eventually it makes sense to see motors and controllers of all one type. For now, however, both exist and there are many reasons to consider each. One advancement in this industry will be a cost-effective, fully configurable ac or dc drive. The drive can be set up to run either an ac or dc motor. Servos and stepper technology will get better with higher-resolution microprocessors. Another movement in the drive industry promotes locating the drive physically at the motor, allowing for further drive use in homes and other commercial markets.

Sensors Sensors and feedback devices will continue to get more exact. They must, as controllers get more exact. Some sensors will have small chips built in to perform predelivery conditioning of the output signal. New types of sensors (i.e., biological, chemical/electric, etc.) will also emerge.

Machine vision This industry continues to get stronger in its edge-detection and image-processing technologies. This trend will continue. More integration of machine-vision systems into other industries and markets will also drive future developments.

Robotics Even though this industry has seen a reduction in suppliers, it will continue to play a major role in factories around the world. Neural systems and fuzzy logic, along with artificial intelligence, will be worked into the already sophisticated controls for these machines. New, lightweight materials will also be incorporated into the mechanical portion of robots.

Communications The biggest advancements in the future will come in this discipline and are already happening. Wireless systems, wide area networking (WANs), and the general need to move data all over the world are driving this industry's developments. More employees can and will work from home in the future, as they become linked with the plant or office.

Total Quality Management This philosophy and its complete implementation is just beginning. With so many standards being set and having to be met, TQM and SQC will continue to parallel all growth in the electronics industry.

Safety This issue will never go away, even as machines and robots are used instead of humans. Too much in the way of liability and escalating insurance costs will ensure that safety standards are frequently adhered to.

As we move into the twenty-first century, the future looks good for automation and electronics. The need is there, along with the expertise, to develop the technologies. Growth will also occur in three related areas: jobs, training, and standards. New technologies create new opportunities. From new opportunities come new jobs. These jobs will be in the manufacturing sector and the service sector, but will require high skill levels. Technical people need to be educated well, both practically and in theory, which leads into training and retraining. Individuals, even if educated and if well-trained, can never stop learning. Training and retraining will be industrial musts for the future. Too much is changing too fast!

Last, new sets of standards will continue to emerge for all disciplines. Although it seems as though we have enough now, new technology and those areas in which standardization is deficient will need to come onboard. Standards are often hard to implement but are fully necessary. Automation and electronics is an industry unto itself. They were made by industry to serve and supply industry. The concept has been with us for ages. Not until the last two or three decades have we been able to take advantage of it. It is now time to exploit the technology for the betterment of all humankind!

A

Common acronyms

ac alternating current
A/D analog to digital
AFD adjustable-frequency drive
AGV automatic guided vehicle
AI artificial intelligence
ALU arithmetic logic unit
AOTF acousto-optic tunable filter
ASCII American Standard Code for Information Interchange
ASD adjustable-speed drive
ATG automatic tank gauge
BASIC beginner's all-purpose symbolic instruction code
BCD bit or binary code decimal
BIL basic impulse level
BJT bipolar junction transfer
bps bits per second
CAD computer-aided drafting
CADD computer-aided drafting and design
CAE computer-aided engineering
CAM computer-aided manufacturing
CASE computer-aided software engineering
CFC chlorofluorocarbon
CGA color graphics array
CHEMFET chemical field-effect transistor
CIM computer integrated manufacturing
CIP clean in place
CMOS complementary metal-oxide semiconductor
CNC computerized numerical control
COBOL common business-oriented language
CPU central processing unit
CRT cathode ray tube
CT current transformer

D/A digital to analog
DAS data acquisition system
dc direct current
DCS distributed control system
DDE dynamic data exchange
DMA direct memory access
DNC direct numerical control
DOS disk operating system
DP differential pressure
DPDT double-pole double throw
dpm digital panel meter
DRAM dynamic random access memory
EEPROM electrically erasable programmable read-only memory
EGA enhanced graphics array
emf electromotive force
EMI electromagnetic interference
EPROM erasable programmable read-only memory
FET field-effect transistor
FLA full load amperage
FLC full load current
GPIB General Purpose Interface Bus
gpm gallons per minute
GTO gate turn off
HCFC hydrochlorofluorocarbon
hp horsepower
HVAC heating ventilating and air conditioning
IC integrated circuit
ID inside diameter
IGBT insulated gate bipolar transistor
I/O input/output
I/P current to pressure
IR infrared or current resistance (drop)
JIT just in time (manufacturing)
kVA kilovolt amps
kVAR kilovolt amps, reactive
LAN local area network
LCD liquid crystal display
LCL lower control limit
LED light-emitting diode
LSI large-scale integration
LVDT linear variation differential transformer
MAP manufacturing applications protocol

Common acronyms

MCC motor control center or metal clad cable
MG motor generator
MIPS million instructions per second
MIS manufacturing information system
MMI man-machine interface
MOSFET metal-oxide semiconductor field-effect transistor
MOV metal-oxide varistor
MRP manufacturing resource planning
MTBF mean time between failures
MTTD mean time to detect
MTTF mean time to fail
NC numerical control or normally closed
NO normally open
OCR object character recognition
OD outside diameter
OEM original equipment manufacturer
PB proportional band
pc personal computer or programmable controller
PCB printed circuit board
PD positive displacement
PE professional engineer
pf power factor
P/I pressure to current
PID proportional integral derivative loop
PIV peak impulse voltage
PLC programmable logic controller
POC price of conformance
PONC price of nonconformance
POS point of sale
PROM programmable read-only memory
PT potential transformer
PWM pulse width modulated
QA quality assurance
QC quality control
R&D research and development
RAM random-access memory
RF radio frequency
RFI radio frequency interference
RGB red green blue
rh relative humidity
rms root mean squared
ROI return on investment

ROM read-only memory
rpm revolutions per minute
RS-232 return signal-232
RTD resistance temperature detector
SCADA supervisory control and data acquisition
SCR silicone-controlled rectifier
SG specific gravity
SMD surface-mount devices
SPC statistical process control
SPDT single-pole double throw
SQC statistical quality control
SRAM static random-access memory
T/C thermocouple
THD total harmonic distortion
TOP technical office protocol
TTL transistor-transistor logic
UCL upper control limit
UHF ultra-high frequency
UPC universal product code
UPS uninterruptable power supply
UV ultraviolet
VDM video display monitor
VFD variable-frequency drive
VGA video graphics array
VHF very-high frequency
VLF very-low frequency
VLSI very-large scale integration
VME virtual memory executive
VSD variable-speed drive
WAN Wide Area Network
WIP work in progress

For acronyms of organizations (ASME, IEEE, NEMA, etc.), refer to Chapter 2.

B

English to metric conversions

Area conversion constants

One Square Millimeter = 0.00155 Square Inches
One Square Centimeter = 0.155 Square Inches
One Square Meter = 10.76387 Square Feet
One Square Meter = 1.19599 Square Yards
One Hectare = 2.47104 Acres
One Square Kilometer = 247.104 Acres
One Square Kilometer = 0.3861 Square Miles
One Square Inch = 645.163 Square Millimeters
One Square Inch = 6.45163 Square Centimeters
One Square Foot = 0.0929 Square Meters
One Square Yard = 0.83613 Square Meters
One Acre = 0.40469 Hectares
One Acre = 0.0040469 Square Kilometers
One Square Mile = 2.5899 Square Kilometers

Weight conversion constants

One Gram = 0.03527 Ounces (Avoirdupois)
One Gram = 0.033818 Fluid Ounces (Water)
One Kilogram = 35.27 Ounces (Avoirdupois)
One Kilogram = 2.20462 Pounds (Avoirdupois)
One Metric Ton (1000 Kilograms) = 1.10231 Net Tons (2000 Pounds)
One Ounce (Avoirdupois) = 28.35 Grams
One Fluid Ounce (Water) = 29.57 Grams
One Ounce (Avoirdupois) = 0.02835 Kilograms
One Pound (Avoirdupois) = 0.45359 Kilograms
One Net Ton (2000 Pounds) = 0.90719 Tons (1000 Kilograms)

Weight

10 Milligrams = 1 Centigram
10 Centigrams = 1 Decigram
10 Decigrams = 1 Gram
10 Grams = 1 Decagram
10 Decagrams = 1 Hectogram
10 Hectograms = 1 Kilogram
1000 Kilograms = 1 (Metric) Ton

Length conversion constants

One Millimeter = 0.039370 Inches
One Centimeter = 0.3937 Inches
One Decimeter = 3.937 Inches
One Meter = 39.370 Inches
One Meter = 1.09361 Yards
One Meter = 3.2808 Feet
One Kilometer = 3,280.8 Feet
One Kilometer = 0.62137 Statute Miles
One Inch = 25.4001 Millimeters
One Inch = 2.54 Centimeters
One Inch = 0.254 Decimeters
One Inch = 0.0254 Meters
One Foot = 0.30480 Meters
One Yard = 0.91440 Meters
One Foot = 0.0003048 Kilometers
One Statute Mile = 1.60935 Kilometers

Length

10 Millimeters = 1 Centimeter
10 Centimeters = 1 Decimeter
10 Decimeters = 1 Meter
1000 Meters = 1 Kilometer

Volume conversion constants

One Cubic Centimeter = 0.033818 Fluid Ounces
One Cubic Centimeter = 0.061023 Cubic Inches
One Liter = 61.023 Cubic Inches
One Liter = 1.05668 Quarts

English to metric conversions

One Liter = 0.26417 Gallons
One Liter = 0.035317 Cubic Feet
One Cubic Meter = 264.17 Gallons
One Cubic Meter = 35.317 Cubic Feet
One Cubic Meter = 1.308 Cubic Yards
One Fluid Ounce = 29.57 Cubic Centimeters
One Cubic Inch = 16.387 Cubic Centimeters
One Cubic Inch = 0.016387 Liters
One Quart = 0.94636 Liters
One Gallon = 3.78543 Liters
One Cubic Foot = 28.316 Liters
One Gallon = 0.00378543 Cubic Meters
One Cubic Foot = 0.028316 Cubic Meters
One Cubic Yard = 0.7645 Cubic Meters

Index

A
ac, 12
ac drive, 176-190, 227
  across-the-line bypass scheme, **181**
  advantages/benefits, 180-181
  analog, 191-192
  braking, 192
  circuitry, 191
  classifications, 180
  coolants, 193
  digital, 191-192
  distances between motor and drive, 195-196
  efficiency, 193
  ground fault protection, 193
  harmonic distortion, 193
  high altitude, 194
  high humidity, 195
  horsepower, 193
  in-rush currents, 192
  input reactors, 196
  main parts of, **180**
  motor quantity, 193
  output contactors, 196
  output reactors, 196
  power factor, 193
  power supply, 192
  selecting, 191-196
  short circuit protection, 193
  speed, 192, **192**
  various names of, 179
  ventilation, 193-194
ac flux vector drive, 196-198
ac motor, **146**, 152-157
  inverter-duty, 154
  life expectancy, 154
  nameplate, **169**
  polyphase, 153
  repulsion, 156
  speed/poles chart, 182
  speed vs. torque curve, **183**, **184**, **185**
  split-phase induction, 155, **155**
  squirrel-cage, 153-154, **153**, **155**
  synchronous, 156
  temperatures for insulation classes, 154
  wound rotor, 156
ac vector technology, 157
acceleration torque, 39
  formula, 40
adjustable speed drive (ASD), 179
affinity laws, 60
air compressor, 62-63
alternating current (ac), 12
alternator, 2
American Institute of Motion Engineers (AIME), xiv, 9, 66, 321
American National Standards Institute (ANSI), 9, 361
American Society for Quality Control (ASQC), 361
American Society of Mechanical Engineers (ASME), 8, 66
American wire gauge (AWG), 26
amperage, 12
amplitude, 16
AND, 71, 81
anode, 23
Arcnet, 263
arithmetic logic unit (ALU), 77
ASCII, 72
  character code equivalents, 73-76
automation, 4-6, 379-380
  definition/description, 1
  future, 384-386
  present activities, 382-384
  service and support, 380-382

Illustrations are in **boldface**.

395

Index

autotransformer, 22
auxiliary blowers, 149, **149**

B

back electromotive force, 211
backlash, 51-52, **52**
backlighting, 292, 297
ball screw, 56-57, **56**
bandwidth, 204
baud rate, 257, **257**
bearings, 238
bevel gear, 51
binary number system, 72
   decimal equivalents, 73, 109
biosensor, 249
bit, 72
blower, applications, 60-61
Boolean algebra, 118-120
brake horsepower (BHP), 37
braking, 55, 210-212
   common busing, 211
   dc injection, 212
   dynamic, 211
   fail-safe, 55
   regenerative, 211
break-away torque, 39
break-down diode, 20
byte, 72

C

cable, 26-28
   coaxial, 255, 270, **270**
   fiberoptic, 255-256
camera, high-speed, 278
capacitor, 19
cathode, 20
CD-ROM, 89, 94, **95**
cell controller, 116-117
central processing unit (CPU), 77, 99, 101-103, 117
   module for PLC, 105, **105**
centrifugal clutch, 54
chain, 53-54
   roller, 53
choke, 21
circuit
   energizing, 19
   relay, 19-20
   parallel, 18
   series, 18
   short, 27-28, **28**
circuit protection, fuses and, 29-30, 222 (*see also* grounding)

closed-loop control, 231-232, **231**
clutch, 54-55
   centrifugal, 54
   eddy-current, 174-175, **175**, 176
   magnetic particle, 54
   over-running, 54
coaxial cable, 255, 270, **270**
collision detection, 125
communications, 251-273, 385 (*see also* networks and networking)
   controller diagram, **252**
   digital bit stream, **253**
   electrical disturbances, 268-270
   high-speed, 253
   isolation vs. nonisolation, 271-272
   machine vision online systems, 284-287
   media, 255-256
   noise, 264-267
   optical isolation, 272, **272**
   parallel, 259-260, **259**
   power fluctuations, 267-268
   protocol, 254
   serial, 257-259, **258**
   transmission distances, 256
   transmission modes, 254
   transmission speeds, 257
commutation notch, 216, **216**
complementary metal-oxide semiconductor (CMOS), 71
computer-aided design (CAD), 88-89, **90**
computer-aided engineering (CAE), 89-90
   hotspotting, 90-91, **91**
computer-aided manufacturing (CAM), 89
computer
   applications, 91-94
   firmware, 82
   future, 94-96
   hand-held, 86-88, **86**, **87**, **88**
   hardware, 78-79
   history, 69-70
   industrial, 69-96, 384
   memory, 79-82
   microprocessing basics, 72-77
   networking, 82-84, 92-93
   peripherals, 82
   software, 78
   storage devices, 79, 89
   system elements, **77**

Index

uploading and downloading information, 84-85
workstations, 85-88, **86**, **95**
condenser, 19
connectivity, 113-115, **114**, **115**
connectors, 26-28
  military style, **27**
constant horsepower, 41, **43**
constant torque, **44**, 221
contact ladder diagram, 122, **122**
control loops, 204, **204**
  closed system, 231-232, **231**
  open system, 230-231, **230**
  programming, 125-130
control systems
  cell controller, 116-117
  distributed, 117-118
  PLC (*see* programmable logic controller)
converter, 190
conveyor, mechanical pieces from motor to, **35**
coulomb, 19
counter EMF (CEMF), 211, 232
coupling, 48, 49-50, **49**
  jaw, 49
  torque-limiting, 50
current source inverter (CSI), 180
cycles, 16

D

dampers, 177
data acquisition, 92-93
dc drive, 177, 198-201, **199**, 227
  digital, 200-201
  efficiency, 201
dc injection braking, 212
dc motor, 157-163, **146**, **158**
  field voltage ratings, 162
  four-quadrant operation, 201, **202**
  parallel, 159
  permanent-magnet speed vs. torque curve, **160**
  replacing with ac machines, 169
  series wound, 159, **161**
  shunt wound, 158-159, **162**
  shunt wound motor curve, **161**
decibel noise scale, 343
decoding, quadrature, 235, **235**
Department of Energy, 170
digital pulse train, **234**
digital revolution, 3

digital tachometer, 225, **226**
diode, 20, **20,** 23
  breakdown, 20
  light emitting, 21, 245
  Zener, 20, **20**
diode rectifier, 186-187, **186**, **187**
direct current (dc), 12
direct drive, 58
distortion
  harmonic, 212, 213, 265-266
  total harmonic, 216-217
distributed control system (DCS), 117-118
downloading, 258
drives, 57-58, 385
  ac, 176-190, 227
  ac flux vector, 196-198
  adjustable speed, 179
  current source, 189, **190**
  dc, 198-201, 227
  direct, 58
  electronic, 221-222
  frequency, 179
  pulse width modulated, 190, **190**
  servo, 202-207
  specialty electronic, 201-210
  spindle, 208-209
  stepper, 207-208, **207**
  tools for repairing, 225
  troubleshooting, 222-226
  variable frequency, 58, 179, 183, 185
  variable speed, 58, 179
  volts-per-hertz, 179
duty cycle, 49
dynamic braking, 211
dynamic random access memory (DRAM), 79

E

eddy-current clutch, 174-175, **175**, 176
efficiency, power transmissions, 64-65
Einstein, Albert, 10-11
electric motors, 47, 145-170, 385
  447T frame, 151, **151**
  ac, **146**, 152-157, 182, **183**
  cooling, 147-149
  dc, **146**, 157-163, **158**, 201, **202**,
  protecting, 149-152
  servomotor, 163-167, **206**
  stepper, 167-168, 208

Electrical Apparatus Service
   Association Inc. (EASA), 9
electrically erasable programmable
   read only memory (EEPROM), 81
electrically programmable read only
   memory (EPROM), 81
electricity, 7-31
   amplitude, 16
   cabling and wiring, 26-28
   electrical formulas, 13-16
   electricity and magnetism, 10-13
   evolution of switch device, **70**
   frequency, 16
   fusing and circuit protection, 29-30
   hardware, 16-21
   standards organizations, 8-10
   symbols, 10, **11**
   tools, 28-29
electromagnetic interference (EMI),
   26, 266
electromotive force, 232
electronic drive, applications, 221-222
Electronic Industries Association
   (EIA), 9
electronic numerical integrator and
   calculator (ENIAC), 70
emergencies, 338-340
encoder disk, 234-235, **234**, **235**
encoders, 233-238, **234**, **235**
   optical, 233
energizing circuit, 19
engagement time, 54
Environmental Protection Agency
   (EPA), 329
equipment, costs vs. productivity, **6**
Ethernet, 263
ETL Testing Laboratories Inc. (ETL),
   9, 329-334, **330**, **331**, **333**

F

Factory Mutual (FM), 327
fail-safe brake, 55
fan, applications, 60-61
feedback devices, 229-250
   dancer scheme, 241-242, **242**
   encoders, 233-238
   laser, 248
   machine vision, 248, 275-302, 385
   mill-duty digital pulse tachometer,
      236-238, **236**, **237**
   resolvers, 238-240, **239**
   servo drive, 205

servomotor, 166
   transducers, 241-242
   ultrasonic, 247-248
fiberoptic cabling, 255-256
field-effect transistor (FET), 25
   metal-oxide semiconductor, 25, 71
field winding, 158
filters, 272-273
finite element analysis, 48
firmware, 82
flash memory, 81
flowcharts, PLC troubleshooting, **135**,
   136, **137**, **138**, **139**, **140**, **141**,
   142, **143**, **144**
flow control, 244
fluid coupling, 177
fluid-based speed variator, 177
fluidized speed variator, 55-56
flux control, 183
flux vector drive, ac, 196-198
flywheel effect, 45, **46**
force, 34
form factor, 162-163
formulas
   acceleration torque, 40
   efficiency, 15
   electrical, 13-16
   electrical power, 14
   horsepower, 14, 36, 39, 59, 60
   inertia, 43
   single-phase amperes, 15
   single-phase efficiency, 15
   single-phase horsepower, 15
   single-phase kilowatts, 15
   single-phase power factor, 15
   speed, 39, 182
   stress, 48
   three-phase amperes, 15
   three-phase efficiency, 15
   three-phase horsepower, 15
   three-phase kilowatts, 15
   three-phase power factor, 15
   three-phase volt-amperes, 15
   torque, 39
   torque requirement of friction, 57
   work equals force times distance,
      34
form wound, 150
Fourier, Baron Jean, 213
Fourier analysis, 13, 213
frequency, 16
frequency drive, 179
friction, 46-47

torque requirement formula, 57
  windage and, 47
full-load amps (FLA), 37
fuses, 29-30, 222

G

gain
  integral, 205
  proportional, 205
gate signal, 24
gate turn-off thyristor (GTO), 23, 24-25
gating, 187
gear, 51-52, **51**, **52**
  bevel, 51
  helical, 52
  miter, 51
  planetary, 52
  spur, 51
  worm, 52
gearbox, 50-51
gear reduction, 48-50
generator, 2
ground fault, 28
grounding, 29, 266-267, **267**

H

hardware
  computer, 78-79
  programmable logic controller, 99-107
harmonic analysis, 214
  data required for, 217-218
harmonic distortion, 212, 213, 215, 256-266
  total, 216-217
harmonic distortion analyzer, 218, **218**
harmonics, 212-218
  commutation notch, 216, **216**
  third, 213
harmonious sound waves, 212
harmony, 212
helical gear, 52
henry, 21
hertz (Hz), 12, 16
horsepower, 14, 35-38, **36**
  constant, 41, **43**
  formula, 15, 36, 39, 59, 60
  torque-horsepower-speed nomogram, 41, **42**
hotspotting, 90-91, **91**

hydraulic systems, 61-62, 173
  components, 61
hyperlinking, 91

I

induction motor, 155-156
  split-phase, 155, **155**
inductor, 21
industrial computers (*see* computers)
Industrial Revolution, 2
inertia, 43-46
  formula, 43
  linear, 44
  reflected, 44-45, **45**
  rotating, 44
inertia matching, 206
information highway (*see* networks and networking)
inlet vanes, 177
input/output, 107-108
  analog and discrete, 107-108, **110**, **111**
  connectivity and, 113-115
inspection, 355
Institute of Electrical and Electronic Engineers (IEEE), 8-9
insulated gate bipolar transistor (IGBT), 25, 180, 188, **189**
insulation classes, temperatures for, 154
integral gain, 205
interference
  electromagnetic, 26, 266
  radio frequency, 26, 218-219, 266
International Electrotechnical Commission (IEC), 329
International Standards Organization (ISO), 361-377
inverter, 179, 188-189
inverter duty motor, 154
ISO-9000, 361-377
  assessment summary report, 369-371, **369**, **370**, **371**
  certificate, **373**
  certification
    benefits, 373-374
    guidelines, 375-376
    marks, 372, **372**
    requirements, 374
  contract review, 363
  corrective/preventive action, 365-366
  data control, 363-364

Index

ISO-9000 *continued*
  design control, 363
  document control, 363-364
  handling and storage, 366
  inspection and testing, 364-365
  inspection control, 365
  internal quality audits, 366
  key functions, 362
  management responsibility, 363
  noncompliance report, 367-368, **368**
  nonconforming product control, 365
  process control, 364
  product control, 364
  product identification/traceability, 364
  purchasing, 364
  quality record control, 366
  quality system, 363
  requirements, 363-367
  servicing, 367
  statistical techniques, 367
  test equipment, 365
  training, 366

K

K-factor element, 23, 215
kilovolt-amps (KVA), 22
Kirchoff's law, 13

L

ladder logic, 120-123, **121**, **122**
lagging, 220
La Place transformation, 13
laser technology, 248
level control, 244
light emitting diode (LED), 21, 245
lighting, 277-278, **278**, 295-296
  back-, 292, 297
  ballasts, 214
  top, 297
line operation, 58
linear inertia, 44
liquid
  gravity of, 60
  pumping applications, 59
local area network (LAN), 260-263, **261**
logic operators, 71, 81, 120
loop control (*see* control loops)
looping, 125-130

M

machine systems
  downtime analysis, 322-324, **323**
  grinding system, 317, **317**
  machining center with individual workcells, **318**
  presses, 318-320, **319**
  robotics, 304-315
  service and warranties, 324-325
  tooling, 317-318
  troubleshooting, 322-324
  web-converting line, 316-317, **316**
machine vision, 248, 275-302, 385
  bad packaging detection system, 286-287, **287**
  backlighting, 292
  black automobile tire, **301**
  coplanarity of several points, **294**
  digital image processing, 279, **280**, 284
  digitized image, 279, **279**
  electrical engineering content, 282
  five disciplines of, 275-284
  fixtures, 297-298
  future, 299-302
  gray-scale edge detection, 280, **280**
  high-speed cameras, 278
  lighting, 277-278, **278**, 295-296
  mechanical content, 282-283
  nondigitized image, 279, **279**
  offline inspections, 293-294
  offline specification sheet, **300**
  offline systems, 288-293, **289**, **290**, **291**
  online systems, 284-287
  PCB inspection system, 285-286, **286**
  resolution, 278
  robotic guidance system, 281-282, **281**
  Ronchi Grid concept, **294**
  tools, 295
  troubleshooting, 298-299
  video imaging, 276, **277**
magmeter, 62, 128, **129**
magnetic flowmeter, 62, 128, **129**
magnetic particle clutch, 54
magnetic pickup sensor, 240
maintenance, power transmissions, 65-66
megahertz (MHz), 71
memory, 79-82
  dynamic random access, 79

Index

electrical programmable read only, 81
electrically erasable programmable read only, 81
flash, 81
hierarchy and evolution, **80**
keeping track of location, 123-125, **124**
random access, 79
read only, 80-81
static random access, 79
metal-oxide semiconductor field-effect transistor (MOSFET), 25, 71
metal-oxide varistor (MOV), 272
metric system, 391-393
metrology, symbols/definitions, 283-284
microprocessor, 3, 71-77
microprocessor industry, 5
microwave technology, 256
mill-duty digital pulse tachometer, 236-238, **236**, **237**
miter gear, 51
mnemonic, 120
modem systems, 84-85
　baud rate, 257, **257**
modular process controller, 126, **126**
modulation
　pulse amplitude, 180
　pulse width, 168, 180
motion control, 3-4, 171-228
　ac drives, 176-190, 227
　braking and regeneration, 210-212
　categories, 172
　dc drives, 198-201, **199**, 227
　definition/description, xiv
　electronic drive applications, 221-222
　harmonics, 212-218
　methods for slowing operation, 176-177
　methods for varying speed, **178**, **179**
　new technology, 175-176
　old technology, 173-175
　power factor, 219-220
　reduced voltage starters, 209-210
　RFI, 218-219
　servo drives, 202-207
　spindle drives, 208-209
　standards, 173
　stepper drives, 207-208, **207**
motor controller, 150, **150**

motor generator set, 174, **174**, 177
motor slip, 182
motor thermostat, 150, **150**
motors, 47-48
　ac, **146**, 152-157, 182, **183**
　brushless, **164**
　cooling, 147-149
　dc, **146**, 157-163, **158**, 201, **202**
　electric, 47, 145-170, 385
　enclosure types, 147-148, **148**
　induction, 155-156
　inverter-duty, 154
　mechanical pieces from motor to conveyor, **35**
　parallel, 159
　protecting, 149-152
　repulsion, 156
　series wound, 159, **161**
　servo-, 163-167, **206**
　shunt wound, 158-159, **161**, **162**
　split-phase induction, 155, **155**
　squirrel-cage, 153-154, **153**, **155**
　stepper, 167-168, 208
　synchronous, 156
　wound rotor, 156
multimeter, 28
multitasking, 117

N

nameplates, 168, **169**
NAND, 71, 81, 120
National Electric Code (NEC), 8, 327
National Electrical Manufacturers Association (NEMA), 8-9, 151
National Fire Protection Agency (NFPA), 8-9, 327
National Institute of Standards and Technology (NIST), 10
National Standards Association (NSA), 10
Nationally Recognized Testing Laboratory (NRTL), 328
networks and networking, 82-84, 92-93, 251, 260-263
　Arcnet, 263
　baud rate, 257, **257**
　client/server scheme, 92, **93**
　connectivity, 113-115, **114**, **115**
　Ethernet, 263
　local area, 260-263, **261**
　programmable logic controllers and, 115-116
　topology, 116

Index

networks and networking *continued*
  transmission distances, 256
  wide area, 263-264, **264**
noise, 63, 264-265 (*see also* harmonic distortion; interference)
  electrical, 264, 266-270
  radiated, 265
  safety considerations, 342-343
  shielding to prevent, 270-271, **271**
  thermal emission, 265
NOR, 71, 81, 120
NOT, 71, 120

O

Occupational Safety and Health Administration (OSHA), 327
Ohm's law, 12-13, **13**
open drip proof (ODP), 147, **148**
open-loop control, 230-231, **230**
operators
  AND, 71, 81
  logic, 71, 81, 120
  NAND, 71, 81, 120
  NOR, 71, 81, 120
  NOT, 71, 120
  OR, 71
optical comparator, 288
optical encoder, 233
optical isolation, 21
optics, 276
OR, 71
oscilloscope, 28
over-running cluth, 54

P

parallel circuits, 18
parallel communications, 259-260, **259**
parallel motor, 159
photoelectric sensors, 245-247, **245**
  reflective/proximity, 246-247, **247**
  retroreflective, 246, **247**
  through-beam, 246, **246**
phototransistor, 21
PID loop controller, 125-130, **127**, 231
pitch, 57
PIV transmission, 177
planetary gear, 52
pneumatic systems, 62-63, 173
polling, 123

power, fluctuations in, 267-268
power factor, 219-220
powers of 10, 107
power supply, 59
power surge, 269
Power Transmission Distributors Association (PTDA), 66
presses, 318-320, **319**
pressure sensing, 243-244
price of conformance (POC), 359
price of nonconformance (PONC), 359
printed circuit board (PCB), 17
  machine vision system inspecting, 285-286, **286**
process torque, 39
programmable logic controller (PLC), 3, 97-107, **104**, **106**, 385,
  applications, 130-133
  cell control, 116-117
  CPU module, 105, **105**
  hardware, 99-107
  hardwired control scheme, **98**
  input/output and connectivity, 113-115
  input/output handling, 107-108
  man-machine interfaces, 113
  modules, 109-113
  networks, 115-116
  operator's manual introduction, **103**
  rack and backplane, **104**
  relay operation, **98**
  software and programming, 118-123
  specifications, **100**, **101**, **102**
  troubleshooting, 133-134, **135**, **136**, **137**, **138**, **139**, **140**, **141**, **142**, **143**, **144**
programming
  loop control, 125-130
  PLC, 118-123
programming languages, 120
proportional gain, 205
proportionately infinitely variable (PIV), 56
protocol, 254
pulley, variable speed, 55, 176
pulse amplitude modulation (PAM), 180
pulse width modulation (PWM), 168, 180
pumping, applications, 59

Index

Q
quadrature decoding, 235, **235**
quality assurance (QA), 356-360
quality control (QC), 354, 356-360
  history, 354-356

R
radiated noise, 265
radio frequency interference (RFI), 26, 218-219, 266
radiowave technology, 256
random access memory (RAM), 79
  dynamic, 79
  static, 79
random wound, 150-151
reactor, 21
read only memory (ROM), 80-81
  electrically erasable programmable, 81
  electrically programmable, 81
rectifier
  diode, 186-187, **186**, **187**
  silicon controlled, 23, 24, 162-163, 187-188, **188**
reduced voltage starter, 209-210
reengineering, 232
reflected inertia, 44-45, **45**
regenerative braking, 211
register, 123
relay circuit, 19-20
relay ladder diagram, 120-122, **121**
repulsion motor, 156
resistance, 12
resolution, 278
resolvers, 238-240
revolutions per minute (rpm), 39
ringing, 205
Robotic Industries Association (RIA), 10
robotics, 304-315, 385
  articulated arm system, 311, **311**
  automated packaging system, **314**
  Cartesian, 312-313, **314**
  circular interpolation, 308, **308**
  current design, 313, 315
  feedback signal vs. vision system signal, 305, **306**
  gantry system, 310-311, **310**
  positioning, 309
  SCARA system, 311-312, **312**, **313**
  work envelope of a robot in one plane, 309-310, **309**
  X-Y Cartesian move to a point, 306, **307**
  X-Y-Z point for 3D moves, 306-307, **307**
roller chain, 53
root mean square (RMS), 14, 23
rotary variable differential transformer (RVDT), 242, **243**
rotating inertia, 44

S
safety, 327-344, 386
  agencies, 327-328
  emergencies and fail-safe, 338-340
  enclosures, 340-342
  environmental protection classifications, 341
  field labeling, 333-335, **335**
  hazardous area definitions of the electrical code, 336
  hazardous locations, 335-338
  industrial noise, 342-343
  intrinsically, 337
  labeling, 328-330
    ETL's classified label, 330, **330**
    ETL's listed label, 330, **330**
  laboratory testing, 330-333
  on-the-job, 343-344
  performance testing, 335
  wiring scheme for analog output intrinsic barrier, 337, **337**
  wiring scheme for discrete input intrinsic barrier, 337, **338**
sawtooth wave, **17**
SCADA, architecture, **131**
semiconductor, 3, 4, 23-26
  complementary metal-oxide, 71
sensors, 229-250, 385 (*see also* feedback devices)
  biosensor, 249
  closed-loop control, 231-232, **231**
  flow and level control, 244
  future, 249-250
  magnetic pickup, 240
  open-loop control, 230-231, **230**
  photoelectric, 245-247, **245**, **246**, **247**
  pressure sensing, 243-244
  speed and position, 232-238
  temperature control, 242-243
  transmitting data, 248-249
serial communications, 257-259, **258**
  handshaking, 259
  RS-232, 258
  RS-422, 258

Index

serial communications *continued*
   RS-485, 258
series circuits, 18
series wound motor, 159
servo drive, 202-207
   control loops, 204, **204**
   feedback devices, 205
   matching components, 206
   stability, 205
servomotor, 163-167, **165**
   brushless, **165**
   connection points, 166
   construction, 164-166
   feedback device, 166
   oscilloscope image, **206**
   rotor, 166
Severance, Frank, xv
shielding, 29, 270-271, **271**
short circuit, 27-28, **28**
shunt wound motor, 158-159
   circuit for, **162**
   curve diagram, **161**
signal, gate, 24
signal conditioners, 272-273
silicon controlled rectifier (SCR), 23, 24, 162-163, 187-188, **188**
   design, 24
sine wave, 186
sintering, 159
Sitkins, Fred, xv
six-step inverter (SSI), 180
Society of Manufacturing Engineers (SME), 10
software, 78
soldering, 27
speed, 39
   ac drive, 192, **192**
   fluidized speed variator, 55-56
   formula, 39, 182
   torque-horsepower-speed nomogram, 41, **42**
   vs. torque curve, 41, **41**
speed of transmission, 257
speed variator, fluid-based, 177
spindle drive, 208-209
split-phase induction motor, 155, **155**
spur gear, 51
square wave, **17**
squirrel-cage motor, 153-154, **153**, **155**
Standards Council of Canada (CSA), 9
standards organizations, 8-10, 66
standing wave condition, 195

starters, reduced voltage, 209-210
static random access memory (SRAM), 79
static-speed control device, 175
statistical process control (SPC), 92, 283, 347-354
   histogram, 350, **350**
   Ishikawa diagram, 357, **358**
   negative kurtosis, 352, **353**
   Pareto analysis, 359, **359**
   positive kurtosis, 352, **353**
   process capability chart, 349-350, **348**
   regression/correlation chart, 349-350, **349**
   scatter plot, 350, **351**
   x and R chart, 350, **351**
statistical quality control (SQC), 283
stepper drive, 207-208, **207**
stepper motor, 167-168, 208
   operation, 167-168, **168**
stiction, 47
stress, 48
   formula, 48
stress analysis, 48
symbols, 10, **11**
synchronous motor, 156
system integrators, 320-325
   downtime analysis, 322-324, **323**
   machine breakdown, 322

T

tachometer, mill-duty digital pulse, 236-238, **236**, **237**
temperature control, 242-243
test equipment, 28-29, 225
theory of relativity, 11
thermal noise emission, 265
thermal overload relay, 150, **150**
thermistor, 242
thermocouple, 242
thermoelectric devices, 242
thermostat, 243
thyratron, 24
thyristor, 23
tools
   electricity and electronics, 28-29
   machine vision repair, 295
topology, 116
torque, 35, 38-40, **38**, 183
   acceleration, 39
   breakaway, 39
   constant, **44**, 221

Index

formula, 39
horsepower-speed nomogram, 41, **42**
process, 39
torque limiter, 50
torque wrench, 40
torsional analysis, 156-157
total harmonic distortion (THD), 216-217
total indicator reading (TIR), 293
total quality management (TQM), 7, 354, 360-361, 385
totally enclosed fan cooled (TEFC), 147, **148**
totally enclosed nonventilated (TENV), 147, **148**
totally enclosed unit cooled (TEUC), **148**
transducer, 241-242
transformer, 12, 21-23, **22**
  rotary variable differential, 242, **243**
  weight comparisons, 23
transistor, 3, 25-26
  insulated gate bipolar, 25, 188, **189**
  metal-oxide semiconductor field-effect, 25, 71
transistor-transistor logic (TTL), 71
triangular wave, **17**
troubleshooting
  drive systems, 222-226
  machine systems, 322-324
  machine vision, 298-299
  PLC, 133-134, **135**, **136**, **137**, **138**, **139**, **140**, **141**, **142**, **143**, **144**

U

Underwriters Laboratories (UL), 10, 328-329
uninterruptible power supply (UPS), 82, 268

V

V-belt, 52-53, **53**
variable frequency drive (VFD), 58, 179, 183, 185

variable pitch fan, 177
variable speed drive (VSD), 58, 179
variable speed pulley, 176
variable voltage inverter (VVI), 180
ventilation
  ac drive, 193-194
  motor, 147-149
video imaging, 276, **277**
  digital, 284, 279, **280**
  digitized, 279, **279**
  nondigitizied, 279, **279**
voltage, 12
voltage spike, 269
volt-amps (VA), 22
volt-ohmmeter, 28
volts-per-hertz drive, 179
volts-per-hertz ratio, 183

W

warranties, 324-325
water, conductivities of, 130
watts, 14
waveforms, 16, **17**
  amplitude, 16
  frequency, 16
  sawtooth, **17**
  sine, 186
  square, **17**
  triangular, **17**
web-converting line, 316-317, **316**
wide area network (WAN), 263-264, **264**
windage and friction, 47
wire, 26-28 (*see also* cable)
  connections, 26-28
wire tracing kit, **29**
work, 34
worm gear, 52
worm gear reducer, 52
wound rotor motor, 156

Z

Zener diode, 20, **20**
zero backlash, 238
zero-crossing, 187

Other Bestsellers of Related Interest

**Programmable Controllers:
Hardware, Software, & Applications, 2nd Edition**
—George L. Batten, Jr.
Provides a solid introduction to programmable controllers—what they are, how they work, and how to select, set up, and use them on the job. Readers learn how to: evaluate hardware configurations; determine software requirements; choose a programming language; control central processing units, peripheral devices, and input/output interfaces; and more. With this in-depth guide, engineers will be able to maximize productivity at the lowest possible cost.
0-07-004214-4 $39.00 Hard

**Power Electronics: Devices,
Drivers, Applications, and Passive Components, 2nd Edition**
—B. W. Williams
The essential text for electronics engineers, designers, and graduate students. In this completely revised and updated reference, the author has added an entire new section on passive components. Important additions to the applications section include converter underlap, reversible converters, and standby and uninterruptible supplies. Many more real examples and problems ensure that the reader gains a thorough working knowledge.
0-07-070439-2 $65.00 Hard

**Power Supplies, Switching
Regulators, Inverters, and Converters, 2nd Edition**
—Irving M. Gottlieb
Thousands of circuit designers—pro and amateur alike—have turned here for the knowledge they need to build quality power supplies for today's most demanding applications. Everything from TV sets and radio transmitters to computers and robots. Theory and practice are covered in illustrated detail, plus you'll master basic power supplies, low-voltage logic devices, and new low-power multiple power supplies.
0-07-024007-8 $27.95 Paper

McGraw-Hill Electronic Troubleshooting Handbook
—John D. Lenk

This practical how-to guide shows readers, in step-by-step fashion, how to pinpoint component faults and correct design flaws for virtually every type of electronic circuit. Lenk delivers a wealth of specific troubleshooting tips for popular consumer items such as TV sets, VCR's, camcorders, and CD players. This storehouse of practical information stands alone as a must-have resource.
0-07-037658-1 $39.50 Hard

**McGraw-Hill Electronic
Testing Handbook: Procedures and Techniques**
—John D. Lenk

No other single volume covers such a wide range of tests for answering so many questions encountered in real-world circuit design. Anyone who works in electronics can put this essential book to immediate use because unlike most similar books, this one is completely self-contained—covering both test procedures and test equipment. It also shows how to interpret test results so they form the basis for troubleshooting electronic equipment and for analysis of experimental design circuits.
0-07-037602-6 $45.00 Hard

How to Order

Call 1-800-822-8158
24 hours a day,
7 days a week
in U.S. and Canada

Mail this coupon to:
McGraw-Hill, Inc.
P.O. Box 182067
Columbus, OH 43218-2607

Fax your order to:
614-759-3644

EMAIL
70007.1531@COMPUSERVE.COM
COMPUSERVE: GO MH

Shipping and Handling Charges

Order Amount	Within U.S.	Outside U.S.
Less than $15	$3.50	$5.50
$15.00 - $24.99	$4.00	$6.00
$25.00 - $49.99	$5.00	$7.00
$50.00 - $74.99	$6.00	$8.00
$75.00 - and up	$7.00	$9.00

EASY ORDER FORM— SATISFACTION GUARANTEED

Ship to:
Name _____
Address _____
City/State/Zip _____
Daytime Telephone No. _____

Thank you for your order!

ITEM NO.	QUANTITY	AMT.

Method of Payment:
☐ Check or money order enclosed (payable to McGraw-Hill)

Shipping & Handling charge from chart below	
Subtotal	
Please add applicable state & local sales tax	
TOTAL	

☐   ☐

☐   ☐

Account No. ☐☐☐☐☐☐☐☐☐☐☐☐☐☐

Signature _____ Exp. Date _____
Order invalid without signature

**In a hurry? Call 1-800-822-8158 anytime,
day or night, or visit your local bookstore.**

Key = BC95ZZA